Systems Analysis and Design

for the Small Enterprise

Third Edition

by David Harris

THOMSON

COURSE TECHNOLOGY ™

Systems Analysis and Design for the Small Enterprise, Third Edition
by David Harris

Senior Vice President, Publisher:
Kristen Duerr

Executive Editor:
Jennifer Locke

Product Manager:
Karen Lyons

Development Editor:
Karen Lyons

Production Editor:
Melissa Panagos

Associate Product Manager:
Erin Roberts

Marketing Manager:
Jason Sakos

Editorial Assistant:
Christy Urban

Text Design:
Ann Small

Cover Design:
Rakefet Kenaan

Compositor:
GEX Publishing Services

*QA
76.9
.S88
H3734
2003*

Disclaimer
Course Technology reserves the right to revise this publication and make changes from time to time in its content without notice.

ISBN 0-03-034903-6

Contents

CHAPTER 2
Project Initiation **32**

SECTION II ANALYSIS

CHAPTER 3
Process Modeling 62

CHAPTER 4
Data Modeling 94

CHAPTER 5
Object Modeling 118

SECTION III **DESIGN**

CHAPTER 9
Report and Query Design **238**

SECTION IV **DEVELOPMENT**

SECTION V **IMPLEMENTATION**

DEDICATION

To my lovely wife Ilene, our children, and our growing family of loved ones

Preface

Desktop computing has changed almost everything about computing, including the way students learn about systems analysis and design. Information systems are a necessity in enterprises of all sizes. Students interact with small enterprises every day. It is appropriate to teach them how to analyze, design, develop, and implement a modest information system in a familiar environment. Students can then apply newly developed skills immediately, in their own community, making the transition from classroom to career easier.

This text provides thorough coverage of small-enterprise systems analysis and design that will prepare students for larger-scale projects. The text includes several integrated case studies tailored to the small-enterprise setting, as well as suggestions on how students can select and develop a Portfolio Project inspired by their local community. The Portfolio Project is a modest information system of their own making. In practical terms, the learning experience is linked to the level of difficulty imposed by the assigned Portfolio Project, not just the text.

Student learning experiences are enhanced by the text's practical application of concepts and skills. Students quickly come to anticipate that each element of the text will provide practical skills they can apply to a project similar in scale and scope to the text's Cornucopia Case Study. Systematically, students replicate the text's Cornucopia Case Study from analysis to implementation, completing the course with the Portfolio Project.

This text is perfect for a capstone course in the CIS applications curriculum. Concepts are associated with the day-by-day, task-by-task experience of systems analysis and design: combining and adapting major 4GL, off-the-shelf application products into a coherent, desktop computer-based system.

INNOVATIONS

This text is distinguished in two ways: First, it presents a project-oriented approach to systems analysis and design, within the context of a small enterprise. Second, it provides students an opportunity to learn by doing. No other book in this market offers such practical application of concepts.

The Cornucopia Case Study appears at the end of Chapters 2 through 15, illustrating how analysts proceed through the systems development life cycle to provide a simple information system for a small music store. This integrated case study includes important concepts presented in the textbook.

Portfolio Project assignments, which appear at the end of each chapter, coincide with the activities and tasks described in each chapter and in the Cornucopia Case Study. Students learn how to work as a member of a team to build a modest information system for a small enterprise in their community. When students complete the Portfolio Project, they will have a project binder of their team's work, as well as a practical understanding of how to build an information system.

CHANGES IN THE THIRD EDITION

The third edition responds to feedback from students and faculty who have successfully used the text, reviewer recommendations, and changes in information system technologies. Without compromising the practical orientation of the text, several significant changes are incorporated into this edition.

- The third edition offers complete coverage of the systems development life cycle methodology in 15 chapters, as opposed to the 18 chapters of the second edition.
- The end-of-chapter material is greatly expanded to include questions on chapter learning objectives, short answer questions, activities, discussion questions, and Portfolio Project assignments.
- The third edition remains focused on the hands-on Portfolio Project, which includes 14 assignments and 6 presentation and report deliverables. An added feature, the Portfolio Project binder, provides students with a uniform and systematic record of the information system development process.
- While the text's focus remains on the traditional approach to systems analysis and design, coverage of the object-oriented methodology is expanded into a separate chapter so that students can become familiar with the object model, use case diagrams, and the Unified Modeling Language diagramming models. In addition, the object-oriented methodology remains integrated into each segment of the systems development life cycle.
- Networking is added to the basic information system components, with continued coverage throughout the text, including greatly expanded coverage in Chapter 13.

FEATURES AND PEDAGOGY

- Each chapter opens with a brief *Overview* and *Learning Objectives* and concludes with a *Summary* in order to emphasize the main points students should understand upon completing the chapter.
- *Key terms*, set in boldface and italicized, help students identify important vocabulary specific to systems analysis and design. Key terms are also defined in the Glossary at the end of the text.

- *TechNote* boxes highlight special-feature, highly technical, or elaborative background material.
- *Thinking Critically* sections present students with real-world ethical and situational dilemmas similar to those they will face as analysts in the field. These scenarios are designed to spark discussion and debate and to help students become aware of the wide range of work-place problems that confront systems analysts today.
- *Integrated case studies* are introduced within the five sections of the text. These integrated case studies reinforce material presented in earlier sections and expose students to a broad cross-section of situations in which analysts work.
- The *Cornucopia Case Study* runs through Chapters 2–15, providing a detailed example for students to follow in developing an information system for a small enterprise. All aspects of the analysis, design, and development process, from client relations to file creation, are illustrated in this case. In addition, Visible Analyst is used to show students how a CASE tool can help analysts create an information system.
- Several end-of-chapter features help students review and extend their understanding of the chapter content:

   Test Yourself questions are directed at chapter learning objectives and content. While the answers to some questions can be developed directly from the text, many questions require students to apply their understanding of the text to slightly different situations.

   Activities ask students to revisit key elements and examples in a hands-on fashion. In most cases, these exercises ask students to analyze and propose solutions to problems suggested in the text. The activities require students to use a variety of commonplace desktop computer software.

   Discussion Questions help students explore chapter topics in a practical, and sometimes personal, way. Many of these questions are suitable for classroom discussion, while others require individual responses. Discussion questions require students to develop answers based on a variety of sources: the text, their experience, the Internet, and classroom participation.
- The *Portfolio Project* appears at the end of each chapter and provides students with hands-on application of theories and practices as they methodically build a modest information system for a small enterprise. Each entry begins with a brief summary of the project's expected progress-to-date and the current challenges facing the student. Three appendices support the Portfolio Project: Appendix A covers project management; Appendix B provides information on the Portfolio Project, including detailed specifications for different examples to use; and Appendix C offers instruction on technical writing and presentations.

ORGANIZATION AND APPROACH

This text is organized into five sections that offer complete coverage of the enhanced systems development life cycle methodology in a sequence that parallels the real-world path used by systems analysts. More than 200 figures and illustrations clarify concepts using numerous sample enterprises. Visible Analyst CASE tool coverage is included in most chapters, permitting instructors to use CASE software if they choose.

The overall approach is very straightforward: first, students become familiar with the material by reading, discussing, and answering questions about each chapter; and second, students apply their understanding of the material with the Cornucopia case study and Portfolio Project.

Section 1: Introduction

Chapter 1 defines an information system in terms of its six components, explores the characteristics of a small enterprise, introduces the systems development life cycle methodology, and discusses the nature of the systems analyst's work. Chapter 2 describes the basic information processing requirements, small-enterprise information system problems, and the initial activities associated with the beginning of a system project. At the end of this section, students complete the first Portfolio Project deliverable: the project contract.

Section 2: Analysis

The analysis phase of the systems development life cycle is organized into a series of chapters devoted to various model-building activities. Chapter 3 covers the process model, offering detailed instruction in constructing data flow diagrams. Chapter 4 covers the data model, emphasizing the importance of file and database design represented in the entity-relationship diagram. Chapter 5 covers object modeling, providing a moderate, but complete, introduction into the object model and Unified Modeling Language diagramming techniques. Chapter 6 covers system modeling, providing a collection of broad views of the information system, which include the user's system diagram, menu tree, and system flowchart. When students complete this section, they will have developed a clear understanding of their Portfolio Project information system by constructing the models described in these four chapters.

Section 3: Design

Coverage of the design phase of the systems development life cycle begins with a broad view of system functions and input form design (Chapter 7), followed by chapters devoted to file and database design (Chapter 8), output design (Chapter 9), and process design (Chapter 10). In the first chapter of this section, students complete the second Portfolio Project deliverable: the preliminary design presentation and report.

Section 4: Development

Coverage of the development phase begins with the third Portfolio Project deliverable: the design review presentation and report followed by a discussion on prototyping (Chapter 11). After the user and analyst agree on the new system design, the development phase is necessarily filled with very detailed activities associated with programming. The use of 4GL products reduces the workload considerably but does not eliminate the need for 4GL programming (Chapter 12). Included at the end of this chapter is the fourth Portfolio Project deliverable: the prototype review presentation and report. This section concludes with expanded coverage of networking technologies (Chapter 13). Throughout this section, students will be consumed with building their Portfolio Project information system. The instructive illustrations of the integrated case studies and the Cornucopia case study provide a template students can replicate in their own projects.

Section 5: Implementation

The final section presents several important topics associated with testing, documentation, and training (Chapter 14), followed by the fifth Portfolio Project deliverable: the training session. Chapter 15 covers equally important topics of project conversion methods and the long-term project maintenance and review activities. Finally, students complete the sixth and last Portfolio Project deliverable, the final report.

AVAILABLE SUPPORT

This text includes a rich collection of supplements for students and instructors. The Instructor's Resource Kit contains a detailed Electronic Instructor's Manual, Student Data Files, Figure Files, PowerPoint Presentations, and the Exam View Test Bank.

The Instructor's Manual includes suggestions and strategies for using the text, including:

- sample course outlines
- suggestions for course management
- installation instructions for the Student Data Files
- suggestions for supervising students' Portfolio Project teams
- instructions on how to set up and use Portfolio Project Binders
- solution suggestions for the Portfolio Projects in Appendix B
- methods for developing Portfolio Projects based on local small enterprises
- answers to end-of-chapter questions and suggested solutions to activities

The Student Data Files are easy-to-use spreadsheet templates for reporting project status, project budgets, analyst hours, and cost/benefit analysis. Figure Files allow instructors to create their own presentations using figures taken from the text. PowerPoint Presentations provide chapter-by-chapter lecture presentations, as well as complete slide shows for the Cornucopia Case Study presentation deliverables. Exam View is a testing software package that can be used to create paper, LAN-based, and Internet exams with multiple-choice, true/false, fill-in-the-blank, and essay questions.

In addition, Course Technology presents online courses in WebCT and Blackboard. Adding online content to a course offers students benefits such as additional hands-on work, self-tests, and links. Instructors can contact their local Course Technology representative for more information on how to bring distance learning to a course.

ACKNOWLEDGEMENTS

My students and colleagues at College of the Redwoods made much of this book obvious by their questions and advice about the real world of systems analysis and design. The theme of their remarks and judgments has always been, "Why is this material important, and how can we use it?"

Certainly, I owe a great debt to those computer educators and professionals whose works are cited as references. No one stands alone in a profession as diverse and ever changing as ours.

Many reviewers have spent countless hours plying over chapter outlines, text material, and artwork to help fashion a book that is correct, complete, well organized, and understandable. This is truly arduous work, especially when combined with the demands of their own teaching and professional responsibilities. I hope they can recognize their suggestions and recommendations in the final product. I offer special thanks to the following reviewers:

- Louis Berzai, University of Notre Dame
- Barbara Doyle, Jacksonville University
- Paul Jordan, Southern Wesleyan University
- Karen Nantz, Eastern Illinois University
- Sandy Puras, Georgia State University
- Laurie Schatzberg, University of New Mexico

Without question, the third edition reflects the most welcomed wisdom of Course Technology's Managing Editor, Jennifer Locke, who has nurtured this project for over a year. Her enthusiasm for this hands-on, project-based approach provided the inspiration to refine, shape, and improve a text used successfully for almost 10 years. As any author will confirm, the number of day-to-day creative, technical, and editorial decisions required for a project of this size and scope is formidable. Karen Lyons, serving as both Product Manager and Development Editor, has always been just a keystroke away in assisting with hundreds, if not thousands, of these decisions. Copy Editor Mark Goodin has been exceedingly faithful in applying his talents to the task of smoothing the rough edges from the text. Production Editor Melissa Panagos has guided the manuscripts, copy edits, design elements, style sheets, images, and proofs through a myriad of hands to produce a readable, precise, attractive, and on-time finished product. Although I have worked most closely with the four individuals mentioned above, credit is due to over a dozen Course Technology staff members directly involved with this project. Specifically, I would like to thank the following individuals whose efforts helped to produce this text: Marketing Manager, Jason Sakos; Associate Product Manager, Erin Roberts; Editorial Assistant, Christy Urban; Manufacturing Coordinator, Laura Burns; Word Processor, Jody Huerkamp; and Proofreader, Harry Johnson. Indeed, I am indebted to the entire Course Technology team who helped make this edition a reality. I have the greatest respect for their collective talents and tenacity.

Finally, my lovely wife, Ilene, and our growing family of loved ones continue to inspire my work. I thank them for their love and support.

READ THIS BEFORE YOU BEGIN

To the Student

Data Disks

To complete some of the Portfolio Project assignments in this book, you will need to create a Data Disk. Your instructor will either provide you with a Data Disk or ask you to make your own.

If you are making your own Data Disk, you will need a blank, formatted high-density disk. You will need to copy a set of files from a file server or the Web onto your disk. Your instructor will tell you which computer, drive letter, and folders contain the files you need. You can also download the files by going to *www.course.com*, clicking Student Downloads, and following the instructions on the screen.

The following Student Data Files accompany this text:

- status.xls — used to report weekly project status
- budget.xls — used to report weekly project budget
- hours.xls — used to report weekly project hours
- cost-benefit.xls — used to develop project cost/benefit chart

To the Instructor

The Student Data Files are available on the Instructor's Resource Kit for this title. Follow the instructions in the Help file on the CD-ROM to install the Student Data Files to your network or standalone computer. For information on creating a Data Disk, see the "To the Student" section above. Course Technology grants you a license to copy the Student Data Files to any computer or computer network used by students who have purchased the book.

Introduction

informationsystems
inthe**small**enterprise

When you complete this chapter, you will be able to:

OVERVIEW

This text assumes that you have used a computer to generate a variety of useful output. In other words, you have entered keystrokes on a word processor to create your resume, or numerals into a spreadsheet program to manage your budget, or course grades and schedules into a database to keep track of your college course work. All of these are separate information products that, collectively, may be considered an information system.

This chapter begins with a brief review of the recent technological advances that have brought computerized information systems within reach of every enterprise. It explains how these trends create tremendous opportunities for the entrepreneurial systems analyst within the small enterprise environment.

The chapter introduces the systems development life cycle (SDLC), which is composed of five major phases: analysis, design, development, implementation, and maintenance and review. You will learn about the six components of a computer information system, the role of the analyst, and the role of technology during each phase of the SDLC. Also, you will learn of the cyclical nature of systems development.

The chapter concludes with a description of a portfolio project. This project continues throughout the text, providing you an opportunity to actually build a modest information system for a small enterprise.

- Distinguish the special characteristics and information needs of the small enterprise from those of other enterprises.

- Outline the five phases of the systems development life cycle.

- Identify the six components of an information system.

- Describe the nature of entrepreneurial systems analysis.

- Describe the purpose of CASE tools.

INFORMATION SYSTEMS DEFINED

Formally, an *information system* is defined as follows:

An information system is a well-coordinated collection of resources that gather and transform data into information products and services that help the enterprise perform its designed functions.

Given the variety of enterprises that exist and the assertion that every enterprise has at least one information system of some sort, agreeing on a more specific definition that would fit all circumstances seems unlikely. Some systems are large in scope; others are restricted to a single segment of the operation. Some systems are completely computerized, others totally manual. Some systems provide a detailed, up-to-the-minute record of the most elementary enterprise operation; others provide only simulations of the real world. Some systems cost millions of dollars to design, build, and maintain, yet others can be delivered for far, far less. What, then, do all information systems have in common?

They all require careful planning, design, development, and implementation. They all require monitoring and maintenance to ensure that they continue to meet the needs of the user. And, eventually, when they fail this last requirement, they must be replaced. Yet, an even more distinctive way to define information systems is to focus on what they do.

They all provide some information, service, or product to a user. Therefore, information systems are often characterized by the audience they serve. In a hierarchical organization, the executives, the managers, and the everyday workers have different information needs, so they require different products and services. The bookkeeper needs an accurate record of the financial transactions of the enterprise, the sales department needs insightful forecasts of market trends, and so on.

Figure 1-1 presents an *information system hierarchy* that associates different information systems with these different audiences. A *strategic system* might provide a CEO with an analysis of an enterprise's performance against industry norms. A *management system* might provide managers and supervisors with summarized reports on employee productivity. An *operational system* might provide the inventory clerk access to the status of purchase orders. Often, all three systems exist within a single enterprise, with the major data-gathering activities concentrated at the operational level and the transformation activities designed to continually refine data as they pass to the other systems. In the small enterprise information system the hierarchical model gives way to a spherical model, with no obvious distinctions between audiences and kinds of information systems.

Finally, our definition of information systems is not complete without some consideration of emerging technology. Three trends provide evidence that information systems evolve right along with technology and support the notion that there is a growing demand for the entrepreneurial systems analyst.

FIGURE 1-1 / *Information System Hierarchy*

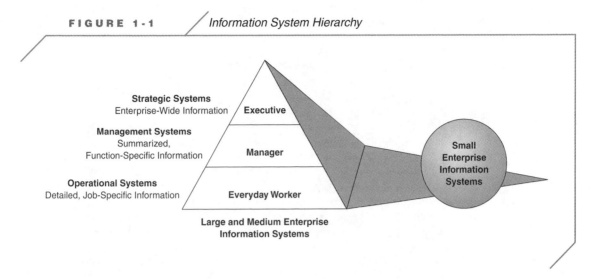

The Emergence of Small Enterprise Computing

When you look at the last 50 years of the millennium, the inevitability of *small enterprise computing* seems obvious. In 1950, computing power was generally confined to large enterprises, government, and education. By the late 1960s, computer information systems were commonplace in medium-size enterprises. The introduction of the microcomputer in 1970 led to an unprecedented string of technological advances, each accompanied by dramatic increases in computer power and dramatic decreases in computer costs. Today, it is difficult to find even the smallest enterprise conducting its affairs without the aid of a computer system of some sort.

The U.S. Small Business Administration reports that small businesses employ half of the private-sector workforce and contribute 50 percent of the private gross domestic product. According to the Cahners In-Stat Group, these enterprises invested more than $57 billion in computer hardware and software in 1998. IDC/LINK (Framingham, MA) adds that small businesses purchase computers at twice the rate of larger businesses.

Gordon Moore, chairman emeritus of Intel Corporation, predicted that microprocessor performance, as measured in millions of instructions executed per second (MIPS), would double every 18 months. Figure 1-2 shows the MIPS rating for the long succession of Intel processors, validating *Moore's Law* as it is sometimes called. Considering that for the past 25 years the price range of a typical small enterprise computer system has remained fairly constant ($2,000-$3,500), it is not surprising that small enterprise computing is flourishing. As a consequence, an ever increasing number of enterprises require expert computer information system services.

FIGURE 1-2 / *Intel Microprocessor Performance Rating in MIPS*

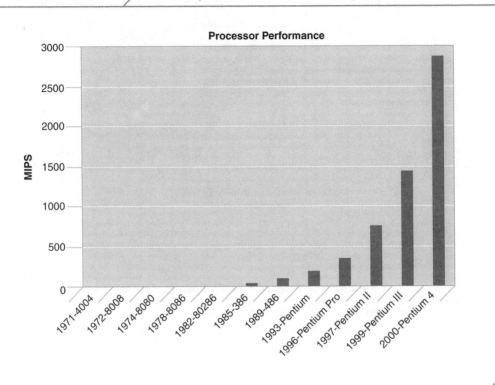

The Internet as a Dominating Technology

Connecting computers together to allow users to share peripherals and information is not a new idea. Users of large and medium-size computer systems have long had this capability. Within the last ten years, even small system users have been able to install a *local area network (LAN)* to connect computers situated in a small geographic area. In short, *telecommunications* has historically been available only to those with access to specialized telecommunications equipment and expertise. The *Internet* removes these requirements, making worldwide access possible, with minimal cost and complexity.

Now every enterprise, regardless of its size, can establish and maintain far-reaching communications, sales, and service areas. The process of collecting, manipulating, and distributing information is no longer constrained by time and distance. The cost to *e-mail* a coworker in the next room is virtually the same as broadcasting a message to a sales force spread over the globe. Every information system should be capable of leveraging this technology. Thus, systems analysts must include careful consideration of Internet technologies in every step of their work.

There are many ways to measure the phenomenal growth of the Internet. In 1981, when IBM legitimized the microcomputer industry with the introduction of the PC, the term "Internet" was used for the first time to describe the connected set of several hundred networked host computers. Today there are an estimated 180 million host computers. Almost every computer in use today is a potential platform for access to this vast and varied array of information. Furthermore, low-cost, high-speed Internet access has expanded the audience to virtually everyone.

The Maturation of Application Software

With each iteration, *off-the-shelf application software* has improved in terms of functionality and ease of use. After 20 years of revision and refinement, such software users are assured of a relatively stable, compatible, and intuitive collection of products in which they can invest precious training time. It is significant that the rate of feature and function changes has leveled off. This permits users to capitalize their information systems in the same fashion that mature financial instruments permit investors to enjoy the benefits of compound interest. Rather than starting completely anew, information systems users can now build on previous knowledge and experience as they migrate from one system to another.

The *software suite* is a natural extension of this process. Microsoft Office combines several products into one package, each with a similar look and feel, but each with distinct uses. As the user becomes familiar with one product, it is easier to learn the next. Further, when new, seemingly unrelated products are introduced, such as Internet browsers or multimedia programs, users have an easier time expanding their software repertoire.

Software developers also benefit from the dramatic advances in processor and software capabilities. Several *computer-assisted systems engineering (CASE)* packages are on the market that specifically address the needs of the systems analyst and information system developer. *Object-oriented application software* further enables the analyst to shorten the process used to analyze, design, and create information systems. As we shall see in later chapters, desktop computer-based software incorporates many features used in contemporary systems analysis and design of medium-scale and large-scale projects.

The Entrepreneurial Systems Analyst

Technological advances in processing capability, network access, and application software eventually demand subtle, though sometimes dramatic, changes in information system design, development, and implementation. Orchestrating this synchronized process usually requires the skills of a computer professional because small enterprise owners, supervisors, and everyday workers seldom have sufficient time or expertise to do this work.

Taken together, these trends create enormous opportunities for the systems analyst. Whereas some small enterprises can afford to hire a full-time computer specialist, most must look outside the organization for help. The *entrepreneurial systems analyst* provides such a service, employing the same sophisticated methods used in large organizations to the small enterprise.

SMALL ENTERPRISE CHARACTERISTICS

The term *small enterprise* is used to describe the diverse activities of many segments of our service and information age. When you hear this term, your first impulse might be to consider only small, single-proprietor businesses such as the neighborhood diner, the baseball card shop, the oil and lube shop, or the storefront retailer. But many, many other enterprises require information in order to continue operating. Local service organizations, community theaters, health volunteers, bed and breakfasts, and reading clubs are merely a few examples. To serve such enterprises properly, you must appreciate the context within which they operate.

For example, the design and implementation of a production control information system for a small manufacturing company that specializes in redwood deck furniture requires an appreciation for woodworking procedures and a sensitivity to the nature of a small enterprise, as well as detailed knowledge of computer systems. Figure 1-3 itemizes some important characteristics of the small enterprise, several of which are discussed in this chapter.

FIGURE 1-3   / *Small Enterprise Characteristics*

1. Owner operated
2. Few employees
3. Few product or service lines
4. Small capital base
5. Low profit margins
6. Low overhead
7. Small geographic physical presence
8. Limited computer expertise
9. Hybrid information systems

Owner Participation

Small enterprises are almost always owner operated or locally controlled. This situation produces a considerable amount of hands-on management. In such an environment, the analyst should remember two things: first, few, if any, intermediaries will be between you and the final decision maker—the owner, director, or chairperson will be keenly interested in what you are doing; second, these people have tremendous demands placed on their time—they will not be able to devote a great deal of time to your questions.

Economic Constraints

The profit margins or expense accounts of small enterprises are likely to be minimal. The costs associated with computer information systems are operating expenses. As such, they reduce profits or service budgets accordingly. You must be prepared to defend such expenditures in terms of their benefits to the enterprise and how they will offset their costs. A section of Appendix A is devoted to the cost/benefit analysis required to make this justification. At this point, acknowledging the small amount of tolerance for cost overruns in small enterprise systems is all that is necessary.

Competitive Pressures

Most small enterprises face stiff competition. Even community service groups must compete for the time of their members, access to the media, and scarce funding. The analyst must consider how the information system affects the organization's ability to compete. The answer to this question greatly influences the user's decision about whether to engage your services as an entrepreneurial systems analyst.

INFORMATION NEEDS OF THE SMALL ENTERPRISE

The small enterprise is much less likely to have a formal, hierarchical organization with distinct information system audiences. In addition to making strategic decisions about products, services, and markets, the owner operator may also supervise, or even perform, the everyday operations. The holistic, or spherical, organizational model requires the analyst to consider the information system in the same way. That is, less emphasis is placed on the classification of the system (strategic, management, operational) and more attention on the need to integrate elements of all three types.

The following discussion identifies many of the common information needs of any size enterprise. This provides a starting point for the needs assessment and problem definition portion of the *systems development life cycle (SDLC)*, which is explained in detail later in this chapter. For the small enterprise, some of these needs are more important than others. Can you argue for those you believe are more critical than others? Can you classify these needs in terms of their strategic, management, and operational information system natures? Can you envision a small enterprise information system with elements of each?

Production and Inventory

Raw materials, goods in progress, and inventory tie up capital and storage space, both of which are in short supply in most small enterprises. However, an information system that reduces such pressures on scarce resources may be difficult to "sell" on these points alone. Other, more immediate problems may provide compelling justification. When a customer or client calls at the busiest time of the day, inquiring about the status of a particular job or whether a certain product is in stock, a production and inventory information subsystem can be instrumental in keeping a valued customer or making a new one. Such a system might help answer such questions as "What is in the shop right now?"; "Who is working on this project?"; or "Do we have any excess capacity to accept a new contract?"

Personnel

Although small enterprises, almost by definition, have few employees, they must maintain records about their personnel. Furthermore, in the likely absence of a personnel specialist, enterprise managers may find themselves devoting many precious hours to record keeping rather than managing. Payroll, benefit plans, training programs, and vacation scheduling are only a few of the obvious information subsystems that can improve productivity, as managers can then devote more time to employee supervision and training, customer relations, or new product development.

Financial Reporting

Each enterprise is accountable for its expenditures and revenues. Financial reporting, such as personnel record keeping, is an overhead expense item. Therefore, information subsystem improvements and time savings do not result in increased revenues. Rather, they free clerical or managerial hours, which can be reassigned to other tasks. An even more important reason to consider computer assistance in this area is that small enterprises are less likely to have the necessary personnel to prudently separate financial functions. A computerized accounting system, although not totally tamper proof, is one way to improve fiduciary integrity.

Marketing, Sales, and Service

Improvements in the marketing, sales, and service information subsystems have a direct effect on revenues. First, they can provide management with reliable information about current performance and historical trends. Second, they can speed the revenue collection process. Both of these are especially important to small enterprises with precarious cash flow dynamics. Prompt billing, sales summaries, and market-targeting statistics can help stretch limited budgets.

Customer Relations

Much of the appeal and strength of a small enterprise is its ability to deliver personal service to its customers. Such service promotes customer loyalty, which is often cited as a bedrock condition for success. The hands-on nature

COMPUTING TERMINOLOGY

Every discipline has a vocabulary and notation that may seem strange to newcomers. With experience, the acronyms become familiar and the new meanings assigned to common household terms seem natural. This text is written with the assumption that the reader is well acquainted with much of the standard computing terminology. Nevertheless, an extensive glossary is included at the end of the text and "TechNotes," such as this one, are used to define terms and expand on technical concepts. For example, this chapter includes several references to various types of software. Because the terms are not fully discussed until later chapters, brief definitions are provided here to help you understand the discussion.

Off-the-shelf application software refers to prepackaged, application-specific computer programs available to users in ready-to-use form. Examples are word processor, spreadsheet, and database programs.

Software suites combine several application software programs into one prepackaged commodity.

Computer-assisted systems engineering (CASE) software is a program that helps the analyst perform many of the charting, design, programming, and project management activities associated with his or her work.

Object-oriented application software is characterized by its focus on predefined information system components, or objects, as they are called. Such software helps the analyst incorporate new and efficient systems analysis and design methods into small enterprise information systems work.

of this kind of service cannot be duplicated by a computer. In fact, the fear that the computer will somehow dehumanize the enterprise is one of the common concerns of potential users. Everyone has been subjected to the personalized, yet generic, holiday newsletters and contest announcements. Such word-processed, mail-merged junk mail is worse than ineffective—it is expensive and sometimes offensive.

Computer-based customer relations should take advantage of what the computer does best, leveraging the faithful memory capabilities of the system. Correlating past customer service and product preferences with current offerings can produce literature that is sensitive to customer differences. A special discount on the anniversary of an account opening or a free consultation to fine-tune a product that is reaching a critical reliability or performance benchmark can be much more effective than a mass mailing. A computerized "tickler" report could serve as a reminder that a personal visit, call, or note is due a particular customer. We do not pretend that impersonal, mechanized aides are substitutes for the real thing, but they do make it easier to deliver the real thing.

THE SYSTEMS DEVELOPMENT LIFE CYCLE

The systems development life cycle (SDLC) is composed of five major phases: analysis, design, development, implementation, and maintenance and review. The first four phases are covered in great detail in the remaining four sections of this text. The maintenance and review phase is summarized in the last chapter. This chapter serves as an introduction to the entire SDLC. You will learn about the five components of a computer information system, the role of the analyst, and the role of technology during each phase of the SDLC. Also, you will learn of the cyclical nature of systems development. Figure 1-4 depicts the SDLC as a circle, with the five phases appearing as distinct wedges. This image will be refined considerably as we move through this chapter.

FIGURE 1-4   / *The SDLC*

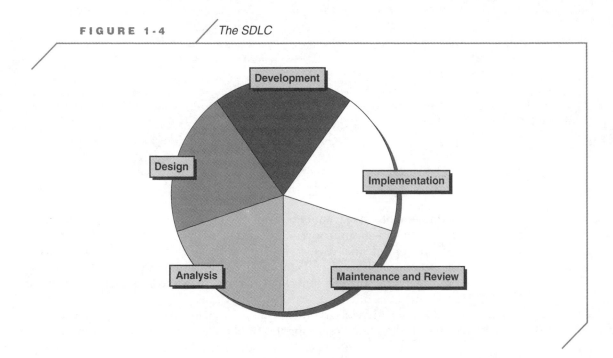

Like any other asset, computer information systems begin to "wear out" almost as soon as they are implemented. This fact is often overlooked in the small enterprise, where "computerizing the operation" is assumed to be a one-time task. Nothing could be further from the truth. Once an organization adopts a computer technology, it is virtually impossible to retreat to a manual operation. Furthermore, as the technology and perceived needs of the user evolve, so must the computer information system.

The Circular Nature of the CIS

Computer information systems should be designed to be flexible, adaptable, compatible, and expandable. In other words, they must be able to change. Identifying the end of one system and the beginning of another is often difficult. In this sense, a circular SDLC is continually operating, looping back on itself. The question often asked is: when must you stop modifying the old system and adopt a completely new system? Figure 1-5 offers one way to view this process.

Functional and Operational Obsolescence

Rarely do computer systems reach the point at which they simply will not operate. The bullet in Figure 1-5 serves as a common terminus to illustrate the point at which a system may evolve from a maintenance and review phase to another analysis phase, thus providing upgrade possibilities to prolong the system's useful life and delay its inevitable functional obsolescence.

Operational obsolescence occurs when a system stops working altogether. This is common with a lawnmower or an automobile. Computer systems, on the other hand, are prone to *functional obsolescence*; at some point, they can no longer be modified to meet the perceived needs of the user. In the best situations, the user has foreseen the functional obsolescence and commissioned an analyst to initiate a new SDLC. All too often, however, the system's obsolescence creeps up on the user, perhaps because the user bought into the "one-time task" fallacy of computerization—or perhaps because the system was not designed to grow with the user's needs.

FIGURE 1-5 / *The Circular SDLC*

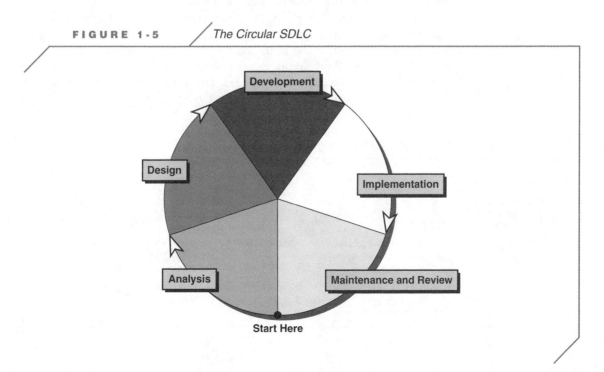

Unplanned Obsolescence

Sometimes information systems suffer from conditions beyond the analyst's control. Such is the case with problems associated with the year 2000, or the Y2K problem, as it was commonly called. This problem arose because many computer systems were built to process only the last two digits of the year. Such systems had problems distinguishing January 1, 2000 from January 1, 1900. Or, when manipulating dates across January 1, 2000, incorrect results would cause payments to be considered late, discounts to be missed, interest computations to be scrambled, and so on.

There were many different responses to the Y2K problem. Many enterprises worked for several years to correct problems with their existing systems. This included developing entirely new systems or modifying old ones. In some cases, entire systems were abandoned because of the cost associated with repairing the problem. Even those enterprises that were already Y2K compliant worried that failures in systems outside their control (suppliers, clients, etc.) might adversely affect their operations.

Although this problem did not just pop up, the situation illustrates that the analyst is often forced to work in an environment not of his or her own making. Large amounts of flexibility, creativity, and patience are required to successfully navigate such uncharted waters.

Reduced Duration of the SDLC

Technological innovation and increased user sophistication have reduced the life span of computer information systems. Whereas a reasonable life expectancy once ranged from four to seven years, you should now plan for a system to last three to five years. Small enterprise systems can be made slightly more durable by overbuying in the hardware and platform component area, but such components inevitably must be replaced. This suggests a "rolling" or recurring SDLC as presented in Figure 1-6. The recurring SDLC often results in entirely new software and hardware to meet evolving user requirements.

FIGURE 1-6   *The Recurring SDLC*

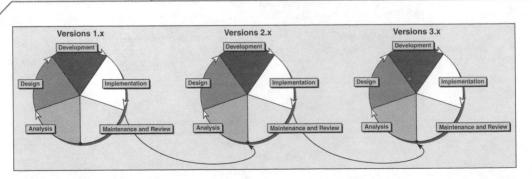

COMPUTER INFORMATION SYSTEM COMPONENTS

The SDLC provides a framework for defining the phases of a computer information system life cycle. Previous sections describe several characteristics of information systems within small enterprises. This section focuses on the information system itself. Exactly what is the nature of the product you are supposed to design, build, and implement?

The six components of a computer information system (CIS) are people, procedures, software, hardware, networks, and data. Take notice: this is not the only way to approach systems analysis. For many years computer professionals focused most of their attention on hardware, software, and data, masking the other elements as a secondary concern. As the profession has matured and users have become more sophisticated, the importance of all six components has become more obvious. Figure 1-7 lists the six components we shall study.

FIGURE 1-7   *Computer Information System Components*

1. People
2. Procedures
3. Software
4. Hardware
5. Networks
6. Data

Networks are a relatively new component of computer information systems and are sometimes referred to as "connectivity" or "geography." These terms are meant to remind us that modern computer information systems are usually connected to other computers via the Internet, or some other network technology, thus extending the reach of the system. Our study includes this important characteristic. You will find the discussion of the Internet, networks, and telecommunications woven into almost every topic.

People

Although technology has replaced a considerable amount of "people work," computer information systems are still designed to help people in some way. In the not too distant past, computer systems were often characterized as "the tail that wags the dog." This meant that people were expected to accommodate the system simply because "that's the way the computer does it." The microcomputer played a significant role in changing this perception. As end users began to effectively bypass centralized computer departments, it became apparent that people could and should be part of the system life cycle, from analysis to implementation. In fact, we shall see that the activities within the SDLC have been significantly modified to accommodate this trend.

Procedures

System procedures, which provide operational instructions for the users of the information system, have a notorious reputation for being the weak link in the usage chain. Many reasons explain this, but for our purposes two observations are paramount. First, programmers hate to write procedures, and second, users hate to pay someone to write procedures. This unfortunate combination resulted in a lot of good systems failing only because nobody knew how to use them. Once again, the PC changed things dramatically. With more users and more competition for their business, computer hardware and software companies were forced to develop user-friendly procedures for every part of the system.

For example, PC operating systems migrated to the graphical user interface (GUI) in response to competition from Macintosh and increased user demand for more understandable operating system procedures. Users expect information systems that are built around interfaces that are understandable. Thus, even the advent of the GUI environment does not eliminate the need for well-written operator and user instructions.

Software

Although software is still formally defined as a series of computer instructions designed to perform a specific task, the practical definition has evolved. It once was safe to refer to all programs as software. Now we must be careful to distinguish among firmware, netware, macros, horizontal software, vertical software, integrated software, custom software, system software, memory-resident software, utility software, and so on. The current practice, especially in small enterprise computing, is to customize off-the-shelf software rather than create computer programs from scratch. Today's analyst must learn how to blend many types of software into one coherent system with a friendly interface.

For example, horizontal software suites, such as Microsoft Office, offer the analyst a foundation upon which to build a small enterprise information system. But, doing so would be impossible without a multitasking operating system such as Windows. Even the vertical software package written specifically to handle the paperwork in a veterinarian's office cannot function without some regard to ROM-based programs (firmware) and disk backup and recovery utility programs.

Hardware

Hardware refers to all the tangible components of the computer and its peripheral devices. The term "hardware platform" is now commonly used to describe the central processing unit and its associated motherboard, memory chips,

SOFTWARE CLASSIFICATIONS

Many types of software are essential to the overall functions of a computer system. Some are not alterable by the analyst beyond the initial decision to include them in the hardware and software purchasing specifications.

Operating system software provides an interface between the user and the hardware, as well as between the application software and the hardware.

Firmware is hardware-specific software that is permanently stored in ROM.

Utility software, such as Norton Utilities, provides diagnostic, repair, and maintenance programs for processor, memory, and disk storage system components.

Application Software

Most computer programs used in today's small enterprise information systems are derived from off-the-shelf software commonly called application software. The analyst adapts this software to create information systems.

Horizontal software: General-purpose software used in a wide variety of situations. Examples are word processor, spreadsheet, and data-base application software.

Vertical software: Specific-purpose software used in a narrowly defined situation. An example is medical billing software.

Turnkey system: Specific-purpose, preinstalled software used in narrowly defined situations. The hardware and software are sold as a package, thus providing the user with access to the information system "at the turn of a key."

Customizable Software

The analyst uses a variety of software to customize the information system for a particular enterprise. This software is classified according to the degree of detail required in the commands.

Third-generation programming language: An English-like syntax used to instruct the computer on how to perform detailed tasks. Examples are Pascal, C, COBOL, and BASIC.

Fourth-generation programming language (4GL): An English-like language used to instruct the computer on which high-level tasks to perform. An example is Microsoft Visual Basic.

Object-oriented programming language: A programming language composed of high-level commands and large libraries of reusable object definitions and functionality. An example is C++.

4GL products: A catchall term applied to user-friendly software products characterized by their visual interfaces, menu-driven functions, electronic help features, on-screen form and report design tools, and code generators. Examples are word processor, spreadsheet, and database packages.

interface cards, and so on. Although a vast array of such hardware is available, this text concentrates on the dominant microcomputer technology: the Intel-based PC. Today's hardware is fast, reliable, affordable, compact, and mobile. In short, it is accessible to practically every enterprise.

Networks

Today's pervasive telecommunications technology extends the possible reach of any information system. Thus, our consideration of the traditional information system components must include a discussion of networking. By connecting computers, networking technology not only alters the boundaries of the information system, it changes the relationships and responsibilities of its component parts. Although future chapters offer a more detailed treatment of these topics, at this point it is important to define some common networking terminology.

A *peer-to-peer network* is the simplest computer network, in which two or more computers are connected to each other by direct cabling. Each computer can serve as client or server to another computer in order to share information and peripheral devices. A *local area network (LAN)* is a system that designates one computer to act as a dedicated server to one or more client computers. A LAN is restricted to a small geographic area, most commonly a single office or building. A *wide area network (WAN)* is a system that extends the principles of local area networking to a wider geographic area by use of a gateway from the LAN to other networks. The Internet is a worldwide connection of computer networks using standard communication protocols to facilitate data and information exchange. The *World Wide Web (WWW)* is one of the Internet's most popular delivery mechanisms. An *intranet* is an enterprise-wide connection of computer networks similar in function to the Internet, but restricted for use within the enterprise. Intranets provide a relatively easy way to connect differing computer hardware and software environments into a common network with a common user interface.

Data

The terms "data" and "information" can be confusing. For instance, it is probably obvious to you that information should be considered part of the system definition. Data, on the other hand, is a bit more slippery. Typically, we define data as the system input and information as system output. This distinction breaks down, however, once we acknowledge that the output or information from one system may be the input or data to another system. Rather than quibble, our approach is to say that data is an integral part of any computer information system and, as such, should be treated with the same care and attention as the other components. Indeed, effective design of data and information gathering, storing, accessing, and sharing is essential to effective information systems.

The Role of the Analyst

Analysts can serve the enterprise in many ways: they can work as outside consultants or as in-house computer specialists; they can work for one day on one specific portion of the project or work for the entire project; and their work may vary from simple end-user support to complex custom programming. In the small enterprise the analyst often becomes the person responsible for every element of the computer system: hardware selection, software installation, information system development and maintenance, personnel training, and so on.

Analysts set the standard for employee attitudes about software copyright protection, data confidentiality, and information integrity. Collectively, computer ethics are not well understood. Software licensing and duplicating policies vary from one product to another, making end-user education and enforcement difficult to manage. The analyst can offer a localized remedy to this problem by providing guidelines tailored to the particular information system. Throughout this text you will find scenarios, such as the one in the "Software Ownership" Thinking Critically box in this chapter, which are designed to help you confront some of the major ethical issues of the information age, particularly as they relate to your work as an analyst. Often there is no clear-cut answer to the questions posed; nevertheless, you are challenged to critically analyze each scenario and develop a considered opinion.

Above all else, the analyst is a problem solver. You must be well versed in computing and be able to quickly learn how many different enterprises can effectively use computing technology. Another way to describe this role is to divide the SDLC model proportionately to show the relative amount of time the analyst devotes to each phase.

Although Figure 1-8 is not a scientific representation, it serves to emphasize the importance of problem-solving skills, with easily more than half of the SDLC consumed by analysis, design, and development. The lines separating these phases are purposefully blurred to illustrate the recent changes in the way analysts work. Common practice is to move between the phases during such a *blurred SDLC*: rarely does one phase complete before the next begins. The strategy, or methodology, used to complete each phase actually supports this approach to the process.

FIGURE 1-8 / *The Blurred Proportional SDLC*

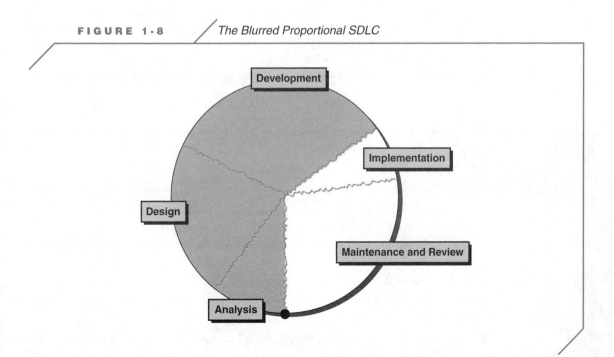

The Analyst as an Agent of Change

Regardless of the benefits—whether dramatic or subtle—of their work, analysts are viewed as agents of change by various people within the organization. Some react enthusiastically to the new information system, while others resist every change you introduce. Certainly, in the short run, your interpersonal skills will influence this situation significantly, but the long-term success of the system depends as much on the user's acceptance of the product as it does on the quality of the product. The discussion of user-friendliness is rooted in this issue.

One way to minimize resistance to change is to make the new system look as much like the old system as possible. User interfaces and procedures can be designed to resemble the familiar look and feel of the old system, even if the hardware and software elements are completely new.

The Analyst as a Problem-Solving Strategist

In rare situations, problem solving can be reduced to pure intuition. Most of the time, however, the successful analyst follows a process that produces a series of incremental solutions that can be folded together to form the final product. The SDLC provides the broad framework for describing the process of solving information system problems. To perform the tasks identified by this framework, the analyst has a toolbox full of *techniques*, which are described in subsequent chapters along with a strategy that explains how and when to use the *tools*. A comprehensive strategy such as this is called a *methodology*.

However, it is hard to envision the analyst as pure strategist, that is, one with no programming experience or responsibilities. Most small enterprise systems require some adaptation of existing application software, which is where programming skills come into play. An introductory course in programming logic and limited experience with a third-generation language such as C is sufficient to prepare the analyst for the *integrated development environments (IDEs)* that

SoftwareOwnership

Over the past three years, D'Amilio's Discount Art Supply of Chicago has grown tremendously. A client survey suggests that smaller, suburban discount facilities would be very profitable. Consequently, D'Amilio's decides to open three new suburban outlets.

Along with this expansion, D'Amilio's contracts you to set up a new computer information system to manage and coordinate all the business-related activities of the new stores. After your analysis, you determine that this project requires you to customize several new software application packages, which D'Amilio's agrees to purchase. Subsequently, you install these packages on both D'Amilio's computer and your own computer.

Under what circumstances or conditions, if any, do you feel it is appropriate to install a client's software on your computer?

accompany many new software packages. IDE offer the programmer sophisticated tools that simplify the programming task. The distinction between the programming experience required now and that required only a few years ago is important: the varied but detailed programming once necessary has given way to knowledge of limited but very high-level programming. Several systems development chapters address this issue in more detail.

The Analyst as a Group Facilitator

Much of the analyst's routine involves working with others. Hardware and software specialists, vendors, other analysts, managers, and end users form a diverse collection of individuals who must be consulted on a regular basis. As with any group, conflicts in perspective and personality create problems that must be resolved. Unfortunately, no easy, foolproof remedy can make this dilemma disappear. However, several theories on how best to deal with group dynamics are available. Most of them begin with a description of the different types of group members and then follow with some advice on how to deal with each. The following offers one such prescription.

Group Member Personas

- **The negative type:** Such people assume nothing will work as planned. Their skepticism cannot be ignored, but it should not be allowed to dominate discussions or discourage the creative risk takers from participating. Some Murphy's Law humor might discharge the negativism and tension that such individuals bring to the group.
- **The positive type:** They are always optimistic about the outcome of a situation, sometimes without good reason. Don't discount the value of positive thinking, but check their work.
- **The know-it-all type:** They never do. But, because they usually know a lot, you need to listen. The group itself will quickly learn to discount their less well-founded contributions.
- **The do-nothing type:** They talk a lot, but accomplish very little. Because they are usually so good at their charade, the group becomes frustrated in its attempts to discipline the individual. Don't let the group waste too much energy trying to correct this behavior. You must take responsibility to bring these people around by assigning them their own tasks and deadlines.
- **The techie type:** Every project needs someone who can solve the particularly technical problems encountered by the group. This person usually talks over everyone's head, so limit his or her time "on stage," but respect and reward the "backstage" miracles he or she performs.

Knowing that group dynamics can sometimes impede team effort and collaboration, you are cautioned against worsening the situation by developing your own superior-know-it-all persona. To frame this issue more positively, consider the substantial and convincing research directed toward quality management. Dr. W. Edward Deming's concept of *total quality management (TQM)* postulates that each team member can and must contribute to the overall success of a project.

TQM substitutes leadership for slogans and exhortations, training and self-improvement for ignorance, performance objectives for quotas, and collaboration for isolation. To become an effective group facilitator, the analyst must adapt the high-minded theory of TQM and the simplistic solutions offered above to each systems project.

THE ROLE OF TECHNOLOGY

Computer technology is obviously the grist from which computer information systems are fashioned. However, the same technology also provides powerful tools to help the analyst do the work. The paper and pencil tools of only a few years ago are rapidly giving way to electronic workbenches that are stuffed with charting utilities, code generators, screen painters, debuggers, and the like. As always, the analyst must be very careful to choose the right tool for the job.

METHODOLOGIES, TOOLS, AND TECHNIQUES

This chapter recalls the evolution of information systems. The methods by which these systems were created have also evolved over the years. The methodology presented in this book is eclectic in that it incorporates what has proven to be the best of several methodologies. Thus, you will learn how to incorporate consistent user involvement in the project. You will learn how to model the system using data flow diagrams, system flowcharts, and entity relationship diagrams. You will learn how to apply object-oriented analysis and design methods. Moreover, you will learn how to build the system by employing both old and new programming practices, user-friendly horizontal software, and electronic workbenches. Although each of these techniques is borrowed from slightly differing methodologies, they combine to form a practical approach to small enterprise systems work. As you study further, you should recognize that the real world often presents challenges that require the analyst to adapt existing tools and methods to new situations.

CASE Tools

Computer-assisted systems engineering (CASE) is an interesting combination of words. It suggests that a computer program can help the analyst to engineer, or create, a computer information system. Sometimes the "S" in CASE stands for "software," which is a more restrictive interpretation. The word "case" communicates an important point, regardless of the exact translation. It can be taken literally to mean a collection of analyst productivity tools organized into a single package, or placed into a common carrying case, if you will. Figure 1-9 associates some of the common CASE tools with the different phases of the SDLC.

FIGURE 1-9   *The SDLC with CASE Tools*

SDLC Phase	CASE Tool Functions
Analysis	Data flow diagramming
	System diagramming
	Entity-relationship diagramming
	Object modeling
	Data dictionary
Design	Input screen design
	Report design
	Structure charting
	Data dictionary
	Prototyping
Development	Code generation
	Data dictionary
	Prototyping
Implementation	Documentation
Maintenance	Reverse engineering

Some CASE tools focus only on a portion of the SDLC, while others cover the SDLC from beginning to end. Collectively, CASE offers the analyst a tool that integrates many of the varied SDLC tasks, providing a common repository of design specifications, system models, and data definitions. Furthermore, CASE provides for consistent documentation at each stage of systems development. The advantages to using CASE tools become particularly obvious in large to moderate-size enterprise system projects, especially when many analysts require access to project materials. With increases in the number of software environments, the frequency of system maintenance requests, and analyst turnover, CASE can be a stabilizing influence in otherwise tumultuous circumstances.

Some CASE packages include *project management* tools. Regardless of its size, every systems project involves an expenditure of resources to achieve a set of objectives. Project management software helps to generate project budgets, progress and completion reports, resource planning and allocation charts, and, in some cases, cost/benefit analysis. Project management tools are discussed in detail in Appendix A.

Introducing Visible Analyst

Visible Analyst is a popular CASE tool that integrates several phases of the systems analysis and design process into one software product. Visible Analyst Student Edition is available for purchase along with this text. To illustrate how a CASE tool works, several elements of a sample project, called Cornucopia, are developed using Visible Analyst. Cornucopia is introduced in the next chapter.

As explained, CASE tools come in many varieties, covering different aspects of the systems project and requiring different levels of user expertise. Visible Analyst can help you develop several different models of an information system. Specifically, it provides facilities for the development of function, data flow, entity-relationship, object/class, state transition, and structure chart models. All of the model specifications are retained in a repository for easy reference.

Figure 1-10 shows the basic Visible Analyst interface. In this illustration, we have chosen CPIA from a list of projects in the Select Project dialog box.

FIGURE 1-10   / *Visible Analyst*

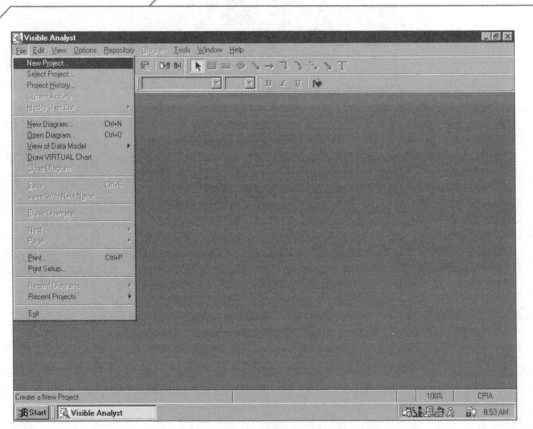

CPIA appears in the lower-right corner of the window and indicates the active project. A basic set of symbol and line tools is displayed.

Each of the menu choices has several options. In all, they supply a complete set of the tools used to construct many of the products required in a systems project. For example, the File option offers choices to create a new diagram or open an existing diagram.

In addition to these tools, Visible Analyst has the ability to verify that the analyst is consistent in the use of project names and definitions. In other words, the analyst can rely on this product not only to provide information about the project, but also to check for some types of mistakes.

Alternatives to CASE Technologies

Sophisticated CASE products cost thousands of dollars to purchase and require a great deal of analyst training and expertise to implement. For small enterprise projects, the benefits of CASE are unlikely to justify the added expense of its use. When a CASE product is not appropriate, the analyst should look for appropriate alternate technologies.

In the absence of a CASE tool, several easy-to-use and inexpensive software substitutes are available. Word processors, paint programs, flowchart and draw utilities, database screen editors, and code generators are but a few. Although such stand-alone packages do not offer the highly integrated, dictionary-styled solutions available with CASE products, they do allow the analyst to create and modify a professional looking product.

THE ENHANCED SDLC

A combination of advances in hardware, software, network technologies, and systems methodologies has changed the way computer information systems are developed. The view of the SDLC presented in Figure 1-11 reflects the new approach. The enhanced SDLC embodies several important concepts. The recurring nature of systems work and the blending of several phases of the SDLC have already been discussed. The influence of the user in the entire process is added, along with descriptive labels to describe the nature of the relationship between the analyst and the user. These labels are discussed in future chapters. Finally, analyst tools, some new and some old, are sprinkled within the perimeter of the SDLC model.

FIGURE 1-11 / *The Enhanced SDLC*

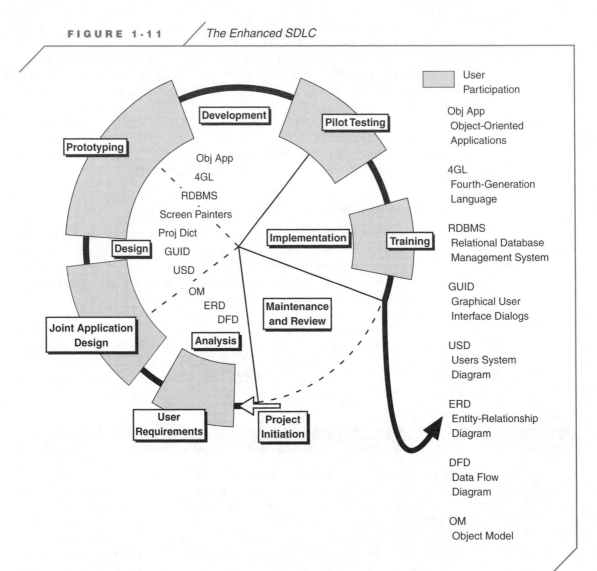

SUMMARY

Knowing the historical underpinnings of your profession is important. This chapter passes over this history lightly, not because it isn't important, but because this text is based on the assumption that you learn best by doing, especially at the beginning. From time to time, the text suggests such exercises or briefly summarizes relevant material. This chapter begins with two important subjects: first, small enterprise, PC-based computing should rely on proven systems theory and techniques; second, the small enterprise approach presents special advantages for our introductory studies.

The chapter continues with a discussion of the systems development life cycle, which serves as the framework we use to study and develop a small information system. This is complemented by a discussion of the strategy used to advance from one element of the framework to another. Both discussions focus closely on the two points emphasized earlier in the chapter.

This chapter introduces several versions of the system development life cycle, beginning with the traditional five-phase model and ending with the enhanced SDLC. Our study will be guided by this last version, which enhances the traditional model by incorporating new computer technologies, analyst techniques, and end-user participation throughout the process.

The chapter also introduces the elements of a computer information system. Six components—people, procedures, hardware, software, networks, and data—make up the system and are addressed as we study each of the phases of the SDLC in the chapters that follow.

Finally, in such a dynamic environment, the role of the analyst expands to include both technical and social issues. The analyst must now possess the skills to handle group interactions, ethical conflicts, and machine-to-human interfaces.

TEST YOURSELF

Chapter Learning Objectives

1. What are the distinguishing special characteristics and information needs of the small enterprise?

2. What are the five phases of the system development life cycle?

3. What are the six components of an information system?

4. What is the nature of entrepreneurial systems analysis?

5. What is the purpose of CASE tools?

Chapter Content

1. An entrepreneurial systems analyst possesses skills that are only applicable to a small enterprise environment. True or False?

2. Computer ethics standards are well established and easily available to systems analysts and computer users. True or False?

3. In terms of computer programming, a systems analyst need only have introductory computer programming experience. True or False?

4. Team building is an important activity of the systems analyst. True or False?

5. An information system is a collection of resources that work together to gather and transform data into information. True or False?

6. All information systems, regardless of their scope, require careful planning, design, development, and implementation. True or False?

7. In every enterprise there are always obvious distinctions between strategic, management, and operational types of information systems. True or False?

8. An information system is composed of people, procedures, software, hardware, data, and networks. True or False?

9. Telecommunications technology extends the possible reach of any information system. True or False?

10. Heavy managerial responsibilities of owner operators, increased competition, and relatively small profit margins associated with small enterprise environments present special problems for the analyst. True or False?

11. It is sometimes necessary to integrate elements of strategic, managerial, and operational information systems into a single small enterprise system. True or False?

12. It is not necessary to differentiate the information needs of a small enterprise from those of a medium-size or large enterprise. True or False?

13. The systems development life cycle is composed of several overlapping phases. True or False?

14. Information systems should be designed so that they can be expanded, revised, and updated. True or False?

15. At some point, obsolescence may force the analyst to initiate an entirely new system development life cycle. True or False?

16. The expected life span of an information system is increasing because of advances in hardware and software technology. True or False?

17. The potential for increased analyst productivity, brought about by advancing technology, is represented in the enhanced SDLC model. True or False?

18. The user plays a very small role in the systems development life cycle of modern computer information systems. True or False?

19. In the small enterprise information system, the information system hierarchy is totally eliminated because small enterprises require much less complicated information systems. True or False?

20. The information system is functionally obsolete when it can no longer be maintained or upgraded to satisfy the changing needs of the user. True or False?

ACTIVITIES

Activity 1

Conduct an informal survey of the existing information system of the next five small enterprises you come in contact with. How many of them are computer based? What types of computers, if any, are involved? Who functions as the information specialist in these enterprises? What type of training does this individual possess?

Activity 2

Review the employment ads in the Sunday newspaper for a metropolitan area. Select five ads that refer to microcomputer systems skills in the job description. Make your best attempt to identify the courses in your college catalog that cover the subjects described in the ad.

Activity 3

Investigate the Internet presence of five small enterprises in your community. How many did you locate on the Internet? How effective is the Internet message they present? Suggest reasons why those who are not on the Internet should be.

Activity 4

Many variations of the SDLC model exist. Use your school library or the Internet to find an alternative to the one presented in this module. Describe the differences and similarities in general terms.

DISCUSSION QUESTIONS

1. Given the discussion on small enterprise information needs, speculate on the information system needs of a small advertising agency. How might its production needs for desktop publishing, presentation graphics, and multimedia hardware and software complicate the information system? What does this suggest about the possible range of computer technology required to satisfy small enterprise computing needs?

2. Explain why "people" costs are often the most expensive of the six elements of a computer information system.

3. What are some of the advantages of focusing your study of systems analysis and design on the small enterprise?

4. What distinguishes an information system from a single application, such as a payroll system or an inventory system?

5. Why is user/client participation so important to the systems development life cycle?

6. Why are information systems more prone to functional obsolescence than operational obsolescence?

7. Why might it be difficult, if not impossible, to rank the six components of an information system in order of importance to the success of the system?

8. Why must a systems analyst possess people skills as well as technical skills?

9. What distinguishes a methodology from a tool or a technique?

PortfolioProject

This text presents a methodology based on well-established systems analysis procedures, while sensitive to the growing importance of small enterprise computing and increased functionality of 4GL software. The small enterprise project approach requires you to apply sound computing fundamentals to situations that you will find very contemporary and perhaps close to your own real-world experience.

Portfolio project assignments run throughout the text, coinciding with the activities and tasks described in each chapter. You will work as a member of a team to build a modest information system for a small enterprise. There are 14 team assignments designed to help your team create six project deliverables: project contact, preliminary presentation, design review, prototype, training, and final report. When you have completed this work you will have a portfolio of your team's work, as well as a practical understanding of how to build an information system.

The Cornucopia case study appears at the end of the remaining chapters. It is a guide to the activities associated with your portfolio project. Three appendices (A—Project Management, B—Student Projects, C—Technical Writing and Presentations) provide further instruction on completing your own information system.

Team Assignment 1: Your Fictional Consulting Firm

It is very likely that you will work as a member of a student team to complete the portfolio project. The first assignment provides an opportunity for team members to get to know one another while working on a task unrelated to the particular small enterprise project assigned to your team. Imagine that your team is actually a small consulting firm. In order to establish your team identity, perform the following tasks:

- Decide on a name for your systems analysis consulting firm.
- Develop a company logo using a graphics package.
- Integrate this company logo into a company letterhead, which you will use on all correspondence with your client. Submit a copy of your stationery to your instructor.
- Create an advertising flyer for your consulting company. Submit a copy of your flyer to your instructor.
- Prepare a Portfolio Project Binder, with tabs for 14 team assignments and each of the six deliverables mentioned above. File a copy of your letterhead stationery and advertising flyer behind the tab labeled "Assignment 1."

projectinitiation

OVERVIEW

Your role as a problem solver is a common theme throughout this text. A first step in the system development life cycle (SDLC) is the development of a clear understanding of the information problems to be addressed. This chapter explains how to sift through a maze of stimuli to focus your attention on the information problems of the enterprise. You will learn how to generate a preliminary sketch of possible solutions and their associated economic consequences. Finally, you will learn how to develop a rudimentary project contract.

- Describe the basic information-processing requirements.

- Identify the common problems associated with the information systems of small enterprises.

- Understand how to become familiar with the existing information system.

- Prepare a feasibility report.

- Use various resources for fact-finding and diagnosis to develop the project contract.

BASIC INFORMATION-PROCESSING REQUIREMENTS

Every enterprise has its own distinguishing characteristics. At first, this presents an overwhelming challenge to the analyst. How can you be expected to understand so many different information systems? How do you determine when modifying an existing system is sufficient, and when to initiate an entirely new SDLC? Although experience has no substitute, some guidelines are available that can help you answer the first SDLC question, "What is the problem?"

To help with this task, we explore several techniques in the fact-finding and diagnosis section of this chapter. But first we examine information system problems in general, particularly in the small-enterprise environment. Such problems can usually be traced to their failure to meet some very basic *information-processing requirements* (Figure 2-1).

FIGURE 2-1   / *Basic Information-Processing Requirements*

1. Information must be relevant.
2. Information must be accurate.
3. Information must be timely.
4. Information must be usable.
5. Information must be affordable.
6. Information must be adaptable.
7. Information must be accessible.

Information Relevancy

The term *information relevancy* describes the usefulness of system outputs. In a fast changing and competitive market, what we judge as information today may become junk tomorrow. For example, a report on term life insurance policies for corporations may be totally useless to an independent insurance agent who specializes in owner-operated enterprises. Likewise, an analysis of last year's teen fashion trends is unlikely to help a small retailer plan for this year's fad. In both cases, the information does not help the user make decisions.

In such situations, printouts go unused, on-demand queries are never demanded, and users never inquire about missing reports. In other words, the information is of no consequence to the user, therefore it is ignored. These are sure signs that the user needs to be consulted about the redesign of the system.

Information Accuracy

Information accuracy is always a concern. However, no system is completely error free. Some errors and inaccuracies are inevitable and therefore do not necessarily point to a critical system problem. During the design phase the analyst and user must establish an *error threshold* to serve as an alarm. When these thresholds are exceeded, the analyst should look first at the data-capturing procedures. Although some processing logic errors persist even after extended product testing, most accuracy problems are caused by a breakdown in the people, procedures, or data elements of a computer system.

If you cannot identify the cause of errors in one of these three components, you have to look at the entire system. Hardware and software problems are particularly difficult to diagnose. Therefore, any extended analysis may have to wait until you understand the system more fully.

Information Timeliness

In most cases, the value of information declines as time passes. Increased wait time and lost product or service opportunities are two common symptoms of this problem. *Information timeliness* can erode as the system ages, or it can suffer from poor design.

One of the great advantages of computers is the speed with which they follow our instructions. Unfortunately, processing speeds have not been matched by input and output speeds, leading to the well-known *I/O bottleneck* as the root cause of most time delays. The solution to this problem usually lies in reducing the human interaction to a minimum by automating input with specialized devices (e.g., bar-code readers, scanners) and replacing hardcopy with softcopy output.

Information Usability

Information packaging is critical to the user. Reports that are difficult to understand frustrate users, require an inordinate amount of training time, and often contribute to misunderstanding. This problem often can be traced to the report content and design specifications. As discussed in Chapter 1, information must be tailored to fit the different information needs of the intended audience. For example, the everyday worker needs a convenient way to enter and retrieve daily transaction data. The supervisor, on the other hand, is likely to need a summary of this data rather than all the detail. The upper-level manager, in turn, looks for information that incorporates the summary data into long-term trends. Although the underlying data is all the same, each user should be served by a different product. Thus, *information usability* is defined in terms of the user's needs.

Information Affordability

Information affordability cannot always be measured in direct dollar expenditures. The time and effort required to gather, process, and disseminate information is taken away from production or service. These *opportunity costs* are equally as real as the actual dollars spent on paper and postage. Although we consider information to be the lifeblood of the modern enterprise, the activities required to supply information are almost always viewed as an overhead cost. That is, unless the enterprise's principal activity is information processing, profitability is inversely related to information costs.

Any system in which information costs are rising faster than productivity and profits is difficult to defend. Allowing for extraordinary initial costs, systems that experience persistent upgrade costs or chronically increasing maintenance costs are probably functionally obsolete. That is, they still run, but they don't run well enough to justify their expense. Such systems are characterized by dramatic mismatches in one or more of the six system components (hardware, software, data, networks, people, or procedures). For example, in an effort to improve usability and timeliness, system software might be upgraded, only to find that the existing hardware can't support the increased processor and video output demands.

Information Adaptability

No one wants an information system that is cast in stone. This is particularly true in an age of accelerated functional obsolescence. Users are highly sensitive to the rapidly changing information technology they see advertised. Users quickly abandon those portions of a system that are of no use to them, turning to informal, ad hoc sources of information.

Information adaptability is often linked directly to system *expandability* and *compatibility*. Unless the system is at an evolutionary terminal point—such as with the task of retrofitting a first- or second-generation microcomputer to handle a contemporary application—adding more hardware or software resources is often the answer to such problems. In custom-designed systems, nonstructured programming practices generally lead to undocumented programs that are difficult to understand, and system changes require more effort.

Information Accessibility

The modern enterprise is characterized by its fast pace, diverse tasks, and dispersed workforce. Information system users must be able to quickly supply or retrieve task-specific information from many locations. One way to address this requirement is to provide multiple transmission modalities for information products. A payroll supervisor may need a printed weekly summary of labor expense, whereas a traveling sales representative may need regular access to an online inventory. Telecommunications technologies, such as the Internet and intranets, provide a possible solution to this requirement of information systems.

Symptom, Problem, and Solution Summary

Despite the risk of oversimplification, a brief summary is in order, including two important reminders. First, information system problems are not found in isolation; most often, a combination of circumstances causes these situations. Second, for a system to experience all of these symptoms at the same time is highly unusual. Thus, a single system could be overly complex, expensive, and confusing, while at the same time it could be accurate, timely, and relevant. Figure 2-2 summarizes the material in this section.

FIGURE 2-2 / *Symptom, Problem, Solution Summary*

Basic Requirement	Symptom	Problem	Solution
Relevancy	The system is not used.	User needs have changed.	Involve the user in the redesign process.
Accuracy	Reports are incomplete or erroneous.	The data input procedures are confusing or too demanding.	Simplify data capture through source document redesign or the use of input automation.
Timeliness	The response time to user requests for information is increasing.	Input and/or output demands exceed the capabilities of the system.	Automate input, upgrade the output and disk storage devices and/or the processor speed.
Usability	Users are confused about how to use the system.	Outputs are inappropriately designed or they are poorly documented.	Redesign the outputs and/or improve the documentation, then retrain the users.
Affordability	System costs are increasing more than user productivity.	One or more of the system elements are mismatched.	Evaluate the system mismatches to see if they can be minimized or if you need to commence a new SDLC.
Adaptability	Users have abandoned some parts of the system.	The system is approaching functional obsolescence.	Upgrade to a more powerful computer platform to allow for software upgrades.
Accessibility	Users must alter work patterns to retrieve information.	The information delivery system does not match work patterns.	Redesign the distribution and retrieval system to include online and on-demand access.

SMALL-ENTERPRISE INFORMATION SYSTEM PROBLEMS

The basic information-processing requirements and potential problem areas presented thus far apply to all enterprises. However, several potential trouble spots occur more often in small-enterprise systems.

Typically, the small enterprise has problems keeping up with the cyclical nature of information processing. The enterprise must consistently commit a great deal of time and energy to capture, edit, input, process, output, and evaluate data. Unless this is a well-articulated part of the job description, personnel who are responsible for such work are often neglectful. This is especially true if they perceive these information system activities as busy work. Again, we can first look to observable symptoms that alert the analyst to an underlying problem (Figure 2-3).

Two of the most common information system deficiencies that cause the difficulties listed in Figure 2-3 are *source data input inefficiencies* and breakdowns in subsystem integration.

FIGURE 2-3   Small-Enterprise Information System Problem Symptoms

1. Product processing controls are ineffective.
2. Client files are inaccurate and incomplete.
3. Customer correspondence is haphazard.
4. Business tracking and forecasting is spotty.
5. Customer billing systems are not timely.
6. Inventory control procedures are unreliable.

Source Data Input Inefficiencies

The day-to-day pressures of the small enterprise fall on a restricted number of employees. Each person is responsible for several jobs that might be assigned to separate specialists in a larger setting. In some cases, jobs are shared by two or more people. In such settings, problems often occur with data capture, editing, and input. Data may be duplicated, lost, or incomplete.

Also, the degree to which this phase of the information system is automated greatly affects the process. Optical scanners, magnetic-ink character recognition devices, and speech recognition systems can improve data capture speeds and reduce errors.

Regardless of the specifics, source data input inefficiencies are magnified as they cascade through the system. Many of the problems listed in Figure 2-3 can be traced to system input points.

Breakdowns in Subsystem Integration

As discussed in Chapter 1, most information systems are really a collection of subsystems, with each subsystem having its own input and output points. If these subsystems are not carefully coordinated, the larger system as a whole does not perform properly. *Subsystem integration breakdowns* provide fertile ground for the analyst to explore when searching for the root causes of small-enterprise information system problems.

For example, using the billing subsystem's customer database as input to the correspondence subsystem would be a logical, natural design choice. Combined with a word-processing mail-merge macro, the database would provide the detailed information required for specialized customer services, such as sales and service announcements, service contract expiration, and so on. However, if the database does not include service contract expiration dates, frequency of account activity, or service preferences, this data must be collected and maintained separately, which then risks introducing new errors and increasing costs. Another breakdown may occur if the database file format is incompatible with the word processor, which would require complicated, time-consuming, and error-prone conversion procedures.

SoftwarePiracy

Tommas, a small shrimp boat business headquartered in New Orleans, currently uses a software package called Quicken to prepare its financial reports. The bookkeeper's complaints about the slow ink-jet printer, the difficulty in reading the small monitor, and the general sluggishness of the computer's CPU prompt the owner to hire you to upgrade the hardware and the accounting system.

During your preliminary investigation, you ask to see the license for the Quicken software, hoping that the owner can qualify for a discounted price for the software upgrade. You discover that the Quicken software was illegally copied from an associate's disk.

How would you respond to this situation?

ThinkingCritically

FACT-FINDING AND DIAGNOSIS

The preceding section provides general background information on typical system problems. However, some specific activities are worthy of mention because they can help the analyst pinpoint problems, clarify user expectations, and foster agreement on a contract or some other project document that defines what is to be done.

For Example — The CIS Lab and Silhouette Sea Charter

Two considerably different small enterprises will help illustrate the analysis phase of the SDLC: the CIS Lab and Silhouette Sea Charter. One is nonprofit, the other hopes never to become nonprofit. One interacts with several existing computer-based information systems, the other now relies only on computer information from the National Weather Service. One is accessed by an almost entirely different set of end users each semester, the other has only two employees. The one thing they have in common, however, is the need for computerized information services, as reflected in the following memos (Figure 2-4).

Can you identify the basic information system requirements that are not met in each of these situations? Is anything about these particular problems peculiar to the small-enterprise environment? How familiar are you with the information system in your school's computer lab? What about the sea charter business? How can you find the answers to these questions? To answer questions such as these, you need to use one or more of the following means of *fact-finding and diagnosis*.

Industry Research

Considering that the analyst is a specialist in relation to computer function, but a generalist in relation to a computer application in the workplace, your investigation must begin with some self-training on the particular type of enterprise with which you are working. No doubt, your knowledge of the sea charter business is limited. Some *industry research* is required before you can fully appreciate the information you will gather on site.

Local libraries, private industry councils, state and federal labor departments, and small-enterprise professional organizations and publications all provide a rich blend of materials that explain the context in which information systems work. Local, state, and national conventions often sponsor workshops that relate to information system problems and products.

Online Research

To expand your familiarity with the enterprise under study, you can use the Internet to conduct *online research*. For example, popular search engines and directories, such as Yahoo! and Google, present several options to narrow or broaden your search. As a test, open your browser, select your favorite search engine, and type "yacht charters" as a search expression. No doubt you will discover many sites that provide insight into enterprises similar to our Silhouette Sea Charter example.

FIGURE 2-4   /   *For Example ... Memos for Help*

January 1, 2004

Memo for File:

The card system we use for recording CIS student lab hours takes too much time to key into the spreadsheet we developed two years ago. Plus, it is very inaccurate.

Is there a better way to do this?

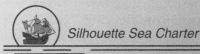

January 1, 2004

Memo for File:

The manual record-keeping system we now employ is no longer adequate to keep up with our expanding business. Our chartered ships, boat captains, and customers are too numerous to allow reliable cross-referencing for scheduling, billing, and payroll.

Let's get a computer in here to help us!

Personal Contacts

Personal contact with the individuals closest to the problem is indispensable during the investigative stage. Three useful techniques directly involve enterprise personnel in the fact-finding process: *personal interviews, questionnaires,* and *on-site observations.* Each method requires careful planning to ensure that these activities are completed within a reasonable amount of time and produce specific answers to the analyst's questions. Even though small-enterprise personnel may work under severe time constraints, they are eager to tell you about their jobs. Without a well-structured plan, the analyst may learn a lot about office politics and personalities and very little about the dynamics of the existing and proposed information system.

The initial contact should be general in nature, with a rough overview of the suspected problem. You are then responsible for the focus of subsequent meetings. As strange as this sounds, do not expect the user to be willing or able to provide all the information you need. Remember, the user's perspective about what is relevant may be much different from yours. The routine or intuitive information-handling details may seem unimportant to the user, but it may be critical to your work. After careful research on your part, you should develop written plans to guide further meetings. Such plans not only serve to structure your meetings, they also provide a tool for documenting your findings.

Documentation Review

Beyond the industry research mentioned earlier, you will need to review the existing information system documentation and procedures material. Although they might not be identified as such, every enterprise has some existing material that describes, or at least illustrates, present information system procedures. These are a prime source for the analyst. During *documentation reviews,* be sure to collect samples of source documents, processing instructions, and reports as you proceed, noting their source or point of reference in case you need to return for more specific information later. Be aware that documentation containing sensitive or proprietary information is likely to require special handling.

Looking at the Six System Components

A broader, more conceptual framework in which to investigate the system is to consider each of the six elements of a computer information system as a source of potential problems. Each of the items listed in Figure 2-3 may be related to deficiencies in one or more of the six components.

For example, the CIS Lab memo complains of inaccuracies in the system, which corresponds to the second item listed in Figure 2-3. This symptom could

TECHNOTE 2-1

FACT-FINDING INTERVIEWS, QUESTIONNAIRES, AND OBSERVATIONS

Interviewing Techniques

1. Interviews should be scheduled for a specific date, time, place, and duration.
2. All participants should receive advance notice, by phone or memo, of these particulars and of the general nature of the interview objectives.
3. Ask the user if significant differences in the level of detail of the desired information exist among different participants. If so, schedule separate interviews.
4. Prepare a list of general and specific questions you want answered during the interview.
5. Allow time for the participants to ask you questions. Remember, communication is a two-way street.
6. Document the answers to your questions and the participants' questions (see point 5), which you may have promised to answer later.

Questionnaire Techniques

1. Direct the questionnaire only to those who can answer the questions. Sometimes multiple questionnaires are required, each designed for a particular audience.
2. Keep the questionnaire short so that the respondent can complete the questions in less than 30 minutes.
3. Use questions that require "yes/no/don't know," scaled (e.g., "rate from 1 to 5"), or one-sentence answers.
4. Test the questionnaire with a small sample and make revisions as required.

On-Site Observation Techniques

1. Plan your on-site observation by listing what you need to find out during your visit.
2. Coordinate your visit with management and on-site personnel.
3. During your observation, keep note taking to a minimum. Afterward, document your findings in detail.
4. Review your findings with the user.

be caused by a poorly written student lab manual, which is part of the procedures component. Or, it might be caused by inadequate lab assistant training on how to use the spreadsheet template, which is part of the people component. The Silhouette Sea Charter memo, on the other hand, identifies systemic problems that most certainly involve a combination of the six components.

The Request for System Services

Obtaining a written statement of what the user wants is so important that it cannot be emphasized enough. If this is not forthcoming from the user, you need to initiate action to generate some document to serve as a starting point for your work. Be careful, however, that all parties agree that the statement accurately represents what is intended. Figure 2-5 illustrates a sample *request for system services* from the CIS Lab.

FIGURE 2-5   /   *Request for System Services*

Date: February 1, 2004
From: T. Foster-CIS Lab Manager
To: M. L. Barnes-Systems Analyst

Description:

At present, students record their lab time-in and time-out on 538 time cards. A separate card is maintained for each student in a lab section. During the week, lab assistants compute and record on the time card the elapsed time for each in/out entry. At the end of each week, lab assistants compute the weekly totals on each card and enter this amount into a cumulative spreadsheet program. Each month, the lab assistants generate a spreadsheet printout for each lab section, which is then posted in the lab for student and teacher reference.

The students and lab assistants make a lot of mistakes during this process. The students and teachers often complain about the long periods between the lab usage spreadsheet postings. Everyone would like to see the lab assistants devote more time to helping students and less time doing clerical work. Can you remedy these problems?

Constraints:

Cost/Budget: Only $1,250 remains in the CIS lab budget, of which
$135 is committed to new printer cartridges and $315 to repair.

Time Frame: The new system should be implemented in time to start the Spring term on April 1, 2004.

Other: The lab has an extra microcomputer system available for this system.

In this case, the lab manager is obviously well acquainted with the standard form and computer information system terminology. Therefore, the college's systems analyst would not need to help in preparing the request. In fact, after a short conversation with the manager, the analyst might simply agree to look into some inexpensive solutions. This represents an ideal situation for the analyst. A well-informed, computer-literate user is easy to work with.

The Silhouette Sea Charter situation illustrates another scenario. The Memo for File makes clear that the owner and the office manager want to make some changes to their information system. But, it is highly unlikely that they have ever dealt with a systems analyst or any of the SDLC terminology. You will need to take an active role in preparing the request form or some other equivalent document.

FEASIBILITY ANALYSIS

Before either the analysts or the user invests a lot of time and money, the analysts must determine, in a very gross or ballpark estimation, whether they can offer a practical solution to the problem. This is called a *feasibility analysis* or *preliminary investigation*. There are four generally accepted elements of feasibility analysis: operational, technical, schedule, and economic. Operational feasibility is probably the most fundamental and obvious requirement. In short, can the analysts envision possible workable designs? Technical feasibility requires the existence of necessary resources (hardware, software, network technologies) and human skill levels (analyst, programmer, and user) to develop and implement the solution. Finally, the schedule and economic feasibility elements require that the project be achievable within agreed upon time and cost constraints.

In the absence of the detailed analysis, to be discussed in Chapters 3 and 4, the feasibility study requires a great deal of courage and faith from all parties. The objective is to give the user some idea about the new system or subsystem: how it might improve the enterprise, how long it will take to implement, how much it will cost, and so forth. Sometimes this analysis is divided into the four separate feasibility categories described above. Regardless of the way the investigation is divided, be careful to emphasize that this feasibility analysis is designed to help the analyst and the user decide whether each party wants to proceed with a full-scale current system analysis.

Remember, the SDLC provides several opportunities for the analysts and user to decide to proceed or quit, to "go or no go," after reviewing the work in progress. The feasibility analysis and subsequent project contract are the first of many specific situations built into the SDLC to foster continual user and analyst interaction.

Build-or-Buy Strategies

Historically, information systems were obtained in one of two ways. They could be built as a collection of customized software, designed and developed by large staffs of corporate systems analysts and computer programmers who relied primarily on third-generation programming languages. Alternatively, complete turnkey systems could be purchased from independent software developers. This strategy freed the user from the worry of managing work in a high-tech, extremely volatile field. It also helped create a software industry that now rivals the hardware industry in terms of its influence on the marketplace.

Today's powerful fourth-generation language (4GL) products provide another option. It is hard to imagine a small-enterprise information system that would need to be built in a from-the-ground-up fashion. Indeed, most of this text provides instruction on how to adapt existing horizontal software to fit a particular need.

Of course, during the past decade independent software developers have greatly expanded the vertical software industry. Almost every small enterprise can look to trade associations, professional publications, and independent software companies to find existing solutions to their problems. The advantages and disadvantages of vertical software are detailed in Figure 2-6.

FIGURE 2-6   /   *Advantages and Disadvantages of Vertical Software*

Advantages
1. It is available immediately.
2. It has a verifiable track record.
3. It is generally tailored to the enterprise.
4. It has a fixed price.

Disadvantages
1. It cannot be easily modified.
2. The user must rely on long-distance assistance.
3. It may not address all the user's problems.
4. It may include features the user doesn't need.

The *build-or-buy* decision is not an all-or-nothing proposition. The eventual mix of vertical, horizontal, and custom software will certainly have an impact on the work of the analyst. Each of these puzzle parts must be integrated with the existing procedures to produce a seamless information system.

Returning to the two samples for a moment, you can argue for two different build-or-buy recommendations. Our investigation of the CIS Lab student-timekeeping system reveals that we were mistaken in our earlier speculation about problems with the lab instructions or the lab assistant training. In fact, we find that the only way to remedy the problem is to automate the process. Common sense tells us that the budget constraint for the CIS Lab project precludes almost any custom-built solution. The analyst therefore is motivated to find an existing vertical software product that fits the budget and solves as many of the problems as possible. Succeeding chapters demonstrate that this is indeed what happens.

On the other hand, we were correct about the broad, systemwide scope of Silhouette Sea Charter's problems. Fortunately, it can operate under a much more generous budget and a slightly longer time frame for implementation. This allows us to entertain the possibility of building its information system from several horizontal software products.

Cost and Delivery Parameters

As discussed in Chapter 1, the small enterprise is particularly sensitive to economic constraints. Minor cost overruns and project completion delays are magnified in a small enterprise, where operating margins are applied to small

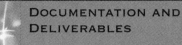

TECHNOTE 2-2

DOCUMENTATION AND DELIVERABLES

Throughout the SDLC, the analyst must systematically collect, organize, and store numerous project documents. This collection is called **project documentation**. Some of these materials are saved for later reference and some are delivered to the user. While Chapter 14 covers project documentation in some detail, this TechNote describes the steps required to begin the assembly process and identifies the project contract as the first project deliverable.

In order to document the activities associated with the preliminary investigation and feasibility study of the SDLC the analyst must:

1. Include the Request for Services to provide an initial definition of the problem.

2. Prepare a summary of the fact-finding efforts, including industry research, interviews, and any documentation the user has of the existing system. The detailed reference documents should also be included or at least identified.

3. Prepare a summary of any research concerning the mixture of available software solutions.

4. Assemble the detailed calculations and justifications that support the recommendations and measurable goals, including baseline measurements, that appear in the project contract.

5. Include the project contract to provide the initial parameters that guide the work to come.

The project contract is the end result of the work described in the documentation discussed above. It is called a **project deliverable** because it is a tangible product, delivered to the user. The other deliverables produced during the SDLC are described in conjunction with the Cornucopia Case and portfolio project.

numbers. A $1,000 mistake is much worse on a $10,000 project than on a $100,000 project.

The user must supply the initial guidelines for both cost and product delivery dates. The analyst, in turn, must evaluate these constraints and offer reasonable alternatives if they are unrealistic. Together, the user and the analyst can formulate the first set of *cost and delivery parameters*.

The Feasibility Report and Project Contract

Throughout the project the analyst generates numerous products, such as those described next. These products are called *project deliverables*. Ultimately, the deliverables are assembled into an organized collection of materials called *project documentation*.

At the completion of the preliminary investigation, the analyst should prepare a report on the project. In most cases, the report is really a collection of documents that details the initial problem definition, the findings of the feasibility analysis, and the analyst's preliminary recommendations.

The *feasibility report* serves as a common point of reference for formal or informal negotiations between the analyst and the client. The resulting *project contract* should establish a clear agreement between the user and the analyst about the work to be done. This is not to say that modifications can't occur along the way—it merely sets a benchmark from which to start. This agreement must include those items listed in Figure 2-7.

FIGURE 2-7   *The Initial Project Contract*

1. **Problem Summary**
   State or list the problems to be addressed.
2. **Scope**
   Define the boundaries of the work to be performed.
3. **Constraints**
   Detail the time frame and cost targets for the project.
4. **Objectives**
   Identify measureable goals of the new system.

The *problem summary* should be a concise statement that includes both the user's original request and the analyst's preliminary findings. As discussed earlier, the analyst may need to sift through a variety of symptoms in order to identify the root cause of the problem. The user must understand such new perspectives and agree to your interpretation.

It is very important that you and the user agree on the boundaries of your work, or *project scope*. For example, the user who wants a computerized customer billing system must be advised of the implications that such a project may have on the existing data capture, master file maintenance, and overall accounting procedures of the enterprise. Agreeing beforehand on the areas that will change and the areas that will be left alone is much better than reworking the project contract later on.

Cost and delivery date are not the only *project constraints* of concern. How much access will the analyst have to current system resources such as personnel, machines, and data files? What are the confidentiality restrictions, if any? What are the tolerance levels for errors in the system? How sophisticated are the intended users of the system? Will the system operate under any unusual physical or environmental conditions? These examples should alert the analyst to look for any circumstance that may place special demands on the design.

Finally, the *project objectives* must be stated in measurable terms so that they are clearly understood by both the analyst and the user. To say that the new system "will improve" present performance is not enough. You must quantify such claims in order to avoid misunderstanding about the worth of your product. In addition, such measures will provide critical feedback during the maintenance and review portions of the SDLC. In order to evaluate product performance, the analyst must establish *baseline measurements* of the existing system if they do not already exist.

THE CORNUCOPIA CASE

Each step of the SDLC is illustrated by the Cornucopia Case, which appears in a series at the end of the remaining chapters. Based on an actual small-enterprise case, the series reinforces the concepts presented in each chapter and serves as a model for your own parallel case study.

Background

Cornucopia is a small music store located in the "Old Town" section of the business district. Its marketing signature is the personal attention it provides its patrons. Coupled with an extensive classical musical inventory of compact discs, cassettes, records, and videos is a modest collection of books on classical music. The customers can avail themselves of a wealth of knowledge about the composers, performing artists, and particular recordings by engaging either the owner or one of the two sales clerks, who seem to always make time to discuss what is of obvious interest to them. All this commerce and conversation takes place amid a few potted plants, beautiful classical background music, and complimentary apple juice or wine.

The last thing the enterprise needs is an expensive, impersonal, complicated computer information system. Yet, the following letter describes some problems that the owner would like to solve (Figure 2-8).

Notice that the owner did not prepare a "request for system services." This is not unusual. Small-enterprise users are not bound by bureaucratic procedures. Indeed, they relish their independence. The analysts will simply work within this context by conducting a careful preliminary investigation and preparing a good project contract.

After two meetings with the owner, during which the analysts were able to observe the business operation firsthand, they confirm that the owner's original letter is accurate in its description of the present system. They also determine that the owner is indeed serious about this project and is willing to invest in a modest feasibility study. The cost of the study will either be absorbed into the project contract or priced not to exceed $200 if the project is aborted.

Feasibility Analysis

The first challenge to the analysts is to determine, in the most general sense, whether they can offer a solution to the explicit or implicit problems the owner has identified. (*Note:* At this point, none of the information needs is unusual or poses particularly difficult technical problems. These conditions are by design, so that you can concentrate on the teaching points of the SDLC. Other complications may arise in the future, but for now, it is assumed that a popular desktop computer platform, along with several off-the-shelf software packages, can provide the raw resources for the solution.)

FIGURE 2-8 / *Cornucopia Initial Problem Statement*

   "Without music, life would be a mistake."

January 1, 2004

M & M Computer Consultants
1130 Avenue of the Giants
Myers Flat, CA 95554

Dear Sir or Madam:

You were referred to me by one of your former clients, who recommended that my business could benefit from a computer information system. Let me tell you a little about how we operate and what we have in mind.

To begin with, we would like to improve our customer record-keeping procedures. All transactions involve cash, personal check, or credit card. Thus, this procedure does not involve accounts receivable. However, we do keep a partial list of customer names and addresses on an old computer, which is located at my home. This information is taken from personal checks or at the request of customers who wish to be placed on the mailing list. It is used to create mailing labels for periodic promotional material.

Second, we would like to improve our reordering system. At present it relies entirely on the observation of the employees. When they notice a particular shortage, they make a note of it on a spiral pad next to the cash register. Periodically, I place orders, via telephone, to the appropriate wholesaler and note the date of the order on the pad. Special orders for disks, cassettes, videos, and books follow the same procedure, except that they are recorded on separate, color-coded sheets of paper. As shipments are received, these notes are scratched off.

Third, we would like to develop a quarterly newsletter for our customers. This would contain information on subjects of interest to our customers, as well as provide advertising for the enterprise.

Fourth, we would like to have information about the sales trends of the business. For example, what products generate the most sales, the most profits, etc.

I hope that you can help us. Please call me at 445-9876 so that we can arrange a meeting to discuss this project in more detail.

Sincerely

Margaret Height

A combination of word-processing, spreadsheet, and database software can be tailored to meet all the information needs itemized so far. A rough budget of $10,000 includes $4,000 for hardware purchases, $2,000 for software purchases, and \$4,000 for labor (80 hours at \$50 per hour). This budget reflects a major premise of this text, that is, desktop computer-based information systems, relying on well-integrated horizontal software, are well within the economic reach of the small enterprise. It also provides a point of reference for future comparison, as a detailed budget is prepared and actual expenses begin to accumulate.

The next challenge to the analysts is to determine the broad constraints of the project. How much time and money is the owner willing to devote to the project? Are those constraints realistic? Is the owner flexible about considering modifications in either the constraints or the project objectives? The analysts' experience will generally guide these negotiations, unless the owner has predetermined, rigid budgets and time frames in mind. The analyst must remember that despite their enthusiasm about systems work, it is in everyone's interest to discontinue the project if all parties cannot agree on what is feasible and practical.

The Cornucopia owner agrees to the analysts' rough budget estimate on two conditions: the project must be completed within four months, and the system must return financial rewards equal to or exceeding the cost within two or three years. The analysts can agree on these terms for the following reasons.

First, the estimate of 80 total labor hours assumes that those hours will be spread out over many weeks, with many interruptions and other small projects intervening from time to time. Four months provides adequate flexibility on this point. Second, the preliminary estimate is that the new system will not add a significant amount of work to the current staff duties. Further, the analysts expect that improved customer communications, combined with fewer lost sales due to out-of-stock problems, will increase sales. A modest estimate of only three additional sales per day, at a net profit of \$5 per sale, generates \$3,900 per year (260 days $\times$ \$15). At this rate, a \$10,000 system is paid for in a little more than two years.

The Feasibility Report

It is essential that all this analysis be documented. The *feasibility report* is the term used to describe this collection of notes, samples, and documents. It can be very formal, or it can be nothing more than a bulging manila folder. For Cornucopia, the analysts choose to prepare a formal report for submission to the owner. This choice accomplishes two things: it reassures the owner that the project contract to follow is well supported by careful analysis; also, it adds an important element of discipline by forcing the analysts to organize, summarize, and analyze the findings, even for projects that never make it beyond this stage. Figure 2-9 provides a snapshot of this report.

Project Contract

The analysts can now develop a written statement that defines the problem and the constraints as agreed to by the analysts and the owner. Although there is no formal request for system services in this case, the original letter from Cornucopia

FIGURE 2-9 / *Cornucopia Feasibility Report*

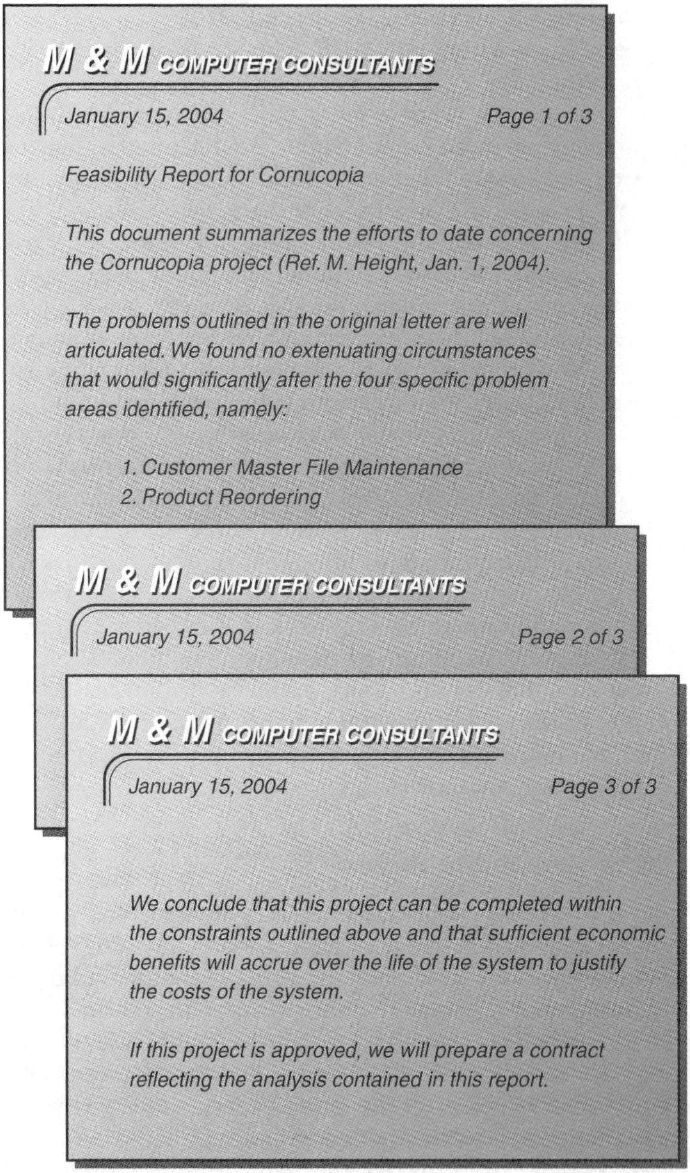

M & M *COMPUTER CONSULTANTS*

January 15, 2004   *Page 1 of 3*

Feasibility Report for Cornucopia

This document summarizes the efforts to date concerning the Cornucopia project (Ref. M. Height, Jan. 1, 2004).

The problems outlined in the original letter are well articulated. We found no extenuating circumstances that would significantly after the four specific problem areas identified, namely:

> *1. Customer Master File Maintenance*
> *2. Product Reordering*

M & M *COMPUTER CONSULTANTS*

January 15, 2004   *Page 2 of 3*

M & M *COMPUTER CONSULTANTS*

January 15, 2004   *Page 3 of 3*

We conclude that this project can be completed within the constraints outlined above and that sufficient economic benefits will accrue over the life of the system to justify the costs of the system.

If this project is approved, we will prepare a contract reflecting the analysis contained in this report.

and the feasibility report provide ample information with which to construct the project contract (Figure 2-10). Notice that some items in this contract are new to the discussion—a brief explanation follows, with more explanation in later chapters.

FIGURE 2-10   / *Cornucopia Information System Project Contract*

Problem Summary:

Cornucopia is a small music store specializing in classical records, cassette tapes, compact discs, and videos. They want to improve their computer information system in four areas:

1. Customer record keeping
2. Product reordering
3. Customer communications
4. Sales trend analysis

Project Scope:

The new information system will include a computerized point-of-sale and inventory subsystem. The point-of-sale portion will include customer account information to facilitate the correspondence segment of the information system.

The system will include easy access to the Internet and a Web page designed specifically for Cornucopia.

Project Constraints:

Cost: The full price of the new system will not exceed $10,000. The product will be under warranty for 30 days from the point of implementation. Thereafter, the ongoing maintenance costs will not exceed 1% per month of the original system costs.

Delivery Date: The new system will be fully operational within four months of the date of this contract.

Other: The cost of the first 12 hours of personnel training is included in the total cost of the system. Additional training time will be billed at the rate of $50 per hour. Cornucopia will bear the cost of the initial customer and CD master file creation.

Objectives:

M & M Computer Consultants will deliver a computer information system that provides users with detailed procedures on how to apply hardware and software resources that address the information needs itemized above. In addition, the new system will be designed to:

1. . . . *not* increase the time it takes to complete a normal sale.
2. . . . *add no more than* five hours per week to the time required to perform customer and CD master file maintenance.
3. . . . *reduce* compact disc reordering time by at least 50%.
4. . . . *increase* repeat customer sales by 5% by the end of the first year of operation.
5. . . . *reduce* "out-of-stock" and "over-stock" situations by 50%, over a two year period of time.

Although not specifically mentioned by the client, an automated point-of-sale and inventory system is an obvious solution to the reordering problem and may also provide useful information for a customer correspondence subsystem. In addition, an Internet presence will certainly facilitate customer communication and may enhance sales.

Information system maintenance and review is a concern to both the user and the analysts. The user does not want to be without some protection against system malfunction. Nor does any user want to go to the Yellow Pages and start all over with a new analyst every time the system might need a slight improvement. At the same time, the analysts want to develop a good reputation. As any successful entrepreneur will confirm, the after-sale service is just as important as the sale itself. Therefore, the analysts agree to provide regular maintenance service beyond the standard 30-day product warranty period. Extraordinary services, such as a major system upgrade or the initiation of a new SDLC, would require a new contract.

User training is discussed in detail in Chapter 14, but the user manuals and system documentation should provide adequate reference beyond that time. Naturally, some ongoing training and user assistance will occur during the regular monthly maintenance visits.

Finally, the objectives section of the contract itemizes several measurable goals while promising to deliver a system that addresses all four areas itemized in the problem summary. The specificity of these goals may seem highly speculative. However, they are based on a combination of the analysts' experience, observation of the enterprise operations, and discussions with the owner. As such, they represent an honest attempt to quantify the performance standards for the project.

Cornucopia with Visible Analyst

System projects generate a multitude of documents that are useful because they describe each phase of the SDLC and each of the five system components. You have already encountered three of these documents for Cornucopia: the request for services, the feasibility report, and the project contract.

Ideally, the analysts begin recording and tracking the project within the CASE environment as soon as the project begins. In this example, select the New Project option in the File menu and follow the directions (Figure 2-11). Visible Analyst stores information about the project in a database. The Btrieve database engine is shipped with Visible Analyst, but you may also use others.

Under Rules and ERD Notation you are required to select from several standard options. This illustrates an important point: there are alternative symbol and rule sets that apply to the various models you will learn about. The best advice is to be consistent. Your instructor will advise you about which standard to follow.

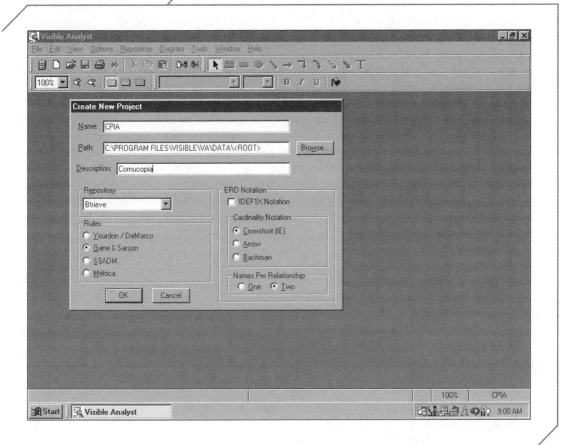

SUMMARY

The analyst's first task is to define the information system problems to be addressed. One way to do this is simply to ask the user to prepare a request for system services. If the user cannot answer your questions in sufficient detail, the analyst can look for symptoms that point to the problem areas. In addition, the analyst's own industry research, personal contacts, and documentation review should provide information about the situation.

Once the problem is defined, a rough idea of what the solution will look like must be developed so that the user and analyst can agree on whether to proceed. At this point, the feasibility analysis is designed to protect both parties from making unrealistic commitments of time and money. From this exercise comes the initial project contract, which serves as a guide for the project through the analysis phase of the SDLC.

TEST YOURSELF

Chapter Learning Objectives

1. What are the basic information-processing requirements?

2. Which common information system problems are associated with small enterprises?

3. How does the analyst discover how the existing information system works?

4. What are the key elements of a feasibility report?

5. What are some resources available to help you diagnose a system problem?

Chapter Content

1. Match the number of the basic information-processing requirement that is not met to each of the situations below:
   1. Relevancy: Information products must serve some useful purpose.
   2. Accuracy: Error rates must be held below an error threshold.
   3. Timeliness: Information products must be available within an acceptable time frame.
   4. Usability: Information must be presented to users in an understandable format.
   5. Affordability: The information system benefits must exceed information system costs.
   6. Adaptability: Information systems should be modifiable to meet changing information processing needs.
   7. Accessibility: Users must be able to interact with the information system with minimal inconvenience.
   ____ a. Reports arrive too late to help managers schedule the next day's work assignments.
   ____ b. Field representatives cannot connect to the customer database from remote locations.
   ____ c. Users regularly confuse detail and summary statistics that appear on a report.
   ____ d. A report of customer demographics is never accessed.
   ____ e. Customer e-mail addresses are habitually incomplete or misspelled.
   ____ f. Software upgrades cannot be loaded because of hardware upgrade limitations.
   ____ g. Sales representatives spend too much time keyboarding customer information.

2. What is the first SDLC question addressed by the analyst?
   a. What are the project constraints?
   b. Who are the intended users of the system?
   c. What is the problem with the current information system?
   d. What type of computer hardware is required?

3. The analyst relies upon _____ participation throughout the SDLC.
   a. owner
   b. user
   c. programmer
   d. all of the above

4. An example of a "source data input inefficiency" is:
   a. Source data may be lost, duplicated, or incomplete.
   b. Data capture procedures may be slow or unreliable.
   c. Automated data capture devices may be incompatible with data processing devices.
   d. all of the above

5. An example of a "breakdown in subsystem integration" is:
   a. Customer mailing addresses are maintained in a duplicate file because customer database information is unavailable to the correspondence subsystem.
   b. Daily sales totals must be reentered into a revenue projection spreadsheet because the sales transaction processing system does not provide summary totals.
   c. An advisor must maintain a separate database of students' course work because he or she does not have access to the transcript subsystem.
   d. all of the above

6. The analyst should be somewhat familiar with the activities of an enterprise before conducting an on-site fact-finding effort. True or False?

7. Online industry research is a sure way to learn about the information needs of an enterprise. True or False?

8. To save time and improve the usefulness of your findings, you should carefully plan your personal interviews, questionnaires, and on-site observations. True or False?

9. When reviewing the documentation of the existing information system, the analysts should look for samples of source documents, processing instructions, and reports. True or False?

10. During a feasibility analysis, the analyst must determine whether he or she can deliver a practical solution to the problem. True or False?

11. The SDLC provides only one opportunity, the feasibility analysis, for the analyst and user to decide whether to proceed with the project or abandon the project. True or False?

12. One possible benefit of performing industry research is that the analyst may discover a preexisting, purchasable solution to the information system needs of an enterprise. True or False?

13. Cost and delivery parameters are established jointly by the analyst and user. True or False?

14. The feasibility report follows a government-approved, standardized format to present a summary of the initial problem analysis and definition plus the analyst's preliminary recommendations. True or False?

15. The feasibility report helps the analyst and client establish a clear agreement about the work to be done. True or False?

ACTIVITIES

Activity 1

Review the Memo for File describing the CIS Lab's information system problem (Figure 2-4). Prepare an answer to the following questions:

1. Which basic information system requirements are not being accommodated by the present card system?

2. Is there anything about these problems that is peculiar to a college or small-enterprise environment?

3. How does your college computer lab keep track of student lab hours?

Activity 2

After reviewing the Memo for File describing Silhouette Sea Charter's problem with its manual record-keeping system (Figure 2-4), prepare a Request for Services for submission to a systems analyst.

Activity 3

Based on your own industry research, personal contacts, documentation review, and Internet research, develop a one-page background description of a small enterprise of your choice.

Activity 4

Using software catalogs, trade magazines, industry-specific professional literature, or the Internet, investigate one type of vertical software. Prepare a one-page analysis of the product or products.

DISCUSSION QUESTIONS

1. Describe a personal experience in which one or more of the basic information-processing requirements have been violated.

2. Using one of the basic information-processing requirements, describe how a small enterprise is particularly sensitive to computer information system performance.

3. Describe the relationship between a problem symptom and the problem itself.

4. Based on personal observation, describe one symptom of a problem with a small-enterprise information system and speculate on the underlying cause of the problem.

5. Describe how well the syllabus to this course serves as a "contract" between the teacher and student.

6. If risk is defined as the potential for financial loss, evaluate how the following risks might affect the analyst and the client:
   a. The project might not be completed within the time constraints agreed to in the contract.
   b. The project might cost more than expected.

   c. The information system might not meet the functional specifications.
   d. The enterprise employees or its customers—or both—might be very unhappy with the new procedures.

7. Log on to the Internet, select your favorite search engine, and type "yacht charter." How many sites did the search engine locate? Visit two or three sites. How might this experience help inform the systems analyst working in the Silhouette Sea Charter example?

8. Given your inexperience in systems analysis work, how would you respond to a friend or client who asks, "How long will it take you to set up my information system?"

9. Why should the analyst spend time studying the existing information system when he or she intends to replace it with a new system?

10. What should the analyst do if there is no preexisting vertical software or turnkey system for the small enterprise under study?

PortfolioProject

Team Assignment 2: Project Initiation

Read the project packet distributed to you. Your project packet may come from Appendix B or from another source. As part of the "fact-finding and diagnosis" activity, you will no doubt have many questions about your client's enterprise and the information system request. In order to formally initiate the project, you must complete the following tasks:

1. Prepare an initial response to your client (use your letterhead stationery) to acknowledge receipt of the project request and your interest in working on the project. In your letter, be sure to include a restatement of the client's problem, your project milestone timetable, and your firm's billing policy. Submit a copy of your letter.
2. Prepare a set of initial questions regarding the systems, data, procedures, and so forth, related to the project. These questions should be forwarded with a cover letter to the client as soon as possible. In most cases your instructor will act as a surrogate client. You should receive a reply (oral, written, e-mail, etc.) to your questions in a timely fashion. For the first few weeks of the project you should continue with the "Q & A" correspondence until you have a firm grasp of what your user wants. Submit a copy of your initial questions.
3. File a copy of these two letters (response and initial questions) in the Portfolio Project Binder behind the tab labeled "Assignment 2."

Project Deliverable: Project Contract

Although you may feel it is premature to agree to anything in writing at this early stage in your portfolio project, you must make an honest attempt to outline the broad parameters of your work. This exercise provides a good benchmark for you to measure your successes and failures as you proceed through the project. The intent is to build your experience so that you can improve with each project you attempt in the future. In order to establish a clear agreement between you and your client, you must complete the following tasks:

1. Prepare a project contract with the following content:
    a. Problem summary
    b. Project scope (boundaries of the information system)
    c. Project constraints (cost, delivery date)
    d. Objectives (measurable goals)
2. Prepare a cover letter to transmit the contract to your client. Submit a copy of your cover letter and contract.
3. File a copy of your cover letter and contract in the Portfolio Project Binder behind the tab labeled "Project Contract."

Analysis

process**modeling**

When you complete this chapter, you will be able to:

OVERVIEW

The problem definition developed in Chapter 2 focuses on the areas within the information system that should be improved or expanded upon. Before beginning to solve these problems, the analyst must fully understand the current information system. Why, you might ask, should I learn about a system I intend to replace? First, you may find that the current system can be modified rather than replaced. Second, your study of the current system will further clarify the problems and needs of the user, while almost certainly suggesting design elements that you can incorporate into your solutions.

Many modeling tools are available that can help describe how the six components of an information system work together. We begin with the data flow diagram (DFD) because it corresponds so well with the way systems operate in the real world. This should make it easier for the analyst to create the series of abstractions necessary to complete the SDLC tasks that lie ahead. *Data flow diagrams (DFDs)* are used to model the people, procedures, and data components of the CIS.

This chapter explains how to develop a series of DFDs, beginning with the context diagram and ending with low-level, task-specific diagrams. You will learn that this is an iterative process, as you continually decompose your models to show more detail.

- Describe why the data flow diagram is called a process model.

- Recognize and describe an abstraction.

- Identify the four elements of a data flow diagram.

- Construct a data flow diagram.

- Decompose a data flow diagram.

MODELING THE SYSTEM

Before we launch into the specifics of data flow diagrams, we need to touch on the *modeling methodologies* presented in this text. Each has a different focus and serves a different purpose, but all of them require the analyst to create an abstraction of the information system.

Process Models

The data flow diagram (DFD) is called a *process model* because it directs the reader's focus to those processes in the system that transform data into information. Also, in depicting the flow of data into, out of, and between processes, the data flow diagram shows many of the important relationships between the data/information, procedures, and people components of the system. Figure 3-1 illustrates this point.

FIGURE 3-1   / *DFD and the Six CIS Components*

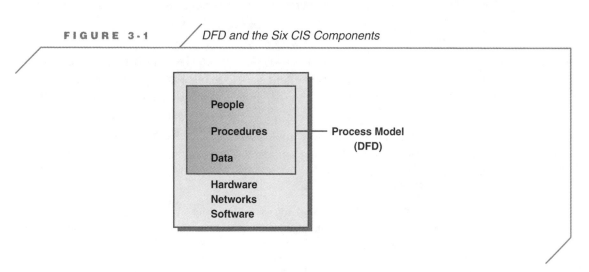

Data Models

As we shall see, data flow diagrams provide for data holding areas that can be accessed via a process. In keeping with their abstract nature, these file storage areas are referred to only by a description of their content or purpose. As such, the DFD does not provide sufficient detail about file characteristics or file relationships to satisfy all the analyst's needs. *Data models*, such as the *entity-relationship diagram (ERD)* described in Chapter 4, provide this detail.

Object Models

Object modeling is a methodology in which the focus is on enterprise objects rather than processes or data. The key to understanding this model is in the definition of an *enterprise object*. Briefly, an enterprise object is an entity defined

ABSTRACTIONS AND COMPUTER INFORMATION SYSTEMS

In the context of computer information systems, an **abstraction** is defined as a simplified substitute for the real world. Your college transcript is such a substitute. It is supposed to reflect your success in the classroom. Of course, we all accept the fact that the transcript alone does not provide sufficient detail to really judge your college career. But we commonly use the abstraction as a means of introducing the subject of your college "what, where, and when" particulars.

The models described in this text are also abstractions. They remove some of the detail that we might include when fully describing the information system. This simplification makes it easier to focus on the different components of the system. In other words, we are forced to conceptualize the component, which helps us to identify the crucial elements *and* their relationship to the overall system.

To illustrate the value of abstraction, imagine you asked your mechanic why your car makes a funny noise. A typical response might be, "That noise is caused by a metal guide on the brake mechanism that rubs against the wheel-rim casing when the brake pad wears down to the replacement point." For most of us, this information is too detailed for us to understand. What we need is a few words that will describe the situation in a nutshell. In this case, translating the mechanic's response to "brake check" abstracts the essential data manipulation activities, thereby improving our ability to understand the situation.

A computer information system is an abstraction in many ways. Users often complain of the abstract nature of the interfaces, the processing features, or the reporting formats. They want products that are user-friendly. Closing the gap between the user's real world and the abstract world of the information system is a challenge to any analyst. Perhaps your experience with the abstract modeling processes presented in this text will help you appreciate this point even further.

by its data characteristics (data) and the functions (processes) it can perform. Thus, the data and process components of an information system are married together into enterprise objects. The object model is of particular interest because of the object-oriented nature of the software products used to construct and implement a small-enterprise information system. Chapter 5 presents a detailed discussion on this model.

System Models

You may have noticed that neither the process model nor the data model includes explicit reference to the hardware, network, or software components of the system. This is not a shortcoming of those models, because they serve other purposes. Various *system models* are designed to document these remaining three components as well as to portray the system in a manner that users can easily understand. The importance of system models in fostering effective and well-informed user participation in all phases of the SDLC cannot be overemphasized. Chapter 6 discusses this point in great detail.

THE DATA FLOW DIAGRAM

The first encounter with an information system can be overwhelming: the analyst is presented with an assortment of source documents, forms, reports, and anecdotal impressions, none of which seems related. In the broadest sense, you may already know what the system is supposed to do, but now you must develop an orderly, systematic picture, or model, of how everything fits together. The DFD presents a picture of what the people and procedures do to transform data into information. The DFD is called a process model because of its focus on these transformations. Thus, data is said to flow through the system in a way that can be illustrated by a popular framework called *input-process-output (IPO)*.

The image in Figure 3-2 suggests that computer information systems are in some ways like a factory. They take raw materials (input data) and transform them (process) into useful products (output information). The DFD methodology provides a formal method to record the analyst's understanding of the numerous inputs, processes, and outputs of the system under study.

FIGURE 3-2   *Input-Processing-Output*

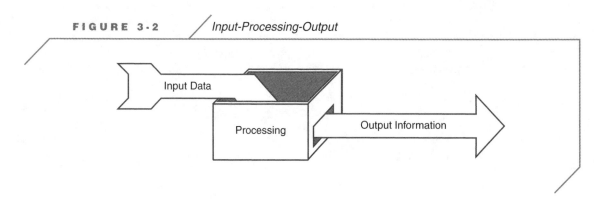

Standard Notation

Data flow diagrams use four distinct symbols, illustrated in Figure 3-3. An *external entity* is either the originator of data or the receiver of information. These data *sources* and *sinks*, as they are also called, are external to the system under study and are not part of the data transformation being modeled. The tendency, perhaps, is to think of external entities as people, such as a customer who orders a product or the manager who evaluates the weekly sales report. But an organization, or even another information system, can also be described as an external entity, as long as it serves as an originator of data (a *source*) or the receiver of information (a *sink*).

A *process* is a series of steps that manipulates data. Collecting, sorting, selecting, summarizing, analyzing, and reporting are common operations associated with data manipulations. The description of the process often reflects the level of abstraction in the diagram. For example, process number one, in Figure 3-3, is labeled Intake, which is very general or abstract. A process labeled Verify Account or Validate Signature is much more detailed, but less abstract. Notice that each of these labels avoids suggesting *how* the process is performed. Rather, they simply identify *what* is to be performed.

FIGURE 3-3   */Illustrated DFD Symbols*

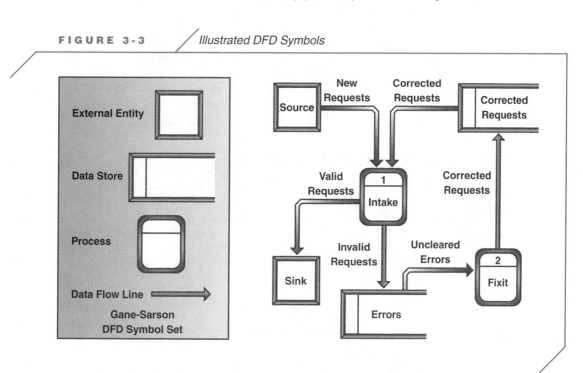

A *data store* is a place to keep data for later reference. The names associated with data stores are also abstractions. They do not describe the detailed characteristics of the store, such as how it is organized or accessed. That information is itemized in other models.

A *data flow* is the means by which data and information are transported from one place to another. The flows are like a delivery service, moving data from an external entity to a process, between processes, or from a process to a data store. Sometimes data flows are called pathways because of the connecting lines that represent the flows on the diagram. You must remember, however, that this, too, is an abstraction. There are no lines painted on the floor between an external entity and a process. The descriptive annotations that appear above or below the flow attempt to identify the data or information that is transported. As you will see, some of these labels are very detailed, others are very general.

To demonstrate the usefulness of this model during the analysis phase, the data flow diagram in Figure 3-3 is intentionally very abstract. Looking at this diagram, the reader would be hard-pressed to determine much about the nature of the enterprise or whether the system was implemented with computers, paper documents, or wireless data communications equipment. But this does not prevent our understanding that two distinct operations are at work in the system: an intake process and a "fixit" process. The first obviously evaluates requests and dispatches them to either an error store or an external entity. The second is responsible for correcting request errors and routing them back into the incoming data flow. Knowing merely this, the analyst can begin to look for possible improvements in the system. For example, if the user reports frequent discovery of duplicate requests processed through to the sink, you might examine the fixit process to see whether prolonged delays occur in the error correction procedures. Such delays might cause impatient clients (sources) to initiate another request, thus introducing a duplicate into the system.

Standard Rules for Construction

To make the diagram easier to read, data flow lines should not cross one another; therefore, you may be forced to show the same data store or external entity in more than one place on the diagram. When this occurs, the symbol should include a shaded lower-left corner to alert the reader to search for other similar entries on the diagram. As a further precaution, using the following rules helps reduce the possibility of adding unnecessary elements or omitting necessary ones:

* Data flows should not connect an external entity directly to a data store or to another external entity—there must be an intervening process.
* Data flows should not connect a data store directly to another data store. Again, there must be an intervening process.
* Data stores should have at least one entry data flow and one exit data flow—and can connect only to a process. A data store with only an entry data flow is probably an archive or backup file, which does not need to be included in the diagram.
* Each process must have at least one entry data flow and at least one exit data flow.

Identifying External Entities, Processes, and Data Stores

Real-world information systems do not naturally present themselves in these tidy categories and diagrams. The analyst must digest a great deal of system documentation, as well as notes from personal contacts and observations, to create an accurate set of DFDs. The following techniques will help.

Begin the process by working with the various system descriptions you have gathered. To get you started, Figure 3-4 presents a *"bottom-up"* DFD creation approach, in which you assemble small details into larger subsystems.

FIGURE 3-4   *Bottom-Up DFD Creation*

1. Develop a narrative of the system.
2. Underline the action words.
3. Develop a sequential task list of action words.
4. Eliminate tasks that do not transform data.
5. Identify cohesive tasks.
6. Fit all remaining tasks to a cohesive task.
7. Develop an IPO chart for each cohesive task.

To illustrate the bottom-up method, we return to the Silhouette Sea Charter example introduced in Chapter 2. Through the fact-finding work on this project, we have developed the following narrative of the steps involved in scheduling a charter. Notice that this is only a portion of the system project alluded to in the user's original memo.

Scheduling begins when the customer <u>submits</u> a charter request to the office manager. The office manager needs to know how many passengers are expected, when the customer wants to begin and end the charter, and where the customer wants to go. For each charter, Silhouette provides a boat captain and a vessel. It does not provide passenger food and beverages.

Sometimes the office manager cannot determine whether the request is reasonable. For example, a request to transport four people from San Francisco to Seattle in four days during the dead of winter may not be a good idea. When this happens, the request <u>is forwarded</u> to the owner, who makes this determination, <u>contacts</u> the customer if necessary, and either <u>adjusts</u> the request or <u>rejects</u> it outright. All requests <u>are returned</u> to the office manager for scheduling and filing.

The office manager <u>consults</u> the ship register and availability sheets, which show the dates when the ship's owner is willing to rent the vessel as well as any dates that have already been reserved for other charters. The first available ship that meets the requirements of the charter in terms of ship size, class, rental fee, and so forth <u>is reserved</u>. Finally, the office manager <u>checks</u> the boat captain log sheets to <u>find</u> someone who is qualified and available to take the charter.

As each piece of this puzzle is fitted to another piece, the particulars <u>are entered</u> on a charter specification sheet. When this sheet is complete, the owner <u>contacts</u> the customer to <u>finalize</u> the terms of the charter, after which the office manager <u>prepares</u> a charter contract for <u>signature</u> by the owner and the customer.

Following the steps in Figure 3-4, we first emphasized all the action words in the narrative that transformed or transported data. Next, we develop the following sequential task list (Figure 3-5), grouped by cohesive tasks. A *cohesive task* is defined as a group of activities designed to perform a specific function. Data transporting activities are noted, but not included, in a task.

FIGURE 3-5   Silhouette's Bottom-Up DFD Worksheet

Action Word	Task Number	Task Description
submits	none	not identified
is forwarded	none	not identified
contacts	#1	evaluate special requests
adjusts	#1	evaluate special requests
rejects	#1	evaluate special requests
are returned	none	not identified
consults	#2	schedule charter
is reserved	#2	schedule charter
checks	#2	schedule charter
find	#2	schedule charter
are entered	#2	schedule charter
contacts	#3	complete charter papers
finalize	#3	complete charter papers
prepares	#3	complete charter papers
signature	#3	complete charter papers

At this point, we can consider each cohesive task to be a separate process. The input and output of the cohesive tasks is a data flow, associated with an external entity, a data store, or another process. The best way to see this is to develop an input-process-output (IPO) chart for each task. Our first attempt at the Task #1—Evaluate Special Requests IPO chart is below (Figure 3-6).

FIGURE 3-6   Silhouette's Task #1—Evaluate Special Requests IPO Chart

Input	Processing	Output
request	contacts	notation
	adjusts	
	rejects	

You should notice that the "contacts" activity does not transform or transport data; therefore, it should be eliminated from the processing list. The other two activities are likely to require the owner to make some transforming notation on the request. Figure 3-7 illustrates the segment of Silhouette's DFD that corresponds to this discussion. Notice that the input and output are identified as data flows, with the input request flowing from a process and the output notation flowing to a data store. This identifies two more elements to Silhouette's DFD, namely "charter intake" and "charter requests," thus illustrating how this model is built piece by piece, with lots of review and revision required before a clear picture of the information system emerges.

FIGURE 3-7   /   *Silhouette Sea Charter's Partial DFD Sketch*

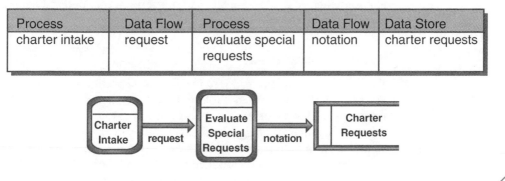

Process	Data Flow	Process	Data Flow	Data Store
charter intake	request	evaluate special requests	notation	charter requests

If, or when, you get bogged down with this approach, consider reversing the process by using a *top-down DFD* creation approach that first identifies the major processes and then breaks them down into subprocesses. Figure 3-8 lists these steps.

FIGURE 3-8   /   *Top-Down DFD Creation*

1. Solicit oral answers to the question "What is the first task of this system?"
2. Continue with "What happens next?"
3. Repeat Step 2 until the response is "We do it all over again," or "We are finished."

Surprisingly, as you record the answers to the questions in Figure 3-8, you will see that they correspond to the cohesive tasks identified in the bottom-up approach. This is reassuring and provides further insight into the way the system works. In practice, you should move freely between the two methods. To complete this example, refer to Figure 3-3. Can you see how that DFD could be modified to illustrate this portion of Silhouette's existing system?

THE CONTEXT DIAGRAM

The data flow diagrams as described are usually preceded by even more abstract diagrams. The *context diagram* shows the entire system as a single process, connected only to the external entities. As the broadest picture of the system, it serves to set the system boundaries and establish those parts of the overall system for which you are responsible. In this respect, the context diagram is closely associated with the scope section of the project contract.

System Boundaries

The CIS Lab example introduced in Chapter 2 fully illustrates the DFD process. You will recall that we were looking for an inexpensive vertical software solution to the lab's problem. Fortunately, many other lab facilities have similar timekeeping needs. This prompted an enterprising programmer to write a small program called, appropriately, the "Time Keeper." The lab manager came across this product at a regional conference and decided to give it a try. Priced at $250, this was not a great risk and has actually turned out quite well. What students and lab personnel now refer to as the "Time Keeper System" (TKSystem) is described in a student handout as follows:

All individuals using the lab facilities are required to have a numbered access card. Separate cards will be issued for each class that has a lab component. Properly enrolled students can pick up their card(s) in the lab during the first week of classes. When you use the lab, the card (appropriate for the class you are working on) should be placed in the holder on top of the computer monitor.

To record your lab usage hours, you are required to "LOGIN" and "LOGOUT" at the TKSystem computer in the lab. The number on your access card is keyed to the course you are enrolled in, so you will only have to key in the access number. From the Main Menu, press F1 to LOGIN or F2 to LOGOUT and follow the screen directions.

Note: If you forget to LOGOUT, the TKSystem will automatically log you out after four hours without recording any of the time to your account. See the lab assistant if you suspect this has happened to you. This also means that if you plan to work in the lab for more than four hours at a time, you must logout before the four-hour limit expires and then login again.

The TKSystem will show you your session time and accumulated total time when you logout.

Unfortunately, no written instructions are available for the lab personnel to maintain the system. As a result, the lab manager now wants the analyst to document the TKSystem and prepare training materials for the lab personnel. Notice that this new request adds to the original request for services (see Figure 2-5), making it much more specific. This additional requirement should be incorporated into the project documentation.

Your own personal experience and common sense might suggest the first sketch of the context diagram for the TKSystem, as illustrated in Figure 3-9.

FIGURE 3-9   *TKSystem Context Diagram*

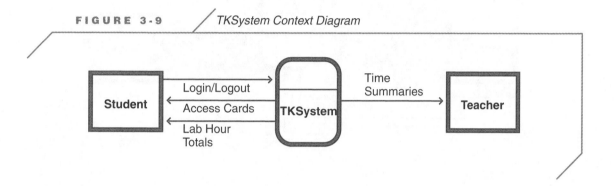

However, your fact-finding review of the lab procedures manual and the TKSystem documentation, interviews with the lab personnel, and general observations reveal an elaborate system involving micro/mini uploading and downloading, file backups, student nonpirating agreements, lab information booklets, class roster reconciliations, lost card procedures, and so on. Apparently, what was first described as a timekeeping system is really much more than that. In other words, the boundaries of the system are wider than we originally thought. Users often misrepresent system boundaries simply because they tend to refer to existing systems in a shorthand or generic, nondescript slang. You must be careful to investigate beyond the superficial descriptions provided to you.

Figure 3-10 illustrates another sketch of the context diagram for the TKSystem. In this version, two new external entities and several new data flows appear, reflecting our revised understanding of the system. The data flows that carry over from the first context diagram are in the shaded area.

This context diagram was constructed by first working from the narrative developed through the fact-finding process (the bottom-up approach) to identify the external entities. Remember, external entities originate or receive data flows, so vocabulary such as "submits," "provides," "delivered," and so on suggest existence of an external entity. When interviewing, keep asking the top-down question, "Then what happens?" over and over. When you hear "... then we give it to ..." or "... so and so sends us ..." you should suspect that an external entity is involved. Once the external entities are defined, the data flows are easier to identify, especially if you have kept track of the nouns or objects associated with the action words and phrases.

F I G U R E 3 - 1 0 / *Revised TKSystem Context Diagram*

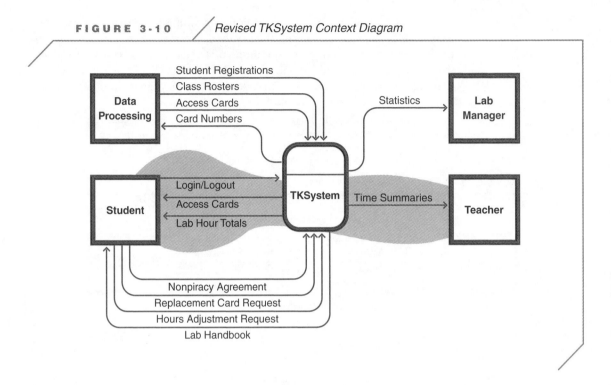

Internal and External Entities

An external entity has been described as something or someone who either generates data or receives information. Our context diagram that follows identifies students, data processing, lab managers, and teachers as entities. Because they are all *outside* the system we are responsible for, they are called *external* entities. This label is difficult to justify when you consider that this is a *student* registration and timekeeping system. It really is a confusing description until you realize that it is intended to remind the analyst that the operations of such entities are beyond his or her control. Therefore, you need not be concerned with the Data Processing Department at all except to note that it receives data and provides information. Likewise, you don't care what the teacher does with the summary reports or where the students keep their access cards.

Another confusing point: does the existence of external entities imply the existence of internal entities? No! Inside the system we are concerned only with processes and stores. It is tempting to identify a person or an object that performs a task, such as "The lab assistant issues access cards to students." But the lab assistant, in this situation, is not considered to be an entity. Because we are interested only in the data transformation that this task involves, we will define this as a subprocess called Issue Access Cards, ignoring in the DFD the person who initiates this action. On the other hand, the lab manager is correctly identified as an external entity as the recipient of a periodic statistical summary from the TKSystem.

LEVEL DECOMPOSITION

After all this discussion, we are still left with a single process and very little detail about what happens inside of it. The context diagram must be decomposed to show this detail. This begins by recognizing that the single process in the context diagram is almost always a collection or series of small, well-defined activities called cohesive tasks, which you should remember from the preceding section.

Identifying Events within a Process

The first step in the decomposition process is to identify the individual events within a process. A subtle, but important, difference exists between a task and an event: an *event* is a benchmark, or plateau, that we use to identify the end of one *task* and the beginning of another. Thus, in the TKSystem, the issuance of a lab card is an event that helps us identify two of the subprocesses: Issue Access Cards and Login/Logout. Once a student obtains a lab access card, he or she can login and logout.

As an exercise, see if you can pick out the inputs and outputs to Issue Access Cards from the following narration prepared by the analyst. Compare your work to the *IPO chart* for the Issue Access Cards task (Figure 3-11).

The lab manager informs the Data Processing Department of the card numbers already issued. The Data Processing Department prints cards for each student registered in a lab course, beginning with the next series of numbers, and sends the cards, along with the printed class rosters and an electronic data file of student registrations, to the lab manager. Thus, for preregistered students, the cards are already printed and in the lab on the first day of class. Students who later add the class must wait 24 hours for this process to catch up. The lab manager inputs the electronic data file to the TKSystem, making it ready to record the hours students spend in the lab. When a registered student uses the lab for the first time, he or she must approach the lab assistants' counter to receive a copy of the lab handbook and fill out a nonpiracy agreement. When the student returns a signed nonpiracy agreement to the counter, the lab assistant checks off the student's name on the class roster and issues the appropriate preprinted lab card. The signed nonpiracy agreements are filed below the counter.

The IPO chart also identifies data flow labels (italics) and the data store label and operation (update, retrieve). From this chart, we can construct a new data flow diagram to show the next level of detail for the Issue Access Cards task (Figure 3-12). Notice that we have labeled this as process number 2 (there are a total of seven processes, as shown in Figure 3-13) and identified three data stores (Agreements, Rosters, and Time).

To test your understanding so far, can you explain why items such as "statistics" and "total hours" are not on this IPO? Why does the task list detail new items, such as "lab account" and "checkoffs"? Should these new items be added to the context diagram? Why don't the data stores appear on the context diagram?

FIGURE 3-11   / *TKSystem Task IPO*

TASK: Issue Access Cards
 INPUTS:
   class rosters (Source: Data Processing)
   access cards for registered students (Source: Data Processing)
   student registration (Source: Data Processing)
   signed nonpiracy agreement (Source: Student)
   name on class rosters (Data Store Retrieval: Roster)
 OUTPUTS:
   access cards (Sink: Student)
   lab handbook (Sink: Student)
   card numbers already issued (Sink: Data Processing)
   checkoff on class rosters (Data Store Update: Roster)
   new lab account (Data Store Update: Time)
   nonpiracy agreement (Data Store Archive: Agreements)

FIGURE 3-12   / *TKSystem Issue Access Cards Task DFD*

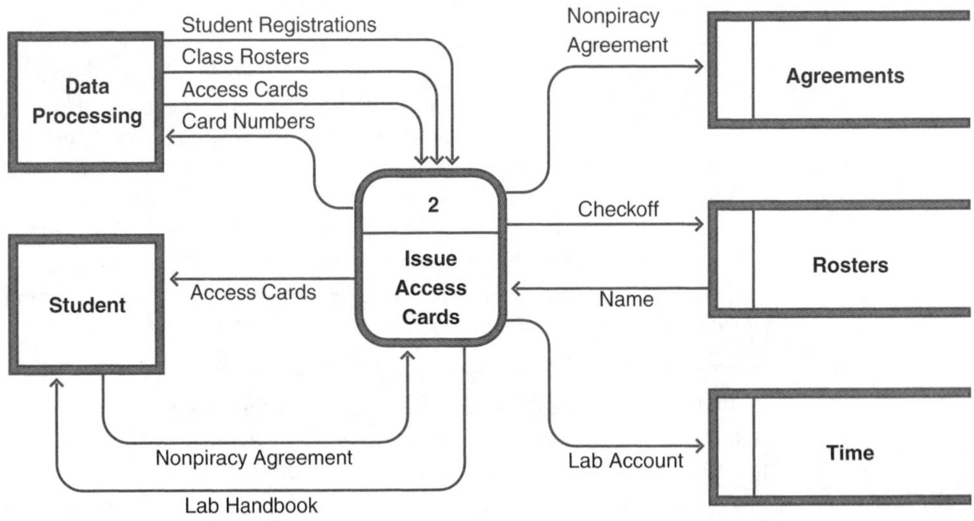

The decomposition continues as we identify other events and other sub-processes. Once again, the bottom-up and top-down development process can be used to identify such events and to develop a list of tasks and their associated

TECHNOTE 3-2

LOGICAL AND PHYSICAL DFDs

Data flow diagrams can be distinguished by the degree to which they identify the way the system is implemented. A **logical DFD** removes all references to the implementation specifics of the system. A **physical DFD** species the real-world objects that are used to make the system work.

The following sequence is recommended when using the two types of diagrams during the analysis and design phases of the SDLC.

Analysis Phase:

1. Develop the physical DFD of the existing system by referring to the real-world objects that are used in the system. For example:
   Add the numbers with a calculator.
2. Abstract the logical DFD of the existing system from Step 1 by removing the implementing detail. For example:
   Compute the total.

Design Phase:

3. Develop the logical DFD of the new system, without regard to the implementing details. For example:
   Accumulate the total.
4. Develop the physical DFD of the new system from Step 3 by assigning real-world objects to perform the processing tasks, contain the data flow data or information items, and retain the data store items. For example:
   Input the numbers with a bar-code scanner.

Ultimately, the analyst uses both approaches; the physical DFD because it is easy to develop a process model that is based on real-world objects and the logical DFD because it is easy to manipulate a process model that is based on abstract objects.

For the most part, the data flow diagrams in this text are physical DFDs. You may find it useful to work with both types at first, without concern for the strict sequencing recommended here. With more experience, you will be able to adopt the formal guidelines more easily.

---

inputs and outputs. Remember, events separate tasks and inputs/outputs identify data flows. Returning to the context diagram for a moment, be aware that every data flow should appear somewhere in the decomposed diagram. By definition, they should all originate or terminate with an external entity (a source or a sink). However, other data flows are internal to a process and, therefore, should be associated with a data store or another process.

Eventually, when all the task IPO charts are completed, the analyst can prepare a *first-level DFD*. A first-level DFD shows the cohesive tasks as separate processes on the data flow diagram. Figure 3-13 presents the TKSystem first-level DFD with seven distinct, cohesive tasks, or subprocesses, to the system. Can you identify the data store archive operations? Should they be included in this chart? Why is the roster data store shown with a shaded corner? What do the numbers in the processes suggest? Notice, also, that although backup data stores are not generally included in the DFD, in this case, Back Up appears as an informational item.

FIGURE 3-13 *TKSystem First-Level DFD*

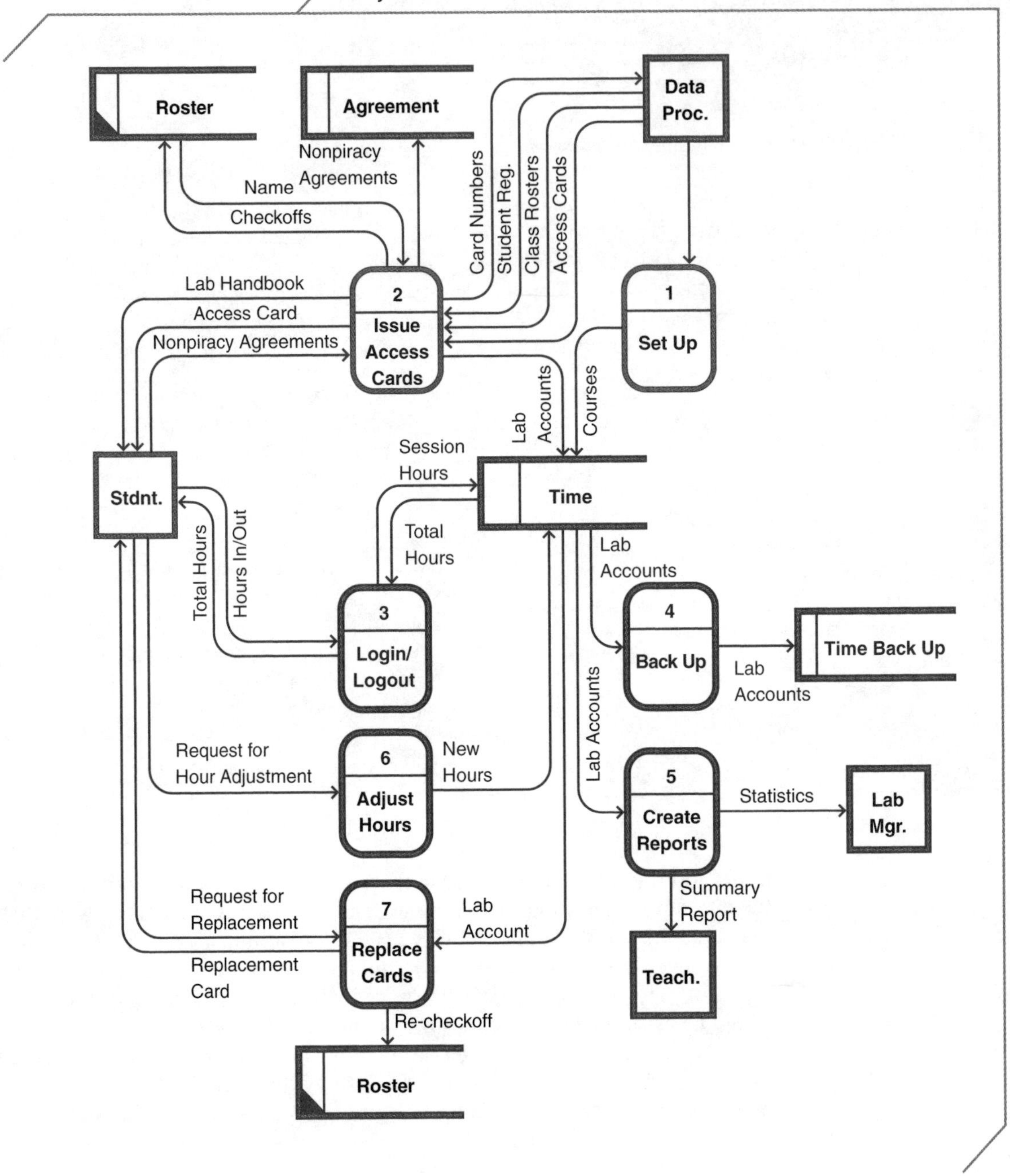

Identifying the Data Flows

Data flows not only connect external entities, processes, and data stores, they also show the data or information that passes from one to the other. Written just above or below the flow line are data, form and report names, or other descriptive labels.

Notice the labels that describe individual data elements (card numbers, total hours) and others that describe a collection of data (roster, summary report). This may seem arbitrary. At this point it is important merely to note, rather than eliminate, the difference in data flow labels. The construction of the current system DFDs is done to help you understand what the system does. Therefore, the labels should be as descriptive as possible.

The analyst must specify all the data items associated with the data stores, external entities, and processes. Many of the data flows presented in the preceding diagrams and task IPO charts are easy to identify, but some are more difficult. The analyst usually begins this task by inspecting the sample forms, reports, and file listings collected during the fact-finding activities. Another, less obvious, method is to simulate the processing steps with some sample data. This is referred to as a *walk-through*.

Figure 3-14 presents a sequentially numbered data flow walk-through for the TKSystem Login/Logout process. Notice that student_id appears on the walk-through but not on the task DFD. Can you explain this data element's purpose in the Login/Logout process? How would you change the task DFD to reflect this new perspective in the Login/Logout process?

Determining When to Stop Decomposition

Although the TKSystem first-level DFD provides much detail, some tasks could require further breakdown to be fully understood. If so, the leveling process should continue with second-level and third-level diagrams, as shown in Figure 3-15. Theoretically, the number of levels you can create is unlimited. But when *DFD decomposition* provides no further insight into what the system does, the analyst should move on to other models.

Note one very important caution with respect to data flow diagrams and small-enterprise systems work: you must keep your work on the DFD in perspective. It is a modeling tool to help you understand what the system does. The end user is rarely interested in the DFD as an end product, especially if it describes a system that you intend to change significantly. Therefore, do not spend an inordinate amount of time trying to "pretty up" the DFD of the existing system. However, you should resist the temptation to skip the DFD even when the existing system functions seem obvious.

You may have wondered about those aspects of the system that the DFD does not reveal. For example, the TKSystem includes procedures to deny lab access to students who refuse to sign the nonpiracy agreement, but this is not evident on the DFD. Also, the DFD does not show that the Data Processing Department interacts with the system for the first ten days of the term, providing updates to the student registration information. To address these and other concerns, the next chapter revisits the other two modeling methodologies mentioned at the beginning of this chapter. The system model illustrates the relationship between data stores and processes in a way that emphasizes the user interfaces, whereas the entity-relationship diagram illustrates the special relationships between data stores.

FIGURE 3-14 / *Data Flow Walk-Through*

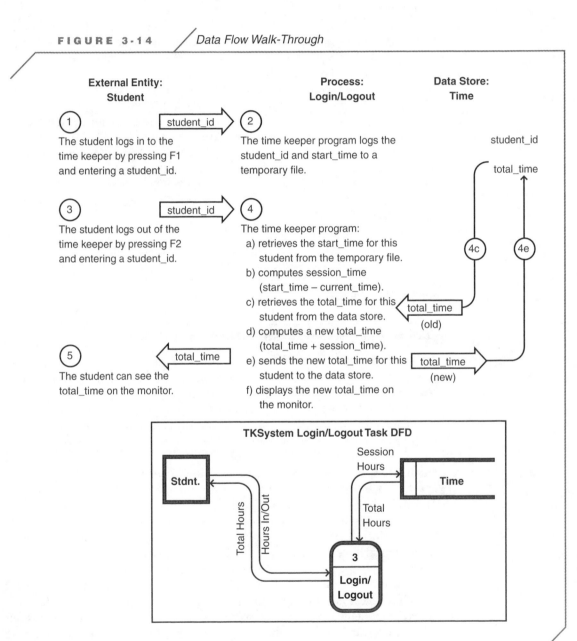

FIGURE 3-15   TKSystem DFD Levels

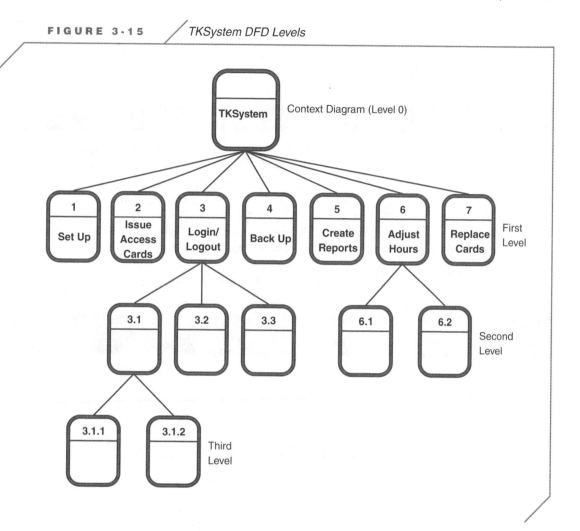

ModelBuildingforPay

This text teaches you how to construct several different models of the information system. Obviously, you will spend a lot of time developing the models. How would you respond to a client who asks why he or she should pay for model building?

ThinkingCritically

These new modeling techniques may, in fact, reveal that we need to return to the DFD decomposition to develop further detail and deeper understanding of the system. Thus, the iterative nature of each of the phases of the SDLC emerges once again.

CASE TOOLS FOR DATA FLOW DIAGRAMMING

Creating a readable data flow diagram is no easy task. Usually several paper-and-pencil sketches are required before you are ready to use a CASE or paintbrush tool. This task presents two specific challenges. First, you need to have an image library of the standard symbols. Second, you need a comprehensive listing of the data stores, data flows, external entities, and processes that are used in the project. These names and their associations with one another are normally contained in a data dictionary. CASE tools provide a means for creating and maintaining dictionaries of this sort. A generic description of CASE tool modeling capabilities appears at the end of the next chapter.

THE CORNUCOPIA CASE

The analysis phase for the project is now well underway. The analysts must first understand the existing system before the new system can be designed. The process model is the first abstraction to develop.

The Context Diagram

The owner's original letter, the fact-finding activities, and the subsequent discussions involving the project contract provide the information necessary to prepare the context diagrams for the existing system (Figure 3-16). Notice that the existing system does not include the owner as an external entity because she never originates input or receives output. All her activities are internal to the system. From the project contract the analysts know that the new system, on the other hand, will produce at least two new information products: newsletter and sales trend reports. Therefore, the owner, in receiving the sales trend reports, becomes an external entity to the new system. Also notice that the sales system is an external entity because it is a separate information system within the enterprise.

FIGURE 3-16 / *Cornucopia Existing System Context Diagram*

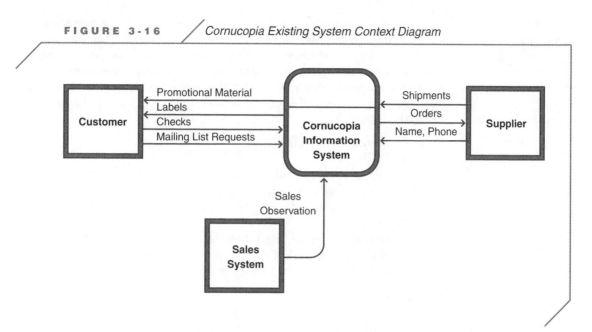

The First-Level DFD

The next step in the analysis process is to develop the first-level DFD of the existing system. Once again, the previous activities provided the insight necessary to identify the specific tasks and task IPO charts that will help the analysts decompose the context diagram. Of the four areas in this project, only the reordering system and the customer record-keeping procedures currently exist. The following narrative is based on the owner's original letter:

> The customer record-keeping procedure does not involve accounts receivable, because all transactions involve cash, check, or credit card. A partial file of customer names and addresses is kept on an old Macintosh computer at the owner's home. The owner _updates this file_ from customers' personal checks or direct requests from customers. Mailing _labels are printed_ from this system. The computer is also used to _create the promotional materials_ that are periodically _mailed_ to customers.
>
> The reordering system relies entirely on the employees' observations. By _observing_ sales activity and/or the inventory directly, they notice stock shortages, which they _make note of_ on a spiral pad next to the cash register. Periodically, the owner _consults a suppliers list_ and then _places orders_, via telephone, to the appropriate wholesaler and, finally, _notes the date_ of the order on the pad. Special orders for discs, cassettes, videos, and books follow the same procedure, except that they are recorded on separate, color-coded sheets of paper. As _shipments are received_, these _notes are scratched off_.

Several action words or phrases are emphasized in the narrative, from which four cohesive tasks emerge, as detailed in the following outline. Note that the customer record-keeping task is simply called customer correspondence and exists completely separate from the other three tasks. In other words, the current system provides no cross-reference between customers and their specific purchases. This fact was confirmed through a last-minute phone call to the owner.

TASK: Customer Correspondence
> INPUTS:
>> personal checks (Source: Customer)
>> customer requests (Source: Customer)
>> customer name, address (Data Store Retrieval: Customer)
>
> OUTPUTS:
>> customer name, address (Data Store Update: Customer)
>> mailing labels (Sink: Customer)
>> promotional material (Sink: Customer)

TASK: Inventory
> INPUTS:
>> observations of inventory (Data Store Retrieval: Inventory)
>> observations of sales (Source: Sales System)
>
> OUTPUTS:
>> notes the shortage (Data Store Update: Orders)

TASK: Order
> INPUTS:
>> notes the order (Data Store Retrieval: Orders)
>> consults supplier list (Data Store Retrieval: Supplier)
>> supplier changes (Source: Supplier)
>
> OUTPUTS:
>> places orders (Sink: Supplier)
>> notes the date (Data Store Update: Orders)
>> supplier changes (Data Store Update: Supplier)

TASK: Receive
> INPUTS:
>> shipments (Source: Supplier)
>
> OUTPUTS:
>> scratch off order (Data Store Update: Orders)
>> shipments (Data Store Update: Inventory)

Figure 3-17 presents the first-level DFD of the existing system. Although this diagram could be decomposed further, doing so would not add significantly to the analysts understanding of the system. One interesting observation about this process is that not only does it help the analysts understand the current system, but it also stimulates some early thinking about the design of the new system. For example, the four data stores are possible database candidates. Also, as previously noted, the current reordering system does not interact with the customers or the sales transactions. However, such interaction is likely in the new system, given the requirement to add customer communications and sales trend analysis.

FIGURE 3-17 / *Cornucopia Existing System First-Level DFD*

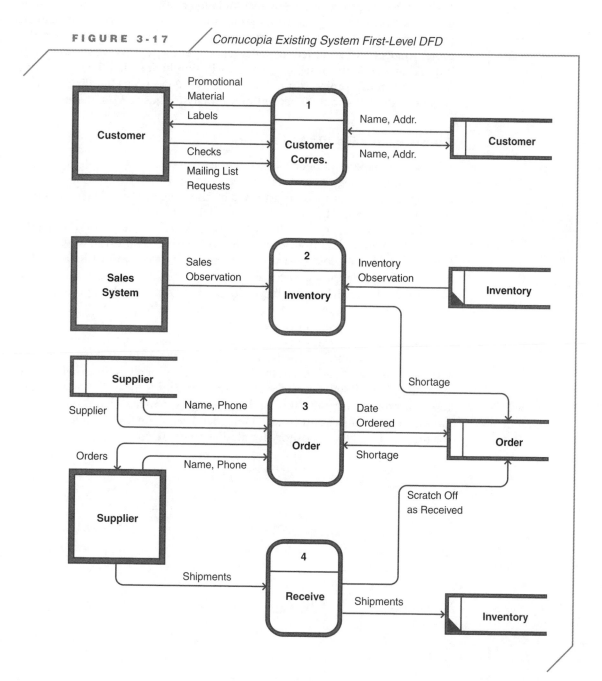

Time and Money

The Cornucopia project is already accumulating costs, consuming analyst time, and making some progress toward meeting its goals. Project management (Appendix A) includes two important tools used to monitor such data: the project budget and completion status reports. These reports are illustrated in future chapters. At this time, the analysts will simply note that they have spent four hours on this project, for a cost of $200 ($50 per hour). Further, the analysts estimate that about 30 percent of the analysis phase is complete.

Cornucopia with Visible Analyst

After several paper-and-pencil sketches, the analysts turn to Visible Analyst to create a context diagram and a first-level data flow diagram for Cornucopia's existing system. This process is activated by choosing the New Diagram option from the File menu. This opens a new model in the work area.

The Diagram Type drop-down list contains all of the models available. Included are class, data flow, decomposition, entity relationship, state transition, structure chart, and unstructured models. Once the model is selected, a blank work space appears, with access to the symbol and line sets appropriate for the selected model. A tool bar, tailored to the open model, appears in the work area. A single mouse click on a tool gives access to the standard symbols, which we can then place in the work area.

Figures 3-18 and 3-19 illustrate the preceding sequence. This text has made several references to the data dictionary. Perhaps this is obvious, but the analysts must create and maintain dictionary entries for all the external entities, data stores, data flows, processes, and task IPOs. This is the only way the CASE tool can provide accurate information through the many menu options discussed here. At this point, understand that Visible Analyst is automatically making entries in the repository for all dictionary items.

Cornucopia Process Modeling with Visible Analyst

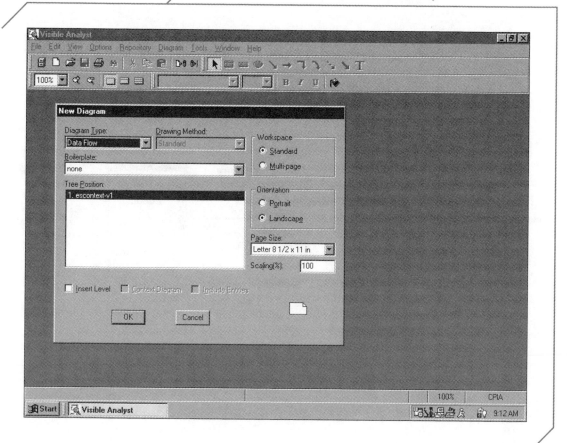

FIGURE 3-19 / *Cornucopia Existing System Context Diagram in Visible Analyst*

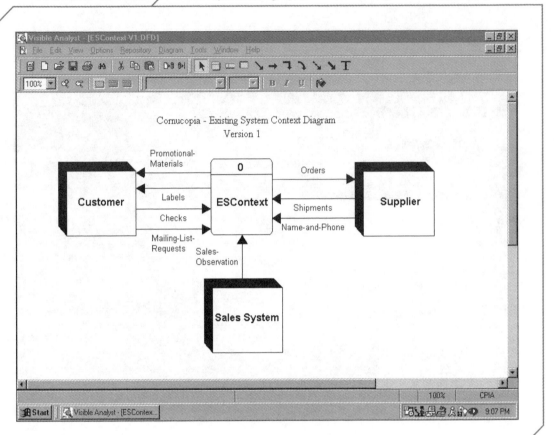

SUMMARY

Data flow diagrams are presented as the first modeling tool used in the SDLC. Such tools are useful because the analyst must strive to develop some understanding of the current system during the earliest stages of a project. This usually begins with a review of the available documentation, interviews, and observations. Soon, the analyst can develop the first context diagram sketch, which reflects the boundaries of the system. Quite simply, this shows the system as a single process, surrounded by external entities. Although, in itself, this is not very descriptive, it does serve as a starting place from which to develop successively more detailed models. For example, as the TKSystem diagrams unfolded, it became apparent that the iterative process was not intended to produce the perfect DFD, but rather to provide a strategy to promote understanding.

The focus of this model is processing. In fact, the question, "What does the system do?" is repeated throughout the modeling effort. This relates to two of the six components of a CIS: people and processing. A third component, data, is addressed by the data flows and data stores of the DFD. The theory is that by concentrating on the processes of the system, the analyst will come to understand each component and their working relations to each other, even though hardware and software considerations remain buried deep within the lowest-level DFDs.

As with any theory, the real world suggests modifications. We identified several shortcomings of the DFD approach. In its preoccupation with process, the model does not completely explain the complex relationships that exist between and among stores, nor does it provide easy access to the system's user interfaces. Both of these concerns are addressed in the next chapter.

TEST YOURSELF

Chapter Learning Objectives

1. Describe the difference between a detailed description and an abstraction.

2. Why is the data flow diagram called a process model?

3. What are the four elements of a data flow diagram (DFD)?

4. Describe the difference between the bottom-up and top-down DFD construction techniques.

5. Describe what it means to decompose a DFD.

Chapter Content

1. Match the model name with its description.

Models

_____ Process model

_____ Data model

_____ System model

_____ Object model

Descriptions

a. This model includes specific reference to hardware and software products.
b. This model unifies all five components of an information system into one diagram.
c. This model depicts enterprise objects, combining data and processes together.
d. This model defines data files and their relationship to one another.
e. This model focuses on the transformation of data into information.

2. Sort the following activities to form an ordered, step-by-step method to decompose a process into a more detailed level of the DFD.

a. Develop and evaluate a detailed narrative of the process, underlining the action words.
b. Identify cohesive tasks and fit all remaining tasks to a cohesive task.
c. Draw and label a new process to describe each cohesive task.
d. Attach and label the data flows to each new process.

e. Develop an IPO chart for each cohesive task.
f. Develop a sequential list of the action words, eliminating tasks that do not transform data.
g. Attach the data flows to external entities and data stores.

3. Using a data flow diagram, the analyst can determine the person responsible for data entry. True or False?

4. By definition, an external entity cannot be associated with the enterprise. True or False?

5. Data flow arrows must always show left to right movement on the data flow diagram. True or False?

6. A data store is a place to keep data for later reference. True or False?

7. Even if two processes pass data directly from one to another, it is incorrect to connect two processes with a data flow line. True or False?

8. Each cohesive task is considered to be a process. True or False?

9. The context diagram is prepared after the first-level data flow diagram. True or False?

10. The context diagram shows all of the external entities, which are described as either a source or sink. True or False?

ACTIVITIES

Activity 1

Using the CIS Lab example as a guide, investigate the student time reporting procedures in your school's computer lab. Prepare the following:
  a. a narrative describing the existing system
  b. a context diagram for the existing system
  c. a series of Task IPO charts that identify the inputs and outputs for each process within the existing system; be sure to include the sources, sinks, and data stores associated with each data flow
  d. the first-level DFD for the existing system

Activity 2

Silhouette Sea Charter's original memo mentions three areas of concern: scheduling, billing, and payroll. This chapter provides a descriptive narrative for the existing scheduling system, along with a rough idea of the task IPO chart.
  a. Develop the context diagram for the existing scheduling system.
  b. Complete the Task IPO chart for the existing scheduling process. Include the sources, sinks, and data stores associated with each input and output.
  c. Develop the first-level DFD for the existing scheduling system.

Activity 3

The first time you registered for a college class you encountered a complex information system. This experience was greatly influenced by your understanding of what you were supposed to do. By our definition, you became an entity, interacting with a process (registration) as you originated data values (student id, name, address, etc.) and received information (your registration confirmation). If the system worked, the process correctly passed this information to a data store (student master file) for later use.
  a. Ask the counseling office or the registrar to explain or provide a diagram that describes this information system (or, even better, perhaps you can get both).
  b. Construct a context diagram to describe this information system.
  c. Construct a first-level DFD to describe this system.
  d. Describe how this system interacts with at least one other information system on campus (e.g., the library, financial aid, or housing).

DISCUSSION QUESTIONS

1. Locate a narrative that explains an information system that you interact with regularly (e.g., the telephone book, the automated teller machine, the syllabus for this class). Identify the processes, external entities, data flows, and data stores. Evaluate the usefulness of this model.

2. Think of a feasible modification to Silhouette Sea Charter's existing procedures for scheduling a charter. Revise Figure 3-3 to reflect your modification.

3. Refer to the revised TKSystem context diagram and the TKSystem task IPO (Figures 3-10 and 3-11) to answer the following questions:
   a. Why doesn't the TKSystem task IPO show all the data elements that appear in the context diagram?
   b. Why do new items appear on the task IPO?
   c. Why don't the data stores appear on the context diagram?

4. Refer to the TKSystem first-level DFD (Figure 3-13) to answer the following questions:
   a. What are the data store archive operations?
   b. Why might the appearance of the data store archive operations be useful to the analyst?
   c. Why is the roster data store shown with a shaded corner?
   d. What do the numbers in the processes suggest?

5. Refer to the Data Flow Walk-Through (Figure 3-14) to answer the following questions:
   a. Can you explain the purpose of the student_id data element in the Login/Logout process?
   b. How would you change the task DFD to reflect the student_id in the Login/Logout process?

6. Try to explain the TKSystem first-level DFD to a friend who is not taking this course. Describe this experience. How would you rate the DFD as a communication device?

7. Explain how the SDLC is an abstraction.

8. All of the models presented in this text are abstractions, or simplifications, of more complex concepts or processes. In what ways is the six-component model (people, processes, data, hardware, software, networks) of an information system an abstraction?

9. In Cornucopia's existing system context diagram, "sales observation" is identified as a data flow. Speculate on what might be a more detailed description of exactly what data is observed and flows into the existing information system under study.

10. After carefully examining all of the figures illustrating Visible Analyst in the first three chapters, speculate on the effort required to learn how to use Visible Analyst and, once learned, the effort required to produce a context diagram in Visible Analyst.

PortfolioProject

Team Assignment 3: Process Modeling

Model building is an important part of the analysis phase of the SDLC. Models help you better understand the information system under study. Your first experience with data flow diagrams may seem very tedious, perhaps frustrating. Frequently, the analyst must revise the diagrams several times before the model accurately represents the way external entities, processes, and data flows work together.

Preparing regular project budget and status reports is an important part of project management. These reports help everyone understand how the project is progressing. In this situation, they provide a valuable reality check for the student analyst. Comparing your actual experience against your initial plans helps you prepare estimates for your next project.

In order to begin the model-building process and develop your first project management reports, you must complete the following tasks:

1. Prepare the existing system context diagram and first-level data flow diagram. Submit a copy of these diagrams.
2. Review the "Project Budgets" and "Status Reports" sections in Appendix A and the "Project Guidelines" section in Appendix B. Insert your data disk and open the budget template (budget.xls) and the project status template (status.xls).
3. Develop a first draft of your week-by-week budget for your project. Your budget should show estimates for the entire project, actual expenditures so far, along with cumulative over/under amounts, by cost category (hardware, software, and labor).
4. Develop a first draft of your week-by-week project status report for your project. Your status report should show the following for each activity: start/stop periods, % complete, status summary.
5. Prepare a cover letter to transmit the budget and status reports to your client. Submit a copy of your cover letter and the two project management reports.
6. File a copy of the context diagram, first-level DFD, cover letter, project budget, and project status in the Portfolio Project Binder behind the tab labeled "Assignment 3."

Note: If your instructor requires you to submit regularly updated budget and status reports, refer to the discussion in Appendix A to help you gather the data needed to prepare the reports.

datamodeling

OVERVIEW

Data models provide an extremely important view of the organization plan for the data that flows into, through, and out of information systems. Unlike other modeling methodologies, data models focus only on data and how its component parts are structured, stored, and related to one another.

In this chapter you will learn how to build on the data flow diagrams to create data models. Data models focus on the relationship between the data stores identified in the DFD. First, you will be refreshed on data fundamentals and then introduced to the notion that file processing in general, and relational database processing in particular, form the heart of information systems.

- Identify data types and structures within a file.

- Distinguish between master and transaction files.

- Define the relationship between two files.

- Construct and normalize an entity-relationship diagram.

DATA FUNDAMENTALS

Figure 4-1 shows that data is a common element in both the DFD and the data model. In fact, of the six system components, data is the only one that appears in all the modeling frameworks. This is not a coincidence, nor should it be surprising. Data manipulation and transformation is, after all, what data processing is all about. Although traces of the data-centered notion of input-processing-output can be found in every phase of the SDLC, data is by no means random, or arbitrary, or disorganized. In fact, the *data model* reflects the analyst's preoccupation with data identification and organization.

FIGURE 4-1   Data Models and the Six CIS Components

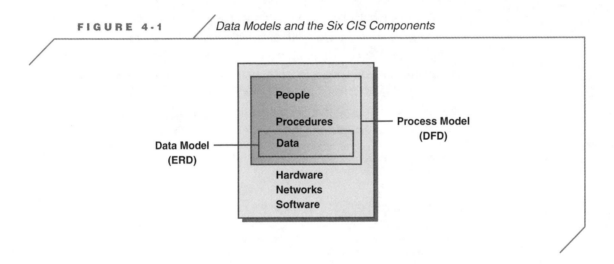

Data Types

Data is defined by three attributes: name, size, and type. Data names provide unique and descriptive labels. Data size determines the amount of space required to hold data values. *Data type* specifies how the computer stores data. Careful declaration of these attributes enables the computer to efficiently store, retrieve, and manipulate data. In addition, the data type classifies and restricts how data can be used.

Primitive data types (e.g., character, numeric, Boolean), which are supported directly by computer hardware, can be combined to create endless complex data types (e.g., date, currency, video). Thus, the analyst is likely to find that Silhouette Sea Charter uses characters to name its boats, digits to record the number of passengers on a charter, and yes/no to indicate if a captain is properly licensed. Furthermore, Silhouette may use dates to indicate the length of a charter, currency amounts for billing, and video to show how a boat performs at sea.

Data Structures

Just as we organize our personal possessions to make their storage, retrieval, and manipulation easier, an enterprise naturally organizes its data for the same reasons. Handwritten lists, boxes of invoices, folders of correspondence, and cash register receipts are examples of data organized in a specific fashion. A *data structure* is a specific organizational strategy for associating individual data elements. Silhouette keeps a copy of each charter contract in a manila folder labeled with the boat's name. Each contract contains a great deal of data associated with a particular charter—the captain, passengers, destination, departure and return dates, and so on. This strategy makes it possible to accurately answer questions about a particular charter or summarize activity for several charters.

Computer-based data structures mimic those of the real world. The analyst soon recognizes the parallels: a tides table can be implemented as a collection of water levels organized into a two-dimensional array of time-of-day and location; a charter's waiting list can be implemented as a queue; and a history of boat maintenance can be implemented as a file. While arrays, queues, and files are data structures familiar to most people, computer programmers must apply special syntax and instructions to use them effectively in a computer information system. The analyst need only have a conceptual understanding of data structures in order to build the data model.

FILE PROCESSING FUNDAMENTALS

Data appears in one of three ways within an information system: as an input data stream, as a file, or as an output data stream. A *data stream* is nothing more than a series of characters that form a command or represent specific data values to or from a program. For example, when you select and activate a program icon in the Windows environment, you send a command data stream to the operating system. Your entry in a word processor's search-and-replace dialog box is an input data stream, while the screen output is an output data stream. On the other hand, the program you activate and the word-processed document you search are examples of files, one an instruction file, the other a data/information file. A *file* is a collection of data that is organized into a specific format.

Master and Transaction Files

Another way to classify files is to consider how file content correlates to events or activities within the enterprise. A *master file* is a collection of data that represents an identifiable person or thing. For example, most enterprises maintain frequently used records about people (customers, employees, suppliers, etc.) and things (equipment, materials, products, etc.). These files are updated regularly to reflect additions, changes, and deletions. Each of the entries in these files is distinguishable by some unique characteristic, such as a special code or identification number. Furthermore, some part of this data is usually a part of every event or activity of the enterprise.

A *transaction file* is a collection of data that represents a particular event or activity of the enterprise. For example, it is very likely that the essence of every sale or service, payroll disbursement, equipment repair, and material usage is recorded for later reference. Notice that each of these events or activities is associated with one or more master files. A single sale event can add to a customer's account balance, increase an employee's commission amount, and deplete a product inventory total. Enterprise transactions usually affect the persons or things of the enterprise.

The time frame in which transactions impact the system is also of some concern to the analyst. Continuing with the sale event described above, it would be wise to change the customer's account balance and the product inventory total very soon, if not immediately, after the sale occurred to prevent the account from going over the limit and the product from being sold out. Such a system is referred to as an *online transaction processing (OLTP)* system. In contrast, a *batch processing* system collects all of the updates together for processing at a later time. There is no need to immediately update the employee's commission amount if payroll disbursements are made every two weeks. Thus, it would be better to collect all of the commission amount updates together for processing just before payroll disbursements are due.

In short, we are concerned with file storage and processing because files preserve the system's electronic history. There are two popular forms of the electronic history: databases and data warehouses. A *database* is a collection of files containing data pertinent to the current and immediate operations of an enterprise. In contrast, a *data warehouse* is a collection of files containing data about the past operations of an enterprise, wherein information workers "mine" the warehouse for information to help managers make strategic decisions.

In a sense, computer information systems function like research librarians. They collect, categorize, summarize, locate, and retrieve data, transforming it into information. Quite apart from their routine input, output, and maintenance functions, library professionals provide an important value-added service when they work to assemble new information from many different, but related, resources. Years of schooling and experience are required for librarians to develop the skills required to do their jobs. Understanding the relationships between and among all their resources is a difficult task. Likewise, in the computer business, it took many years before computer scientists developed the tools to offer the same service to their customers.

Today, computer files are organized in much the same way as is this text: the user (the reader) can proceed sequentially from any page or topic, or the user can access the index to pinpoint a particular subject. The complete text serves as a master file (the book). From the moment the text is published, a variety of proposed additions, deletions, and revisions begin to be assembled into a transaction file. This is then used to update the master file, producing a new edition of the text. At present, the file user has access only to the hardcopy, but it is possible that there is an online version available through a publishers' Web site. In such circumstances it would be possible to capture and process immediate reader comments and questions regarding the text, thus illustrating online transaction processing (OLTP). The next logical question is, "Could such a service allow the user to find out more about a subject by electronically cross-referencing this text to other texts?" The short answer to this question is yes, and the long answer requires a discussion of relational databases.

Figure 4-2 presents the charter contract and two associated files from the Silhouette Sea Charter example. This illustration shows portions of the Customer file and the Charter files. Both files are maintained as paper ledgers, relying on the charter contract as their source for specific values. From the preceding discussion, can you identify which is the master file and which is the transaction file? Can you anticipate any problems in working with these two files to produce a report in answer to a customer request about specific charter dates, boats, and captains during the past three years?

FIGURE 4-2   *Silhouette Sea Charter Files—Part 1*

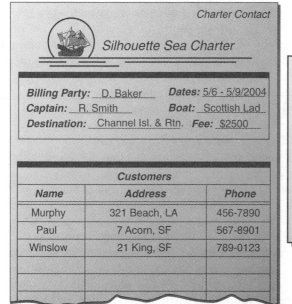

Charter Contact

Silhouette Sea Charter

Billing Party: D. Baker   **Dates:** 5/6 - 5/9/2004
Captain: R. Smith   **Boat:** Scottish Lad
Destination: Channel Isl. & Rtn. **Fee:** $2500

Customers

Name	Address	Phone
Murphy	321 Beach, LA	456-7890
Paul	7 Acorn, SF	567-8901
Winslow	21 King, SF	789-0123

Customer File

Customer	Address	Phone
Calderwood	45 Beale, SF	123-4567
Kirby	34 Gough, SF	234-5678
Murphy	321 Beach, LA	456-7890
Paul	7 Acorn, SF	567-8901
Winslow	876 Palm, LA	678-9012
Winslow	21 King, SF	789-0123
Wolfe	34 Gough, SF	890-1234
Wolfe	34 Gough, SF	890-1234

Charter File

Billing Party	Dates	Captain	Boat	Destination	Fee
G. Frank	2/6 - 2/10/2004	R. Smith	The J.T. Allen	San Diego	$3000
R. Walters	3/1 - 3/1/2004	J. Daley	Mary's Dream	Local	$250
R. Walters	3/1 - 3/1/2004	J. Daley	Mary's Dream	Local	$125
D. Baker	5/6 - 5/9/2004	R. Smith	Scottish Lad	Channel Isl. & Rtn.	$2500

Relational Databases

In your previous studies you have performed many of the standard file-processing activities previously discussed, using either off-the-shelf software or programs you wrote yourself. By now, the advantages of file processing should be obvious to you. It should also be clear that you would not want to keep everything in a single file. Trying to store every systems analysis text in a single master file would be ridiculous. The time required to review the table of contents alone would be prohibitive. By the same token, repeating the same information in each separate text creates duplicate and potentially conflicting information, especially if text revisions are not synchronized. While duplicate and conflicting information is an inconvenience and a source of frustration, humans are very adaptive and forgiving. With a computer system, however, the same situation can be catastrophic. Relational database theory points to the solution.

A *relational database* is a collection of files that are tied together by common fields. Generally, it is used to help with the day-to-day operations of an enterprise. Customers' information may be added to a master file, sales information may be added to a transaction file, and pay vouchers may be generated for employees. Summary reports of product inventory may be printed, inquiries about sales activity may be answered, and supplier purchase orders may be generated. Data associated with each event or activity is captured and filed away for easy access.

Consider the Silhouette Sea Charter example. Figure 4-3 highlights only two of several files involved in the existing information system. You have probably guessed that they also maintain files for captains and boats. (All these files, by the way, would appear as data stores on the DFD models previously discussed.) Figure 4-4 presents our first attempt at a data model for Silhouette Sea Charter's existing system.

The present Customer and Charter file layouts contain several problems. First, the only way to distinguish among customers who have the same name is to look at the address and phone data values—and even that won't work if two customers have the same name, address, and phone. Second, the only way to report on the charters for a particular customer is to scan the entire stack of charter contracts. Third, distinguishing among specific charters is impossible without referring to almost every value in the line entry.

At best, this arrangement makes acquiring a comprehensive view of Silhouette Sea Charter's operations a difficult proposition. To remedy this problem we need to add unique identification or account numbers to the data files. This will provide the ties, or relational connections, between and among the files, which is the first step in creating a relational database. This recommendation illustrates that the study of the current system often leads to design concepts for the new system. Figure 4-4 shows the proposed file layouts. The shaded areas indicate those values that uniquely distinguish each entry in the file. These are called *key fields*. Why might the analyst recommend that the Registration # and SSN fields in the Captain and Boat files be substituted for the Captain and Boat fields in the Charter file?

FIGURE 4-3 / *Silhouette Sea Charter Files—Part 2*

Customer File

Customer	Address	Phone
Calderwood	45 Beale, SF	123-4567
Kirby	34 Gough, SF	234-5678
Murphy	321 Beach, LA	456-7890
Paul	7 Acorn, SF	567-8901
Winslow	876 Palm, LA	678-9012
Winslow	21 King, SF	789-0123
Wolfe	34 Gough, SF	890-1234
Wolfe	34 Gough, SF	890-1234

Charter File

Billing Party	Dates	Captain	Boat	Destination	Fee
G. Frank	2/6 - 2/10/2004	R. Smith	The J.T. Allen	San Diego	$3000
R. Walters	3/1 - 3/1/2004	J. Daley	Mary's Dream	Local	$250
R. Walters	3/1 - 3/1/2004	J. Daley	Mary's Dream	Local	$125
D. Baker	5/6 - 5/9/2004	R. Smith	Scottish Lad	Channel Isl. & Rtn.	$2500

Boat File

Boat	Registration #	Purch. Date	Cost	Length	Sleeps
The J.T. Allen	CA123456789	5/1/73	$35,000	40ft	6
Scottish Lad	CA234567890	2/10/85	$50,000	35ft	6
Mary's Dream	CA345678901	4/13/90	$28,000	25ft	4

Captain File

Captain	SSN	License #	Address	Phone
J. Daley	123-45-6789	Class 1 - B7	27 Portsmouth, SF	321-7654
C. Kirk	234-56-7890	Class 3 - A2	4 Wayward, LA	432-8765
R. Smith	345-67-8901	Class 1 - A1	45 Shaw, Berkeley	543-9876

FIGURE 4-4   / *Silhouette Sea Charter Files—Part 3*

Customer File

CustID	Customer	Address	Phone
	Calderwood	45 Beale, SF	123-4567
	Kirby	34 Gough, SF	234-5678
	Murphy	321 Beach, LA	456-7890
	Paul	7 Acorn, SF	567-8901
	Winslow	876 Palm, LA	678-9012
	Winslow	21 King, SF	789-0123
	Wolfe	34 Gough, SF	890-1234
	Wolfe	34 Gough, SF	890-1234

Charter File

CharterID	Billing Party	Dates	Captain	Boat	Destination	Fee
	G. Frank	2/6 - 2/10/2004	R. Smith	The J.T. Allen	San Diego	$3000
	R. Walters	3/1 - 3/1/2004	J. Daley	Mary's Dream	Local	$250
	R. Walters	3/1 - 3/1/2004	J. Daley	Mary's Dream	Local	$125
	D. Baker	5/6 - 5/9/2004	R. Smith	Scottish Lad	Channel Isl. & Rtn.	$2500

Boat File

Boat	Registration #	Purch. Date	Cost	Length	Sleeps
The J.T. Allen		5/1/73	$35,000	40ft	6
Scottish Lad		2/10/85	$50,000	35ft	6
Mary's Dream		4/13/90	$28,000	25ft	4

Captain File

Captain	SSN	License #	Address	Phone
J. Daley		Class 1 - B7	27 Portsmouth, SF	321-7654
C. Kirk		Class 3 - A2	4 Wayward, LA	432-8765
R. Smith		Class 1 - A1	45 Shaw, Berkeley	543-9876

ENTITY-RELATIONSHIP DIAGRAMS

An *entity-relationship diagram (ERD)* shows data files at rest. That is, the file structures and their relationships provide the focus for the model, rather than the file processing that is the focus of other models. The ERD comprises five ingredients: entities, attributes, relationships, cardinality, and linkages.

In very practical terms, an *entity* is a data file. Or, to put it another way, it is the data store that you encountered in the data flow diagram. Thus, ERDs show the details of the data stores from the DFDs. We already know that data files are really an abstraction of some real-world object or event. The *attributes* of a data file are simply those items selected by the analyst to represent the real world. We have referred to them as fields. Thus, customer name, address, and phone number are some of the attributes of a typical customer file.

File Relationships

Relationship describes how data files are related to one another. We could say that customers "request" charters and thereby describe the relationship between these two entities. *Cardinality* refers to the specific type of relationship the files display in the real world. Because each charter may have more than one customer and, over time, each customer may request many charters, we classify the cardinality as *many-to-many*. In other cases, the cardinality is described as *one-to-one* or *one-to-many*. The fields, or attributes, that the files have in common are called *linkages*. Silhouette Sea Charter's Captain file can be linked to the Charter file by the captain's SSN or some other unique ID value.

Figure 4-5 illustrates the standard entity-relationship diagram symbols, along with the ERD for the proposed Silhouette Sea Charter system. To remove some of the clutter, the attributes of each file do not appear on this illustration. The linkage information is nonstandard in that no symbol is set aside for this piece of important information, even though it appears as an annotation on many entity-relationship diagrams.

The cardinality symbols require some special clarification. The one-to-many relationship means each entry in the Boat file can be associated with one or more entries in the Charter file. This simply allows for the probability that a boat will be used for many different charters. The same is true for captains. The many-to-many relationship between charter and customer reflects the possibility that, over time, a single customer can sign up for many different charters and a single charter can have many customers. This creates a complex linkage between the two files. Figure 4-6 presents two equally unattractive solutions to the implementation problems caused by such a relationship.

The first solution combines the Charter and Customer files into one new file, which results in dangerous and expensive data duplication and redundancy: handling more data than necessary is a waste of machine resources; keeping more than one copy of data floating around the system is an invitation to error. For example, the system must change the number in several places when a customer's phone number changes.

FIGURE 4-5 / *Illustrated ERD Symbols and Silhouette Sea Charter*

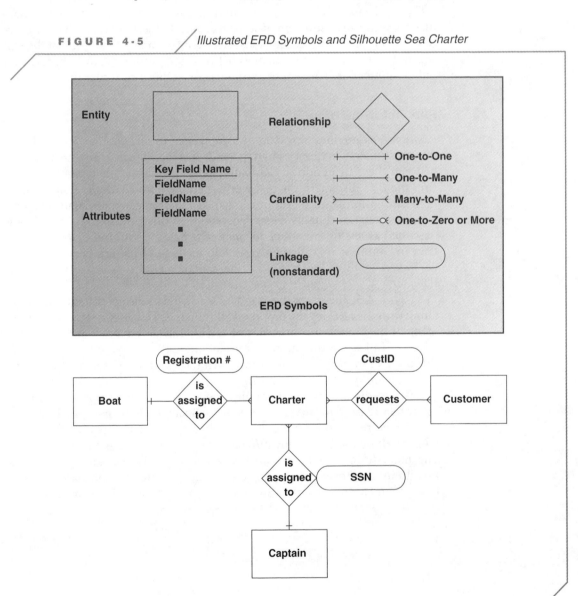

FIGURE 4·6 / *Silhouette Sea Charter Inefficient File Diagrams*

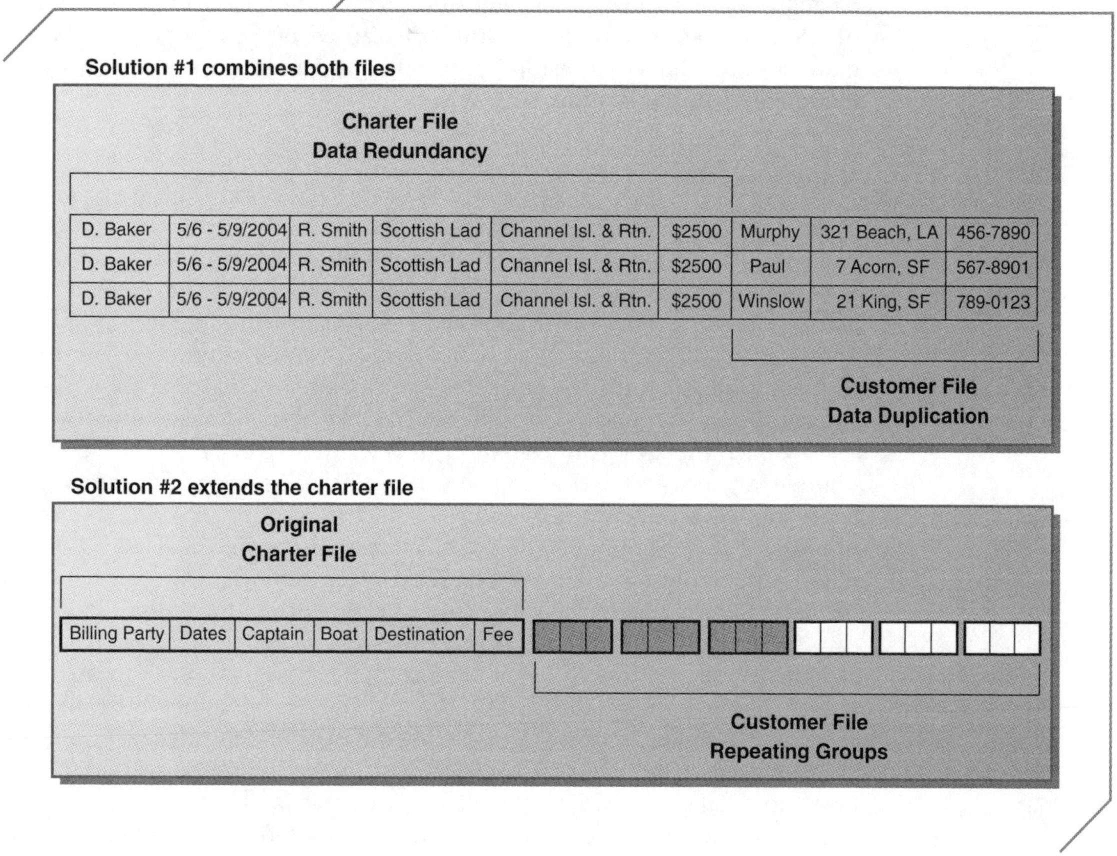

Solution #1 combines both files

Charter File
Data Redundancy

D. Baker	5/6 - 5/9/2004	R. Smith	Scottish Lad	Channel Isl. & Rtn.	$2500	Murphy	321 Beach, LA	456-7890
D. Baker	5/6 - 5/9/2004	R. Smith	Scottish Lad	Channel Isl. & Rtn.	$2500	Paul	7 Acorn, SF	567-8901
D. Baker	5/6 - 5/9/2004	R. Smith	Scottish Lad	Channel Isl. & Rtn.	$2500	Winslow	21 King, SF	789-0123

Customer File
Data Duplication

Solution #2 extends the charter file

Original
Charter File

| Billing Party | Dates | Captain | Boat | Destination | Fee |

Customer File
Repeating Groups

The second solution extends the original Charter file to include separate data fields for as many as six customers, which is the maximum sleeping capacity on any of the existing boats. These fields, which you might name customer1, customer2, customer3, and so on, are collectively called a *repeating group*. This new file design proves to be a poor idea for several reasons. Special processing logic is required to handle the unpredictable number of customer fields containing real data. Furthermore, handling those records containing unnecessary blank characters wastes precious disk space and increases disk access time. Finally, this arrangement requires special processing logic to prevent the office manager from inadvertently booking six customers on a boat with fewer than six sleeping berths. To put this another way, would you knowingly reserve space for empty storage boxes in your closet and then take the time to move the empty boxes to and from your work table every time you needed to access the full boxes?

Relational database theory, in its attempt to protect the integrity of the database and maximize processing efficiency, is very specific about the rules of file design. The remedy to the problems caused by Silhouette Sea Charter's many-to-many file relationship lies in one of the most important concepts of that theory—normalization.

Normalization

Normalization describes the process whereby each file in a database is redesigned so that every attribute within the file is dependent only on the key field of that file. That is to say, every nonkey attribute in the file depends on or is determined by the value in the key field. This provides the technical explanation as to why we must avoid either of the previous solutions to our many-to-many problem with the customers and charters. What follows is a more conversational approach to explain dependency and normalization.

We have already identified Silhouette Sea Charter's Charter file as a transaction file in which each record contains data values associated with a single charter. If Silhouette Sea Charter's conducts 180 charters during the first six months of the year, there should be 180 records in the Charter file. As Figure 4-4 illustrates, each record has a unique Charter ID value, known as the key field. This means that each Charter ID value appears only once in the entire Charter file. Also notice that each record contains the name of the boat, billing party, and captain. These are nonkey fields because these values are likely to appear in many records of the Charter file. Of the 180 charters, a popular boat might be chartered 25 times. Now we can examine the issue of dependency.

Notice that the values associated with boat, billing party, and captain point to a single record in the Boat, Customer, and Captain master files. Looking at these four files on a single page, there is no need for the Charter file to include data values for a boat's purchase date, cost, or length because we can obtain those values by following the Boat linkage established between Charter and Boat. A boat's purchase date is dependent upon a particular boat, not on a particular charter. Charter and Boat are normalized because neither contains

DatabaseAccuracy

Given the need to adhere to strict relational database standards for data integrity, to what extent is the analyst responsible for cleaning up the errors that may appear in the enterprise's existing data files before the new database can be implemented?

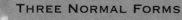

TECHNOTE 4-1

THREE NORMAL FORMS

Formal database design theory outlines a three-step design process that ensures efficiency of the file's key fields, attributes, relationships, cardinality, and linkages. This process is called normalization. The three steps are called normal forms. Thus, during this process you progress from first normal form to third normal form as you rework the database design. Actually, two more normal forms are also used, but they are highly technical and rarely necessary.

First Normal Form

Your database is in first normal form (1NF) if it contains no repeating groups. If you discover a repeating group in a file, you must eliminate the repeating group. This is best accomplished by creating a new file to contain the repeating group information and a field to link the new file to the original file. Silhouette's Customer/Charter file solves the original many-to-many problem and avoids the repeating group problem.

Second Normal Form

Your database is in second normal form (2NF) if it is in first normal form and every attribute within each file is dependent only on the key field of that file. That is to say, every noncrucial attribute in the file depends on or is determined by the value in the key field. This explains why we want the ChartID and CustID attributes in the new Customer/Charter file.

Third Normal Form

Your database is in third normal form (3NF) if it is in second normal form and all the dependent relationships within the files are contained within that file. In practical terms, then, it would be unnecessary and unwise to add the customer name attribute or the charter destination attribute to the Customer/Charter file.

a repeating group and because each nonkey field is dependent upon the key field and not on some other nonkey field.

Just as with the decomposition of the data flow diagram, the ERD may go through several transformations before you settle on a version that accurately represents the data relationships of the system. As mentioned, specific rules govern these relationships. For our purposes, two of these rules are worth repeating: first, no repeating groups should exist within any file; second, all many-to-many relationships should be transformed into two one-to-many relationships.

The normalized version of Silhouette's proposed system is illustrated in Figure 4-7. The addition of a Customer/Charter file reduces the data redundancy in the system and protects the database from the type of invalid entries mentioned previously. Can you identify the attributes in this new file? What is the key field or field combination?

FIGURE 4-7 / *Silhouette Sea Charter Normalized ERD*

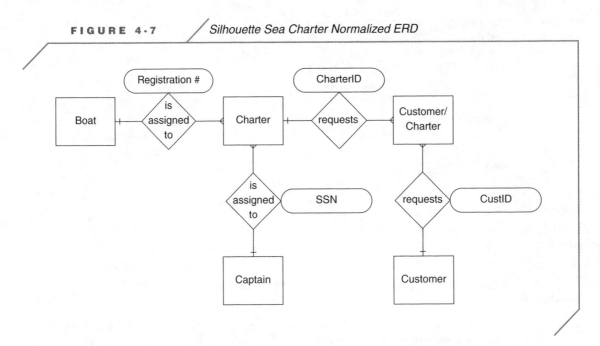

CASE TOOLS FOR DATA MODELING

The diagramming described in this and the preceding chapter can be very tedious and time consuming. CASE tools are particularly well suited to this task because of their image libraries, data dictionaries, chart-building utilities, and error-checking functions. With these tools, the analyst works interactively to define diagram objects (e.g., data stores, processes, entities) for the data dictionary, specify the relationships among the objects, and create the diagrams. Many CASE products also offer tools to verify that models are complete and consistent with one another and that they comply with the standard rules previously described. For example, CASE tools with these capabilities can report on DFD errors involving data stores without an input data flow, or ERD errors involving improperly normalized databases. Because these tools are designed to enforce the basic rules of modular, structured analysis and design, they do more than assist the analyst in diagram construction, they actually help guide the analyst through the methodology.

In some respects, these attributes account for the popularity of CASE tools. Electronically developed, error-checked, and maintained information system models are as appealing to systems analysts as are spreadsheet models to financial analysts.

THE CORNUCOPIA CASE

The analysis of the current system continues with the development of the data model. This not only proves to be interesting but also provides some new insights into the project.

The Data Model

The first-level data flow diagram pictured in the last chapter (Figure 3-17) is a process model, focusing on the processes that transform data into information. This image provides a starting point for the analysts' effort to develop the data model. The DFD identifies four data stores: customer, supplier, order, and inventory. Even though "customer" is the only existing system data store that is presently computerized, all four data stores are part of the information system. Therefore, they are likely to be part of the new system and are good candidates for consideration as entities in the data model. Thus, the analysts' goal is to prepare an entity-relationship diagram using these four data stores as the entities.

Each entity is fully defined by its name, its attributes, its relationship to the other entities, and its cardinality within these relationships. The attributes for each entity, as presently defined by the owner, are detailed in Figure 4-8. This information was gathered partly through the feasibility study and partly through a series of follow-up interviews with the owner.

The analysts can make several interesting observations about these entities and their attributes. First, they must realize that the owner does not have a piece of paper that documents this information. This business enterprise is not concerned about such detailed computer jargon. Only after much coaching do the analysts learn that the owner simply knows which supplier to call for which products. Likewise, the order number attribute is definitely not subject to a systematic, sequential numbering procedure. The order number is nothing more than a handwritten digit on the top of the page. Each day begins a new sequence, starting at 1. Thus, if the owner wanted to find a particular order, she would need to know both the order number and the order date. The UPC (universal product code) is the only attribute that appears in more than one file. Indeed, the paper tablet used for jotting down CD orders has room for as many as three different UPCs. Presumably, if the tablet were larger, the owner could write more than three UPCs on a single order. The UPC-quantity ordered attribute pair is an example of a repeating group.

Supplier phone is a common attribute between the supplier and order entities. The cardinality relationship between supplier and order is one-to-many. That is, for every order there is only one supplier, because no one wants to order the same disc or tape from two suppliers at the same time. What is very likely, however, is that over time several orders will be placed to the same supplier, resulting in that supplier's name or number appearing more than once in the order file.

FIGURE 4-8   / *Cornucopia Existing Entity Attributes*

Entity Name	Customer	Supplier	Order	Inventory
Attribute 1	name	name	order number	UPC
Attribute 2	address	supplier phone	order date	title
Attribute 3	customer phone		supplier phone	artist
Attribute 4			UPC (1)	label
Attribute 5			quantity ordered (1)	price
Attribute 6			UPC (2)	
Attribute 7			quantity ordered (2)	
Attribute 8			UPC (3)	
Attribute 9			quantity ordered (3)	

UPC is a common attribute between the order and inventory entities. The order and inventory files present a many-to-many cardinality relationship. From the inventory side of this relationship, it is possible that a single UPC in the inventory file can be matched to *many* records in the order file. This is true because, over a period of time, Cornucopia may sell several copies of the same CD. From the order side of this relationship, it is possible that a single order can be matched to *many* (up to three anyway) records in the inventory file. This is true because the order tablet has room for three different CDs.

Figure 4-9 shows the first attempt at Cornucopia's existing system ERD. Notice that the customer entity does not interact with any of the other entities, which is consistent with the information on the data flow diagram.

Cornucopia Existing System ERD

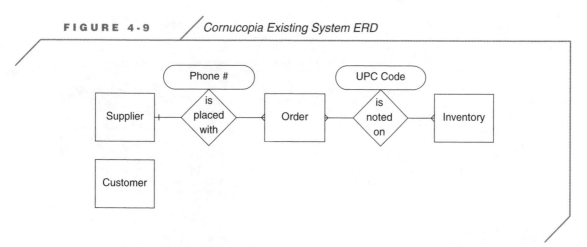

While these file definitions have worked well for the present system, the new system will require more precision. With the understanding that the new system will be based on a relational database, the analysts sketch out an improved set of entity attributes, as shown in Figure 4-10.

One of the most important additions to the entity definitions is the designation of a key field for each entity, shown by the asterisk preceding the attribute name. Furthermore, the addition of the supplier id attribute to the order and inventory entities will make order processing and inventory maintenance much easier, because every order will have its own unique order number. When confronted with an improperly filled order, the owner can easily look up supplier information using the supplier id on the order form. Likewise, when faced with inventory shortages, the sales clerk can quickly determine which supplier to call for assistance. Finally, the definition of a new file (order/inventory) eliminates the many-to-many file relationship between order entity and inventory and the repeating group in order. This text will address several other features of these entity definitions in later chapters.

The analysts' effort to improve the entity attribute definitions and normalize the files, well before the new system design phase formally begins, illustrates how the modeling process helps the analysts develop an understanding of the existing system and may even suggest changes for the system to come.

FIGURE 4-10   / *Cornucopia Improved Entity Attributes*

Entity Name	Customer	Supplier	Order	Order/ Inventory	Inventory
Attribute 1	*customer id	*supplier id	*order number	*order number	*UPC
Attribute 2	first name	name	order date	*UPC	title
Attribute 3	last name	street address	supplier id	quantity ordered	artist
Attribute 4	street address	city		quantity received	label
Attribute 5	city	state			price
Attribute 6	state	zip code			quantity on hand
Attribute 7	zip code	phone			supplier id
Attribute 8	phone	e-mail			
Attribute 9	e-mail	fax number			
Attribute 10	status				

Time and Money

The analysts spent several hours developing the data models for the existing system. The analysts estimate that the analysis phase is now well over 50 percent complete. The time was evenly split between desk work and field research—the analysis phase is not a solitary activity. It actively involves the analyst and the owner. Remember, enterprise personnel are prime sources for information about how the existing system works.

Cornucopia with Visible Analyst

The normalized entity-relationship diagram is illustrated in Figure 4-11. The ERD is subject to a rigor provided by a data dictionary or repository. The entities on this diagram must be named exactly as they are on the data flow diagram. Although the analysts may become frustrated by this obsession with consistency, the CASE tool makes the job much easier by including pick lists of the appropriate dictionary elements.

FIGURE 4-11   *Cornucopia Normalized ERD*

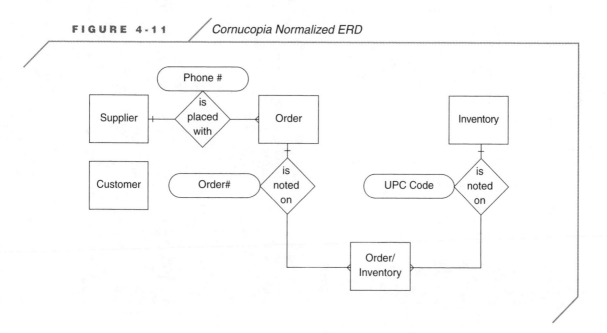

As you might expect, several iterations of the ERD are created before a final version emerges. This demonstrates another advantage to CASE tools: they make it easy to modify your work. The entire SDLC is a process that requires a perpetual sequence of development, review, and modification. Figure 4-11 shows the revised ERD. The final version, as it appears in Figure 4-12, would take many more hours to produce if each intermediate version were created anew.

FIGURE 4-12   *Cornucopia Data Modeling with Visible Analyst*

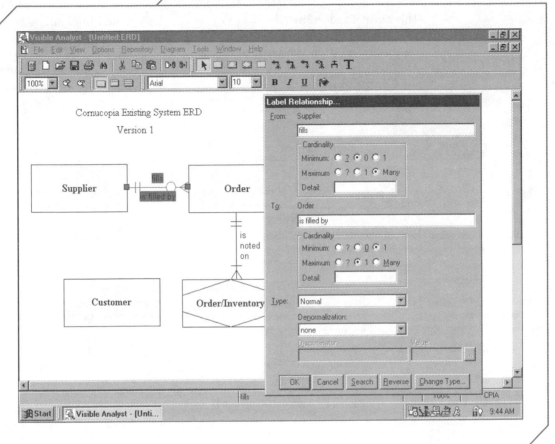

SUMMARY

This chapter provides a review of data types and data structures and then focuses on data files and databases. It distinguishes between master and transaction files—wherein master file records define persons or things and transaction file records describe events that involve those persons or things. To further characterize the nature of these events, this chapter distinguishes between online transaction processing and batch processing. The online transaction processing system presents special challenges due to the immediacy of the capture and processing of data, whereas with batch processing systems, there is less stress

placed on the information system because of the delay between these activities.

All of this prepares the way for a detailed discussion of the data model and the entity-relationship diagram. The analyst must reexamine data stores identified on the DFD, looking for detailed definitions and relationships. This process often leads to further redefinition as the ERD is normalized to eliminate data redundancies.

The ERD complements the DFD and further adds to your understanding of the system under study. Still, there are other views to be developed, as we shall see in the next two chapters.

TEST YOURSELF

Chapter Learning Objectives

1. What are the basic data types and methods of data organization?

2. How are master and transaction files similar and dissimilar?

3. What are the three possible file relationships in a relational system?

4. What does it mean to normalize an entity-relationship diagram?

Chapter Content

Refer to the Silhouette Sea Charter example to answer the following questions:

1. A master file contains record values that describe the characteristics of an object and the record values seldom change. True or False?

2. A transaction file contains record values that describe an event. True or False?

3. In the Silhouette Sea Charter example, the Charter file is a master file and the Customer file is a transaction file. True or False?

4. In the Silhouette Sea Charter example, the Boat file and the Captain file are master files. True or False?

5. In the Silhouette Sea Charter example, the Customer/Charter file is better described as an intersection file rather than a master or transaction file because it serves as a bridge between the Customer and Charter files. True or False?

6. Over a period of time, it is possible that several records in the Charter file will have the same value for Billing Party. True or False?

7. Over a period of time, it is possible that several records in the Customer file will have the same value for Customer. True or False?

8. An entity in an entity-relationship diagram is likely to appear as an external entity in the data flow diagram. True or False?

9. To say that one file is related to another is to say that the files have a common data field. True or False?

10. The beginning analyst must be able to normalize a database into the fifth normal form. True or False?

11. Which of the following fields from the Captain file would you substitute for the Captain field in the Charter file?
    a. SSN
    b. License #
    c. Address
    d. Phone

12. Which of the following fields from the Boat file would you substitute for the Boat field in the Charter file?
    a. Registration #
    b. Purchase Date
    c. Cost
    d. Length

13. Select the situation that exhibits one-to-many cardinality between the files.
    a. Hundreds of students enroll in many classes.
    b. Details of your birth record are kept at the county records office.
    c. A student completes many classes.
    d. Some employees possess a special credential.

14. Select the best group of attributes for the Customer/Charter file.
    a. Billing Party, Dates, Customer
    b. CharterID, CustID
    c. Purchase Date, License #
    d. CharterID, CustID, Registration #, SSN

15. If, over time, the same three customers take two different charters, how many records will this produce in the Customer/Charter file?
    a. three
    b. two
    c. twelve
    d. six

ACTIVITIES

Activity 1

Refer to the process model for the CIS Lab example (Figure 3-13) to identify the data stores and data files in the TKSystem. Answer the following questions:

    a. Which of these files would you classify as master files?

    b. Which of these files would you classify as transaction files?

    c. Describe the relationship between these files.

    d. Normalize these file relationships, adding new files if necessary.

    e. Prepare the entity-relationship diagram for the TKSystem.

Activity 2

Identify one master file and one transaction file in each of the following situations:

    a. Video store customers rent videos.

    b. Students enroll in courses.

    c. Credit card holders charge purchases.

    d. Patients make appointments with doctors.

Activity 3

Consider a library information system with two master files, as defined below:

CardHolder	LibraryBook
*CardNumber	*CatalogNumber
LastName	BookTitle
FirstName	BookAuthor

Define the fields in the transaction file named CheckOut.

DISCUSSION QUESTIONS

1. Describe three computer systems that you interact with as a user. In what ways are the systems friendly or unfriendly? Do you think users participated in the design of these systems? If so, to what extent?

2. Explain why the nonpiracy agreement and roster data stores appear on the TKSystem USD but not on the system flowchart.

3. Explain the correspondence between the data stores and data flows of the DFD and the data files and file attributes of the ERD.

4. In your own words, explain the cardinality of the relationship between the students in this course and the course itself. Draw an entity-relationship diagram to model this relationship.

5. Refer to Figure 4-4 and explain why the analyst might recommend that the Registration # and the SSN fields in the Captain and Boat files be substituted for the Captain and Boat fields in the Charter file.

6. What attributes would you suggest for the Customer/Charter file that appears in Figure 4-7?

7. Given the central importance of relational databases to information systems, to what extent must the analyst be an expert in database procedures?

PortfolioProject

Team Assignment 4: Data Modeling

This assignment builds upon the model building activity in Team Assignment 3. The easiest way to begin building the data model is to study the data stores identified in the process model. Normally, each data store on the DFD is a potential file, or entity, in the entity-relationship diagram. For example, if you identified a customer data store on the DFD, you should identify *customer* as an entity on your ERD. Sometimes a single data store might lead you to define several entities. For example, a casually defined inventory data store might actually require several related database tables to implement.

As with the DFD, the ERD provides another opportunity for you to increase your understanding of the system under study. Your first attempt at the ERD will probably require several revisions as you continue through the analysis phase.

In order to begin the data-modeling process and develop your first entity-relationship diagram of the existing system, you must complete the following tasks:

1. Prepare the existing system entity-relationship diagram. Submit a copy of this diagram to your instructor.
2. File a copy of the existing system entity-relationship diagram in the Portfolio Project Binder behind the tab labeled "Assignment 4."

objectmodeling

When you complete this chapter, you will be able to:

OVERVIEW

Object modeling provides a relatively new view of an information system. Whereas the data flow and entity-relationship diagrams differentiate and isolate data and processing into separate abstractions, the object model combines both elements into a single abstraction. The popularity of this approach is closely associated with the growth in object-oriented programming languages and event-driven 4GL software, both of which are discussed in later chapters. Be aware that in a small-enterprise environment, where off-the-shelf application software provides the information system's processing backbone, object modeling is rare.

This chapter begins with an overview of the object model methodology and rationale for its use. This is followed by a generic description of objects and how to construct the object model and the use case diagram. You will learn about the Unified Modeling Language (UML), which promises to become the standard for this approach to systems work. Finally, you will learn how to use a data dictionary to catalog all of the data elements associated with the project.

- Differentiate between object-oriented and traditional methodologies.

- Identify objects and construct an elementary object model.

- Identify the elements in a use case model.

- Identify the elements in four of the most common UML models.

- Define, construct, and maintain data dictionaries.

OBJECT-ORIENTED METHODOLOGY

Recall from earlier discussions that the SDLC includes several phases. Even though you have not yet studied the design and development phases, you should appreciate the notion that successfully completing these phases is dependent on a successful modeling effort during the analysis phase. Analysis-phase models should help you create design-phase models, which, in turn, should help you develop implementing software.

The first object-oriented programming language appeared in the mid-1960s. Event-driven software with a graphical user interface appeared in the 1980s. *Object-oriented systems analysis and design (OOSAD)* methodology evolved to accommodate these developments. The rationale is simple: if you plan to implement your design with object-oriented development tools, you should use an object-oriented methodology to create the design. Over time, many standardized objects are identified, designed, and developed for reuse in subsequent projects. As object libraries grow, the effort required to build new systems generally diminishes.

Briefly, an *object* is defined by two components: attributes and behaviors. Object attributes are similar to the fields that make up a record in a data file. Thus, a car object has attributes for make, model, year, value, and so on. Object behaviors are similar to the processes that manipulate data files. The car object has a behavior that permits the user to change the value attribute of the car. Behaviors are slightly more complex in that they are composed of executable programming instructions, called methods, which respond to events spawned by users or other objects. The owner of a car initiates a request to depreciate the value of the car. You might consider an object to be a self-contained, preassembled, pre-programmed, slightly customizable worker robot. An information system built on object-oriented technology is composed of a collection of objects that work together to satisfy the system's information-processing objectives. Naturally, objects are the primary focus of OOSAD. The object model and the use case diagram are two very important components of this methodology.

OBJECT MODELS

The *object model* is another way to extract details from a complex information system. In this model, the analysts refocus their attention on the data and processing components with a new perspective, as shown in Figure 5-1. As stated previously, objects embody both data and processing.

FIGURE 5·1 / *The Object Model and the Six CIS Components*

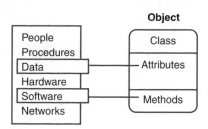

The analyst must first identify the objects and then determine how they work together to produce information. This activity is called *object-oriented analysis (OOA)*. As with the other models we have discussed, analysis helps us understand the system and prepares us for the object-oriented design activities that follow. TechNote 5-1 presents a synopsis of object-oriented terminology.

As the object-oriented methodology has matured, a consensus has developed about its nature and techniques. We will concentrate on those elements that are commonly used and that can be practically applied in the small-enterprise setting. You might characterize this as a gentle, practical introduction to OOSAD.

Object Identification

By definition, a *class* embodies *data attributes* and performs functions or methods. The class provides a blueprint from which one or more specific object *instances*, or occurrences, of the class are created. The class communicates with other classes through *messages*. A message contains service requests and responses. For example, a company's job announcement for a systems analyst position might request information on specific data attributes (name, address, phone, years of experience, professional references, etc.) and list the services the applicant is expected to perform. This defines a class named Systems Analyst. Each applicant is an instance of this class, with specific data attribute values and specific skills to perform the itemized services. Each instance of a class is an object. Once hired, the company requests services and receives responses from the analyst via e-mail or some other messaging system. In short, each instance of the class is an object that knows how to behave as a systems analyst.

TECHNOTE 5-1

ANATOMY OF AN OBJECT

If you have experience with an object-oriented programming language, then the objects and the defining terms below are familiar to you. Even if you haven't had such experience, objects are commonplace and easily recognizable in application software products. Clip art in word-processed documents, charts in spreadsheets, command buttons in Windows dialog boxes, and sound clips in Web pages are examples of objects. By right-clicking the mouse on these objects, you gain access to information about the object; sometimes, you can even manipulate the object's appearance and functionality. The following terms describe objects and how they operate.

Class is an abstraction of a real-world person, place, thing, or activity. There are two parts to the abstraction: **data attributes** and **methods**. Data attributes for the Teacher class might include name, title, department, and discipline. Methods are services performed by anyone who is a member of the class. Teachers lecture, advise, grade, or compose e-mail.

Instance is a specific person, place, thing, or activity. Instances are created from a class blueprint. Each instance is distinguished by the specific values assigned to its data attributes. Furthermore, each instance can perform the services defined in the object class.

Encapsulation describes the enveloping, membrane-like environment in which an object exists. The object is made functional by the methods associated with its class. Methods are normally of two types: methods that perform input/output services, so objects can communicate with other objects that exist outside the membrane; and methods that perform behavior services, so that the object can manipulate data inside the membrane. The object operates like an airplane's black box, wherein its contents are a mystery, but its functions are well known.

Messages are the means by which objects communicate with each other. For example, a teacher instance might use his or her e-mail method to assign homework questions. In turn, a student instance, after receiving such a message, might return answers via his or her e-mail method. The teacher now invokes a behavior method to grade the submissions.

Polymorphism allows for the possibility that a method may behave differently when encapsulated in different classes. Thus, we may obtain different responses from the same method, simply by altering the way we invoke the method. For example, the teacher's grade method behaves one way when invoked by a pass-fail student and another way when invoked by a letter-grade student.

Inheritance imprints class definitions on new generations of the class. This reuse of existing components proves to be a great time saver. For example, the Teacher object class could be used to spawn subclasses called elementary, junior high, and high school. These new classes automatically inherit the attributes and methods of the Teacher class, leaving only subclass-specific attributes and methods to be defined.

Class relationships describe the way classes interact with one another, the extent to which inheritance is employed, and the cardinality between classes. These interactive associations have very practical consequences when it comes time to design and implement the information system. For example, consider that, over time, a single Teacher class instance may teach several courses and a single Course class instance may be taught by several teachers. Understanding this, and the requirements of the implementing database software, the analyst is compelled to define a third class, called Schedule, to record each intersection of the teacher and course instances.

In keeping with our gentle introduction to OOSAD, we begin the *class identification process* with the data model. Recall that the entity-relationship diagram presents data files and their relationship to one another. Consider Silhouette Sea Charter's normalized ERD in Chapter 4 (Figure 4-7), with its Boat, Captain, and Charter files. To identify classes, we must satisfy the definition requirements by defining both data attributes and class functions. The data attributes requirement can be defined rather quickly by looking at the file structures (fields), but the functional definitions require more deliberate thought. Notice that the ERD suggests file functionality through the relationship description. In this situation, boats and captains are assigned to charters. In other words, to create a charter, the charter file must be able to assign a boat and a captain. Thus, the charter object now embodies data attributes (charter id, billing party, dates, etc.) and performs functions (assigning boats and captains). The Boat and Captain master files are classes as well. Their attributes are derived from their respective file structures, while their functions include the creation, update, and deletion of master file records, or instances, as we shall now refer to them. The object messaging system is implemented through object interfaces, more commonly known as graphical user interfaces or screen forms. Thus, the class called Boat includes a service (interface) that permits the user to create a new boat record (instance).

Although our discussion has ventured slightly into topics covered more completely in the design chapters, you can see the close association between the data model, the object model, and the database software used to implement information systems.

It is very important to note that an analyst would not develop all three of the models (DFD, ERD, OM) discussed to this point. We have simply used the DFD and ERD models as a springboard to introduce the object model. If the analyst chooses the OOSAD methodology, there is no need to develop the DFD or ERD. Indeed, there are well-developed practices for identifying classes and class relationships, as well as a widely accepted object-oriented modeling sequence.

Object Relationships

Classes relate and interact with one another in many ways. The Boat, Captain, and Charter classes have a *collaborative relationship*, wherein one class provides information to another class. In this case, Boat and Captain supply information (i.e., respond to service request messages) to Charter through an object interface. Figure 5-2 presents Silhouette Sea Charter's object model (OM). Notice that the model is quite similar to the ERD. One difference is the symbol set used to express the cardinality of the class relationships. Another difference is the identification of some special object class relationships.

FIGURE 5-2 / *Silhouette Sea Charter's Object Model*

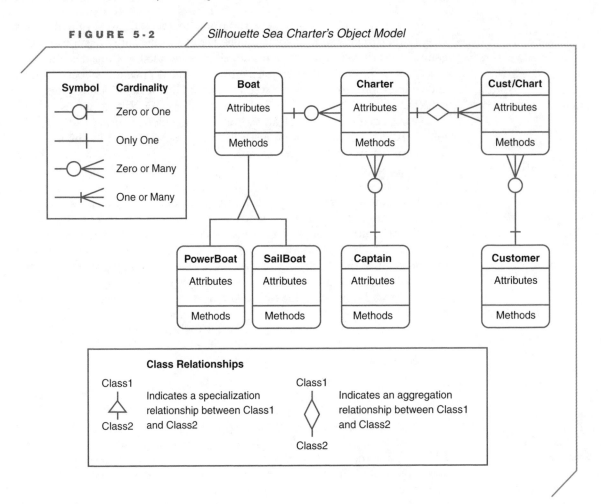

Classes defined as subsets of another class benefit from a principle called *inheritance*. Just as a child inherits characteristics from its parents, a subclass inherits attributes and methods from its superclass. For example, Silhouette Sea Charter could define SailBoat and PowerBoat as subclasses of Boat. This "is a" relationship, referred to as *specialization*, means that each SailBoat and PowerBoat is a subset of Boat. Subsets inherit all of Boat's attributes and methods, plus, they may have their own specialized attributes and methods not shared by Boat or other subsets of Boat.

Finally, there is the "has a" relationship, referred to as *aggregation*. This relationship exists when an instance of one class is composed of one or more instances of another object class. Cust/Chart is related to Charter in this way, indicating that one or more customers can sign up for the same charter.

USE CASE MODELING

Use case modeling provides another way to identify objects. It requires the analyst to develop a series or set of task scenarios. Each scenario is focused on a system event, describing the person who initiates the event and the service performed by the system. A use case is literally a detailed description of what happens when someone or something interacts with an information system. For example, customers must interact with Silhouette Sea Charter's information system to reserve a charter. This event sets in motion a series of activities designed to complete the reservation process. This use case might be titled "reserve charter." Other use cases might be "create new customer," "add new boat," and "pay boat captain."

It is probable that use cases need to interact with one another. Each pay period the pay boat captain use case must get information from reserve charter in order to determine which boat captains to pay. The meticulous and sometimes tedious task of defining each use case and the relationships among use cases constitutes use case modeling. Eventually, the nouns that appear in use case titles provide a basis for object identification. In other words, Silhouette Sea Charter's reserve charter use case implies the existence of a charter object that is an instance of the Charter class. Furthermore, when this class is actually implemented during the development phase, we are likely to find that each instance of Charter is an object that knows how to behave as a charter.

UNIFIED MODELING LANGUAGE

Within the last few years, the *Unified Modeling Language (UML)* has emerged as the generally accepted modeling standard for object-oriented methodologies and techniques. As with any modeling approach, the UML provides a means by which the analyst can represent a complex system as a set of simple diagrams, charts, and narratives. The UML makes it easier for the analyst to transition from one phase to the next because each element in the modeling set is part of the same methodology. Thus, there is a natural continuum of work, with consistent terminology, symbolism, and point of view.

UML Diagrams

There are nine UML modeling diagrams, as identified in Figure 5-3. We will look at four mainstream diagrams that are either similar in function to traditional diagrams (DFD, ERD, data flow walk-through) or particularly instructive about the way an object behaves in an event-driven application.

FIGURE 5-3   Nine UML Modeling Diagrams

1. **Class diagram**
   This diagram shows the existence of classes and their relationship in the logical view of a system.
2. **Object diagram**
   This is a special variation of the class diagram showing specific instances of classes.
3. **Use case diagram**
   This diagram shows the systems' use cases and which actors interact with them.
4. **Interactive (Sequence) diagram**
   This is an interactive type diagram that shows the objects in the system and how they interact via time-ordered messages.
5. **Interactive (Collaboration) diagram**
   This interactive type of diagram also shows the objects in the system and how they interact via messages, but it emphasizes the object relationships rather than the timeline dimension of the sequence diagram.
6. **Statechart diagram**
   This diagram shows the state space of a given context, the events that cause a transition from one state to another, and the actions that result.
7. **Activity diagram**
   This special type of statechart diagram emphasizes the flow of control between objects.
8. **Component diagram**
   This diagram shows the dependency between all of the software components in the system.
9. **Deployment diagram**
   This diagram shows the configuration of runtime processing elements.

- **Use case diagram:** The use case diagram's focus is on the functions (uses) and users (actors) of the information system. This model corresponds to the context diagram and first-level data flow diagram in that uses are similar to processes and actors are similar to external entities.
- **Class diagram:** A class diagram shows the set of classes and class relationships associated with a use case. This is the most common object-oriented model. It closely corresponds to the entity-relationship diagram, wherein the object class somewhat resembles an entity.
- **Interaction diagrams:** The interaction diagrams, which are classified as either sequence or collaboration diagrams, show how several objects work together to provide services. In its different forms, these models document both the nature and sequence of messaging between collaborative objects. Sequence diagrams are more detailed than collaboration diagrams and are generally preferred by systems analysts.

- **Statechart diagram:** The statechart diagram shows how objects react to a variety of events and stimulations from other objects. By identifying these reactions, the analyst sets the stage for the design of object behaviors.

Silhouette Sea Charter's UML Diagrams

To illustrate, we will discuss how these four diagrams apply to Silhouette Sea Charters. Recall that the traditional approach begins with the process model, where the analyst first develops a context diagram and then a first-level data flow diagram. The UML approach is similar in that its focus is on the processes, or uses, of the system. Each major function of the system is considered a use case. Each use case interacts with people or systems outside of the system under study. This notion fits neatly with the external entities that appear in the context diagram. Thus, as illustrated in Figure 5-4, the use case diagram is very much like our familiar context diagram. In this image, actors are similar to external entities, use cases are similar to processes, and associations are similar to flow lines.

FIGURE 5-4   */Silhouette Sea Charter's UML Use Case Diagram*

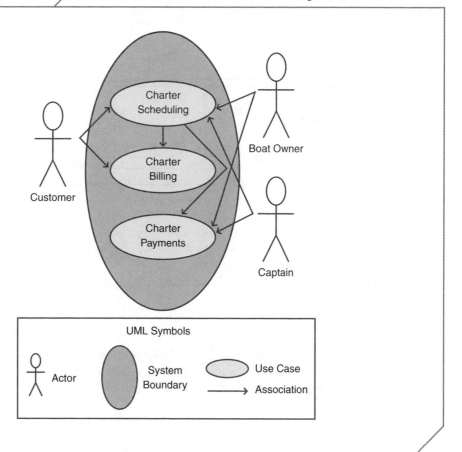

The next UML model is the class diagram, depicted in Figure 5-5. It is no coincidence that the class diagram looks very similar to the OM. Indeed, the OM is a forerunner of the class diagram. There are three differences to note. First, the UML class is represented by a simple rectangle rather than a rounded rectangle. Second, the cardinality notation is numerical rather than graphical. Third, UML supports many-to-many class relationships, as shown between Charter and Customer. It is important to note that the class diagram illustrated here is related to a single use case (scheduling) in the use case diagram, but in practice, class diagrams may be larger in scope.

FIGURE 5-5   / *Silhouette Sea Charter's UML Class Diagram*

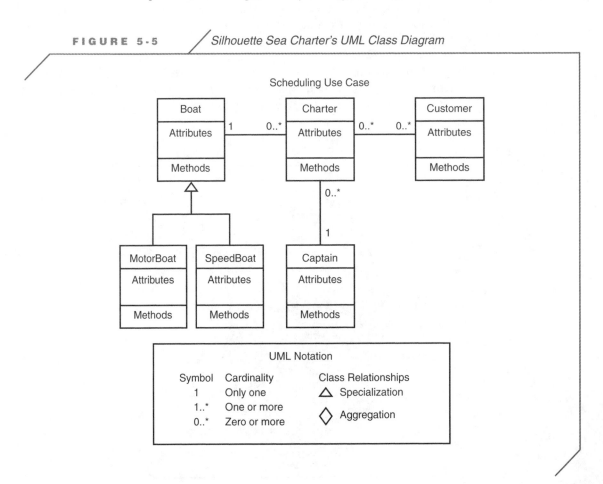

The UML interactive sequence diagram in Figure 5-6 shows how classes work together by passing messages between one another. Remember that a class embodies both data attributes and methods. In other words, the class object can actually perform services and manipulate data. Thus, the Boat class can respond to an inquiry message from the Charter class by finding an available boat to fulfill a particular charter request. The interactive diagram shows the connecting messages between classes.

FIGURE 5-6   *Silhouette Sea Charter's UML Interactive Sequence Diagram*

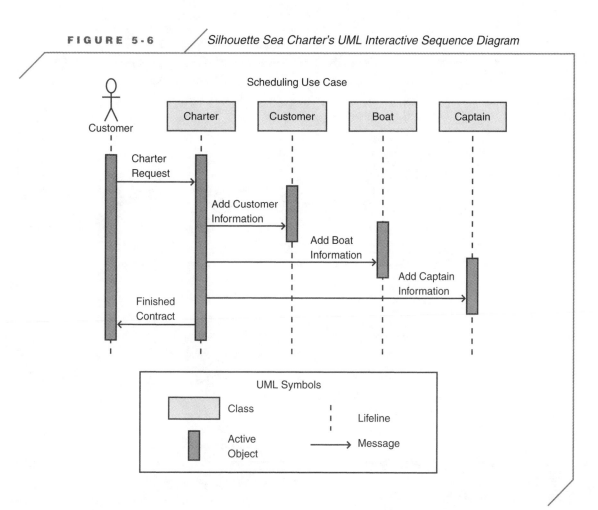

Sometimes the operations performed by a class are so complex that it is necessary to develop a statechart model to help describe the process. This model shows how the class attributes change as events occur during the process. For example, Figure 5-7 shows how the Charter class goes through several states in order to process a charter request. In fact, the transitions from one state to the next correspond to the messages identified in the interactive sequence diagram.

FIGURE 5-7   /   *Silhouette Sea Charter's UML Statechart Diagram*

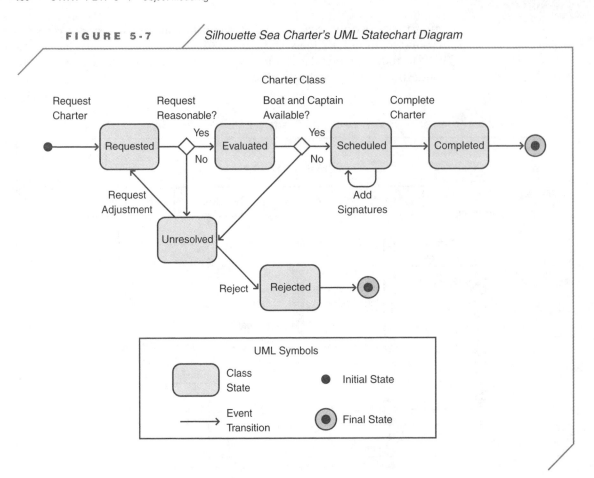

InflatableLabor**Charges**

Your firm employs four analysts, who are assigned to work on many different projects at the same time. One of your projects involves the Redevelopment Agency in Lexington, Massachusetts, where you are designing and developing an information system to help the agency manage a small business park.

As the project leader, one of your duties is to validate the analyst hours charged to the Redevelopment Agency project. You discover that, while the project is on schedule and within its budget constraints, your most talented analyst has consistently inflated the labor hours charged to the project. The analyst reports that this extra time was spent learning about object-oriented methods that might apply to this project. However, none of these methods were used on the project.

What action should you take?

Although this abbreviated example may not show it well, the UML uses similar syntax, symbols, and notation in all of its diagrams. Furthermore, when the diagrams are taken as a whole, they model every phase of the SDLC. These notable features give substance to the claim that the UML is a unified, consistent, and very efficient modeling approach. With all of these models, the focus is on objects: their distinguishing characteristics, behaviors, and relationships; how and by whom they are used; how they interact with one another to provide complex services; in what order they perform actions; and how they respond to events. Taken together, these models provide a blueprint for constructing the information system with object-oriented software products.

THE DATA DICTIONARY

The analysis phase and modeling process creates an avalanche of data descriptions. It is impossible to keep track of all the names, definitions, associations, and existing or proposed uses of this data without a systematic procedure. A *data dictionary*, much like a standard language dictionary, is an ordered catalog of terms and their definitions, but, in this case, of data element names and their definitive characteristics. Every data element should be included so that data names are not duplicated or assigned one name in one part of the system and another name in another part of the system. Using the standard dictionary entry form and symbol set, as illustrated in Figure 5-8, promotes consistency.

This chapter has already discussed some of the data entries required in the dictionary. The data stores and data flows of the DFD are the first items entered. This, incidentally, should make the construction of the other models easier because the definitions for the counterpart files and attributes already exist. As the design for the new system unfolds, the analyst adds entries and maintains an up-to-date data dictionary.

FIGURE 5-8   /   *Illustrated Data Dictionary Form and Symbols*

Form Symbols		Form Entries	
Multiple entries are permitted.	{ }	Element Name:	Uniquely names and describes the data element.
This is not a required entry.	[ ]	Type:	Describes the data as a data flow, a data store, or an independent data stream.
This is the key field or attribute.	Underline	Description:	Provides narrative comments about the data element.
The data structure is composed of the following attributes.	=	Content:	Defines the attributes for data stores.
		Usage Cross-Reference:	Lists other uses or names for the data element.
This symbol separates the attributes.	+	Storage Reference:	Provides the disk filename and path.

Data Dictionary Entry Form

Element Name: *customer*   Type: *data store*

Description: *customer master file*

Contents: customer = *custid +*
  name +
  address +
  phone

Usage Cross-Reference: *DFD, USD, ERD,*
  System Flowchart

Storage Reference: *customer.dbf*

CASE TOOLS FOR OBJECT MODELING

CASE products have kept pace with the evolution of the object-oriented programming languages, 4GL software, and OOSAD methodologies, including the new UML diagramming models. The latest version of Visible Analyst provides a tool to evolve entities from the data model into classes in the object model.

THE CORNUCOPIA CASE

Given the constraints imposed on this sample project, there is little practical value in creating the object model. The analysts do not intend on using an object-oriented programming language to implement Cornucopia's new information system. Even the objectlike features of our implementing 4GL software do not require object identification or use case modeling. Nevertheless, this does provide another opportunity to demonstrate how an analyst might make the transition from the data model to the object model.

The Object Model

The object model (OM) is easily derived from the ERD. Figure 5-9 shows Cornucopia's existing system OM, with five classes and annotations describing the types and inheritance of the class associations. The data attributes are the same as those identified for the data model, with bold type used to signify attributes that are used to communicate between classes. In the Order class, notice that the originating class name is used as a prefix to avoid confusion about naming. The use of standard methods, such as create instance, reflects a very casual analysis of the services provided by the class objects. More specific methods will be identified during the design activities.

The aggregate relationship between Order and Order Line means that a single order instance must have one or more corresponding Order Line instances. This reasoning follows the normalization rules that apply to relational databases. This is important to remember because eventually the analysts will use a relational database to implement their design.

FIGURE 5-9 / *Cornucopia Existing System Object Model*

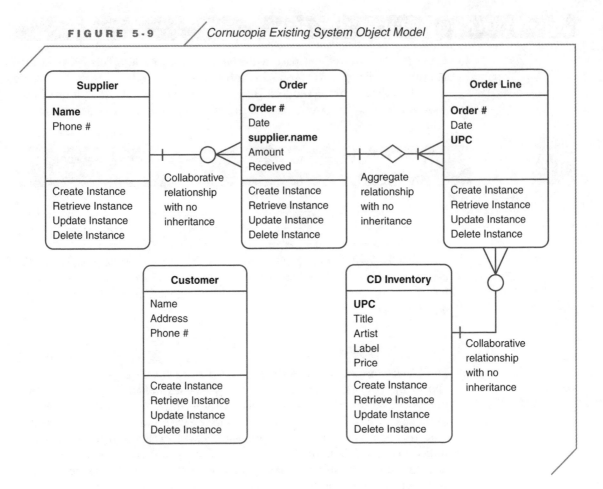

Cornucopia with Visible Analyst

Cornucopia's Customer class definition appears in Figure 5-10. This dialog box is available from the Repository menu. Visible Analyst's repository automatically catalogs every data element, every file, every diagram, and every graphical item entered for the project. As illustrated by the numerous form tabs, text boxes, and control buttons on this form, this CASE tool provides a complete data dictionary, with many ways to define and describe every item in the project.

FIGURE 5-10 *Cornucopia Object Definition with Visible Analyst*

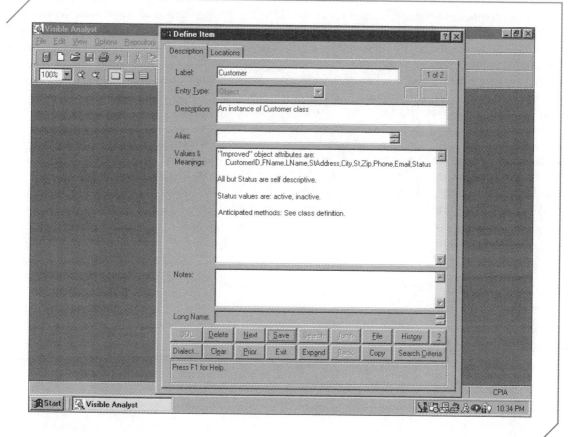

SUMMARY

This chapter presents the object model as an alternative approach to systems analysis and design. Object-oriented technologies are known for their reusable libraries of objects, sensitivity to the increasingly event-driven nature of information systems, and user-friendly graphical user interfaces. Even though the small-enterprise analyst may not engage in object modeling to the extent described in this chapter, object-oriented technologies are important because of the analyst's extensive use of off-the-shelf 4GL software, which is object based. Furthermore, an analyst's responsibilities often shift to medium and large-scale enterprises, where object-oriented programming is rapidly becoming the norm. For these reasons, the systems analyst is well advised to become familiar with the object model, use case diagrams, and the Unified Modeling Language diagramming models. This task is made easier by the parallels between the data model and the object model, as well as between the process model and the use case diagram.

The next chapter presents a broader view of the information system through several system models.

TEST YOURSELF

Chapter Learning Objectives

1. What are the basic similarities and differences between the object-oriented and traditional methodologies?

2. How are objects, classes, and object relationships depicted in the object model?

3. How does use case modeling help the analyst identify objects?

4. How do use case, class, interactive sequence, and statechart diagrams model an information system?

5. What is the primary purpose of the data dictionary?

Chapter Content

1. Select the data flow diagram component that is closely related to entities on the ERD and classes on the OM.
   a. external entity
   b. process
   c. data store
   d. flow line

2. How does the OM compare to the UML?
   a. There is no comparison; they are two totally unrelated models.
   b. They are similar; the UML is much more complete, providing a unified set of models for the analyst to use.
   c. The UML is a nonstandard modeling concept.
   d. The OM is the only accepted object-modeling standard.

3. Which of the following statements correctly expresses the way class definitions are used in an object-oriented environment?
   a. The class blueprint is used to create an instance of the class.
   b. The class definition includes data attributes and methods, but has nothing to do with objects.
   c. Only graphic images can be defined as objects, which limits the usefulness of class definitions.
   d. Object-oriented environments are only found in scientific applications, which limits the usefulness of classes and the identification of enterprise objects.

4. The object identification process can be shortened by first creating which of the following products?
   a. the project contract
   b. the data flow diagram
   c. the object model
   d. the entity-relationship diagram

5. Which of the following is an example of the specialization relationship between classes?
   a. A teacher has 20 students enrolled in her first-grade class.
   b. A single purchase order has eight individual items listed.
   c. A special-teams player is a type of football player.
   d. A student completes about 30 courses as an undergraduate.

6. Which of the following is an example of the aggregate relationship between classes?
   a. A teacher has 20 students enrolled in her first-grade class.
   b. A single purchase order has eight individual items listed.
   c. A special-teams player is a type of football player.
   d. A student completes about 30 courses as an undergraduate.

7. Which of the following is *not* an example of the collaborative relationship between classes?
   a. A teacher has 20 students enrolled in her first-grade class.
   b. A single purchase order has eight individual items listed.
   c. A special-teams player is a type of football player.
   d. A student completes about 30 courses as an undergraduate.

8. Class methods can be considered software because they are really small programs designed to perform a specific service. True or False?

9. The only way for a class to inherit attributes and methods from another class is to copy and paste all of the associated code from one class to another. True or False?

10. Object relationships include an expression of the cardinality between classes because the implementing software imposes specific cardinality rules. True or False?

ACTIVITIES

Activity 1

Visit Visible Systems Corporation's Web site (*www.visible.com*) to learn about the latest features of Visible Analyst. Prepare a one-page summary of your findings.

Activity 2

Review the entity-relationship diagram you prepared for the TKSystem at the end of Chapter 4 (Activity 1). Prepare an object model for the existing TKSystem.

Activity 3

Visit Rational Software Corporation's Web site (*www.rational.com*) to learn about the latest features of UML. Prepare a one-page summary of your findings.

DISCUSSION QUESTIONS

1. How do the attributes and behaviors of the Windows desktop reflect object-oriented technology?

2. What are the similarities and differences in the OM and the ERD?

3. What are the similarities and differences in the use case diagram and the DFD?

4. What are the advantages to developing an ERD before attempting to develop the OM?

5. In what ways are the OM and the UML class diagram similar and dissimilar?

6. Why is it important to use a data dictionary beginning with the very first model you develop?

PortfolioProject

Team Assignment 5: Object Modeling

As stated numerous times in this chapter, a small-enterprise project is not normally a candidate for object-oriented methodologies. Nevertheless, this exercise provides an opportunity for you to become familiar with the object model and the UML class diagram. In order to develop an object model from the data model you completed in the last chapter, you must complete the following tasks:

1. Review the existing system entity-relationship diagram you prepared for Team Assignment 4, giving consideration to the potential for each identity to be defined as a class.
2. Prepare the UML class diagram for the existing system. Submit a copy of this diagram to your instructor.
3. File a copy of the existing system UML class diagram in the Portfolio Project Binder behind the tab labeled "Assignment 5."

systemmodeling

OVERVIEW

System models provide a variety of holistic, physical views of the information system. They help users, analysts, and programmers develop real-world associations to counteract any confusion or apprehension that results from the system's logical framework, which is generally hidden from view. Each of the system models presents a slightly different combination of the six components of a computer information system.

In this chapter, you will learn how to create three system models: the user's system diagram, the menu tree, and the system flowchart. Combined with the process, data, and object models, these system models can help you understand how an existing information system works, and they will undoubtedly suggest ideas for how a new information system should work. As a precursor to the design phase, we will briefly discuss new systems in this context.

Although a detailed discussion of project management appears in Appendix A, the Cornucopia case includes practical examples of several important project management tools. Project budgets, status reports, task and resource management, and disk directories are implemented with everyday software products.

- Create a user-friendly model of how the system works.

- Create a user-friendly model of functions and services the system provides.

- Create a system flowchart.

- Develop several elementary project management tools.

THE USER'S SYSTEM DIAGRAM

The models described in this chapter provide wide-angle, practical, user-friendly views of the information system. Each model focuses on a slightly different set of components (Figure 6-1). We begin with the *user's system diagram (USD)*, which is an image-based, iconographic diagram that is adapted to suit the analyst's need to model the information system for the user. The user is a key participant in the SDLC. The relationship between analyst and user begins with the request for services and continues throughout the project. Many of the analysis activities described in the preceding chapters involve the user as a source of information for the analyst. To ensure that the user continues to be an effective and well-informed participant, the analyst must maintain regular communications with the user.

FIGURE 6-1   *System Models and the Six CIS Components*

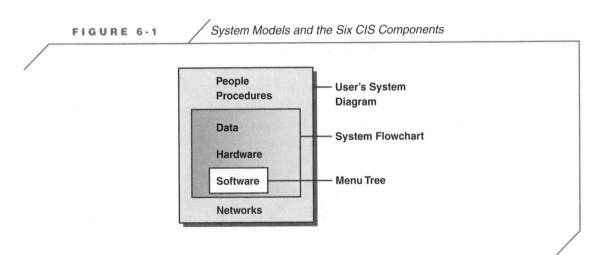

For example, after you develop the context diagram and the data flow diagram for the existing system, you should return to the user to verify your work. However, you must remember that the DFD is primarily intended to help the analyst understand the system. This model may not be very user-friendly, depending on the user's experience and the complexity of the system. Therefore, you need to develop a better communication tool. This USD model is specifically designed to describe the system in terms the user can understand—this begins to build the two-way communications necessary for productive user participation.

Computing and the information system end user have matured to the point that every enterprise, regardless of its size or purpose, must reconsider the way modern information systems are designed, developed, implemented, and maintained. In the Information Age many workers use desktop systems that, with software included, cost between one and two thousand dollars. Yet, these computer systems deliver more processing power than a 1985 IBM 3090 mainframe computer originally valued at millions of dollars.

However, statistics on cost and processing power quantify only one dimension of the Information Age. The end user has undergone similarly dramatic transformations. The fanfare associated with new hardware and software is rivaled by matching increases in end users' computer expertise and sophistication. Corporate information system professionals are racing to develop methods to blend end users' desktop processing solutions into existing enterprise-wide computing systems. It follows that new information systems must also accommodate these new realities.

Contemporary systems analysis and design methodology is based on well-established systems analysis procedures and is sensitive to the emergence of the desktop computer and the end user as elements of principal importance in today's information systems. System modeling provides a good opportunity to develop and maintain an effective relationship between the analyst and user.

The user's system diagram is simply a data flow diagram with (1) familiar, user-friendly images replacing the analyst-friendly symbols and (2) generalized, document-level data descriptions replacing the detailed data element names. The purpose of the USD is twofold. During the analysis phase, it depicts the analyst's understanding of the current system in terms the user can quickly understand. This helps the user validate the analyst's work. During the design phase, the USD increases the likelihood that the user can quickly recognize the critical elements of the proposed information system. This helps the user to suggest changes and confirm acceptance of the new design.

Although the USD is not one of the formal computer information system modeling methods, you will recognize it as a popular form of communication. For example, Figure 6-2 presents a diagram that recently appeared in newspapers across the country. It attempts to show how the cloning process works. The degree to which a complicated process is simplified into familiar icons is a matter of judgment on the part of the author of the article, but the intent is clear. The resulting image is a communication device, designed to help the lay person understand a complex system.

FIGURE 6-2 / *How the Cloning Process Works*

Researchers at the George Washington University Medical Center in Washington have possibly extended a cloning process, known to work on amphibians and some mammals, to humans.

1 *In vitro* fertilization occurs.

2 Fertilized egg develops into young embyro.

3 Egg is taken apart into separate cells or nuclei.

4 An egg is obtained from another woman, and the material containing genetic information is removed through a surgical procedure. Proteins and fats necessary to egg development remain.

5 A nucleus from the first fertilized egg is then implanted into the second unfertilized egg. This process, called nuclear transplantation, imbues the second egg with genetic characteristics of the first.

This process can generate one or more clones with the same characteristics.

Source: "Cloning of Frogs, Mice and other Animals," McKinnell, Robert G.; University of Minnesota Press. New York Times graphic.

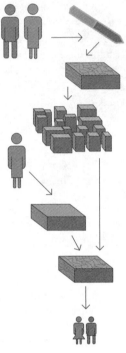

To illustrate further, we will create the user's system diagram for the TKSystem. Remember that the first-level data flow diagram in Chapter 3 (Figure 3-13) identifies seven processes, each of which performs a different function. Each process is complemented by data flows, data stores, and external entities.

In the USD (see Figure 6-3), we substitute icons for the standard symbols in the hope that we can make the diagram more familiar to the user. To make the diagram less complicated, the data flow annotations are replaced by single lines that show only the direction of the flow.

FIGURE 6-3   *TKSystem USD*

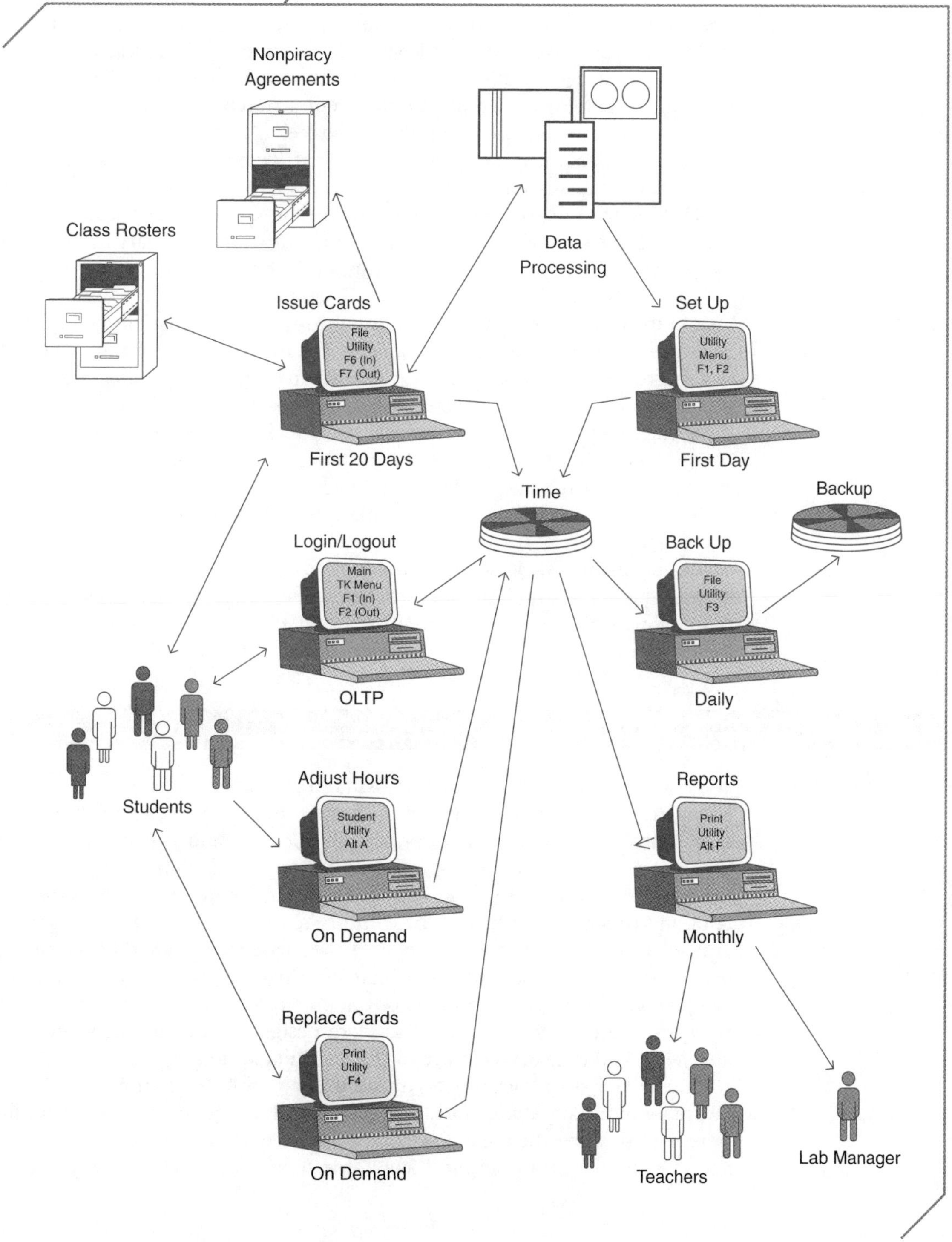

Nonpiracy Agreements

Class Rosters

Data Processing

Issue Cards
File
Utility
F6 (In)
F7 (Out)

First 20 Days

Set Up
Utility
Menu
F1, F2

First Day

Time

Backup

Login/Logout
Main
TK Menu
F1 (In)
F2 (Out)

OLTP

Back Up
File
Utility
F3

Daily

Students

Adjust Hours
Student
Utility
Alt A

On Demand

Reports
Print
Utility
Alt F

Monthly

Replace Cards
Print
Utility
F4

On Demand

Teachers

Lab Manager

In addition, several new pieces of information are included to make the diagram more relevant to the user's everyday experience. First, the computer screens show the menu names and function key options associated with each process. Second, the computer keyboards are titled with a description that informs the reader about how and when the process is invoked. For example, the Login/Logout option employs online transaction processing (OLTP), which allows users immediate access to their time records. The Adjust Hours option, on the other hand, is run only when a student "demands" an adjustment.

Because this is a nonstandard system model, the analyst has great latitude in selecting the images and layout of the diagram. The actual construction of the diagram is easy with a simple paintbrush program. A library of predrawn images, popularly known as clip art, can make the job even easier.

Notice that, by design, this diagram illustrates the system in a physical way. That is, the icons are supposed to remind the reader of familiar physical objects. Sometimes this may suggest a completely machine-based operation, when, in fact, a great deal of nonmachine activities may be involved in a particular process. Nevertheless, the diagram is intended as a communication device for the average computer user, and as such, the USD should be tailored to meet the needs of each situation. In some circumstances it may be best to construct several versions, each with a different audience and purpose. A given enterprise's management, for example, might require a diagram that shows less operational detail, allowing—or requesting—the analyst to emphasize the interfaces to other information systems within the organization. In the TKSystem example, as you may recall, the lab manager wants us to develop documentation and training materials for the system. The USD is well suited as an introduction to the other, perhaps more complicated, documents in such a package.

THE MENU TREE

Today's computer user is perhaps most familiar with the menu tree model. A *menu tree* is a hierarchical display of the operational choices available to the user. Many popular computer application packages and consumer products use a variety of pull-down, pop-up, and button menus to inform the user of the operations available within a specific context. For example, the automated teller machine presents deposit or withdrawal options only after the user has entered a valid password. Not only does the menu tree bridge the analyst and user worlds, but it also adds important information about how the software can be constructed piece by piece, or one menu option at a time. As we shall see later in the TKSystem project, modern design and development methods are well suited to take advantage of this modular approach.

In the TKSystem, each process is part of a hierarchical collection of processes. Figure 6-4 illustrates the menu selections for each of the seven processes in the system. In this way, the menu tree corresponds very closely to the previous DFD and USD models. For example, Issue Cards appears as a rectangular symbol in the data flow diagram, a desktop computer in the user's system diagram, and a series of function key selections in the menu tree.

Notice, however, that the external entities, data flows, and data stores are eliminated from this model. In a way, the menu tree presents an important functional abstraction that is useful to both the analyst and the user. It tells the user exactly how to invoke specific system functions. It tells the analyst a great deal about how the functional software modules fit together. It also provides another tool for effective communication between the analyst and the user.

F I G U R E   6 - 4   */TKSystem Menu Tree*

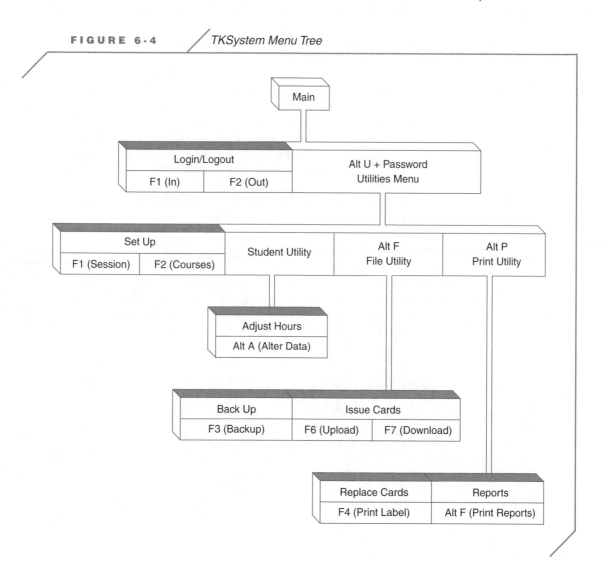

THE SYSTEM FLOWCHART

We now move to a system model that is designed to serve the computer professional more than the user. The *system flowchart* shows the relationships between software and data inputs and outputs, and at the same time, it depicts the real-world devices used to implement the system. Thus, it distinguishes keyboard from scanner input, disk from tape storage, and screen from printer output. This information is useful to the analyst because it isolates the computer programs—and the data files on which they operate—from the rest of the information system components.

For now, though, notice that Figure 6-5 uses the standard system flowchart symbols to illustrate the system flowchart for the TKSystem. In this diagram, the programs are represented as rectangles, the input and output devices are represented by several symbols, and the input and output operations are indicated by the direction of the arrows on the connecting flowlines. There is a close association between this chart and the earlier data flow diagram model, but that may not be apparent at first glance. The three files at the top of the chart are not shown on the DFD because they are really a part of the data-processing system, which is an external entity to the TKSystem. The Time files are shown as data stores on the DFD, while the Report and Summary files are depicted as data flows. The login/logout input stream is also a data flow. The Label file is actually embedded in the DFD's Replace Cards process.

Notice in our sample that the seven processes are collapsed into one program. In reality it is more likely that each process would constitute a separate module within a single controlling program. Nevertheless, this illustration is useful as a system diagram. Later chapters explain other models, such as structure charts and program flowcharts, which address the issue of internal program structure.

FIGURE 6-5 *TKSystem Flowchart with Symbols*

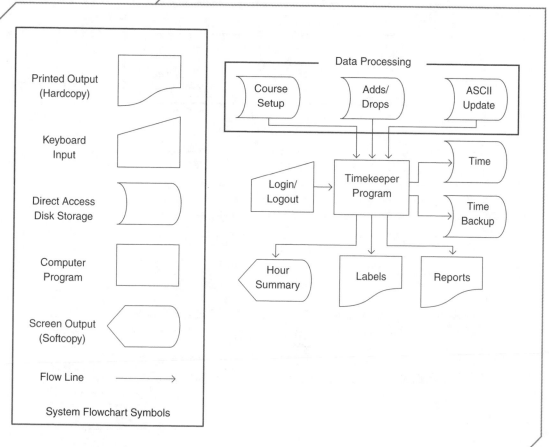

To illustrate a fairly common use of the system flowchart, consider how the TKSystem might interface with another information system. Suppose that the lab manager is required to implement a system to keep track of computer usage, repair, and maintenance. Wouldn't it make sense to look at the TKSystem to see whether some of its information could be used in the new system? Because the system flowchart provides a wide-angle view of the files and their particular implementation, the analyst can use it to find such a file. In this instance, the Report file is a good candidate, as long as its hardcopy implementation can be changed to a direct access disk implementation. The file could then serve as input to the new system (shaded area), as illustrated in Figure 6-6.

FIGURE 6-6   TKSystem Flowchart (Revised)

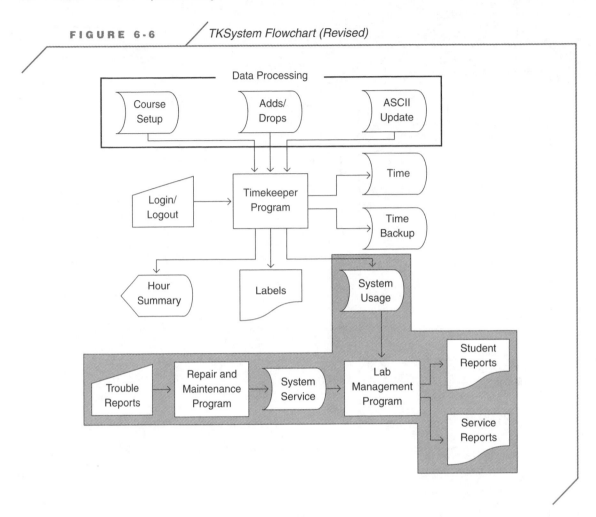

Thus, our system flowchart begins to mature, showing more complex relationships among programs and files. In this case, the output from one system becomes the input to another. You certainly don't want to confuse the TKSystem users with this new development. In this example, you are content merely to change the manager icon on the USD to a system icon of some type, use the revised system flowchart to reflect the broader system, and proceed to develop a much broader lab management information system. Therein lies the true nature of systems work. The task at hand begins to act like a mixture of yeast and water. It may tend to grow and grow, sometimes well beyond the original intent. When this happens, the analyst is wise to retreat to the project contract to refocus attention and energies.

To prepare the system flowchart you can use the DFD to identify some of the obvious data inputs and outputs. But, you will probably need to consult the program specifications in order to complete the chart. For an existing system, these specifications should be found in the system documentation. For a new system, the analyst must develop the specifications as a part of the design phase.

To be sure, the system flowchart is not the perfect abstraction. It does not show all six components of the information system, and it informs only a small segment of the SDLC participants. However, the same could be said of the DFD, and the USD and menu tree models we developed earlier. The point is that this is simply one more tool that the analyst can use to come to terms with the system under study.

NEW SYSTEM DESIGN—A FIRST LOOK

As discussed in Chapter 1, the enhanced SDLC acknowledges the blurred distinction between analysis and design. Indeed, it is natural for the analyst to consider potential design innovations or modifications while investigating and charting the existing system. Figure 1-11 is repeated here as Figure 6-7. The boxed area indicates our approximate position in the process as we proceed in the inevitable changeover from analysis to design.

FIGURE 6-7 / *The Enhanced SDLC (Reprise)*

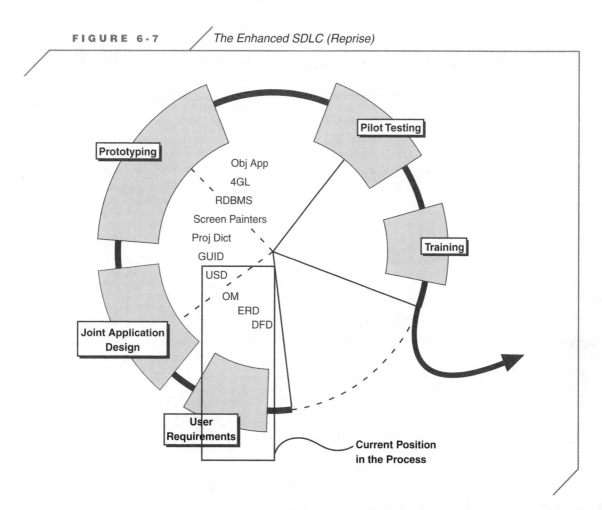

The Proposed New System USD

The modeling work directed at the current system, from DFD to ERD, acquaints the analyst with many facets of the project. One element that is evident through all these modeling efforts is the importance of user interaction in the process. The user is often eager to be a part of the early design work. Because the USD offers the most user-friendly picture of the system, the analyst should attempt to sketch the new system in these terms.

However, this may seem to be a premature exercise. After all, we have barely moved one-fifth of the way through the SDLC. Granting that all projects differ, the inherent system complexity may make this attempt at early design futile for medium- or large-enterprise systems, even with the enthusiastic participation of the user. The small-enterprise system, on the other hand, is much more likely to offer an early opportunity to develop the broad parameters of the new system if you simply follow the path suggested by the project contract.

Remember that the USD is a very high-level view of the system. It probably won't commit the analyst to any specific design detail, yet it pulls the user into the design process very early and provides some point of reference for the project schedule and budget that are to follow.

CASE TOOLS FOR SYSTEM MODELING

CASE tools certainly provide features for creating system flowcharts, but they do not accommodate some of the user-friendly modeling presented in this chapter. Furthermore, proprietary software development environments, such as Microsoft Visual Studio, gain in popularity with each new and improved version. Without detailed corporate partnership agreements, CASE tools cannot support these proprietary products. In addition to continually monitoring the evolution of traditional CASE tool products, the analyst must be aware of alternative model-building software.

ThinkingCritically

Open-Minded Discovery

It is very likely that the analyst and client will explore a lot of ideas for the new system during the early stages of the SDLC. What level of commitment, legal or otherwise, is there to deliver on these ideas?

MODEL-BUILDING SOFTWARE

While CASE tools, such as Visible Analyst, provide sophisticated modeling environments, along with data dictionaries, code generation, error detection, and so on, there are other model-building products on the market.

Microsoft Visio is one example (*www.microsoft.com*). The standard version of Visio offers more than 25 different templates from which to choose. Aside from the standard process, data, and object-modeling shapes and symbols, there are networking, work flow, and office layout templates as well. This makes it easier for the analyst to create user-friendly system models such as the USD and menu tree. In addition, Visio offers project management templates for preparing status charts, schedules, timelines, and task dependency diagrams.

Rational (*www.rational.com*) markets a product called Rational Rose, which is geared to the UML framework. On its Web site it writes, "Software development spans the disciplines of use case and data modeling, architectural modeling, component modeling, code construction and unit testing." It goes on to state, "Our market-leading modeling tools will help you create software that meets requirements, yet remains resilient to change."

Oracle (*www.oracle.com*) offers Oracle Designer, which "provides an intuitive modeling environment." It too supports the usual variety of visual modeling approaches, as well as tools to model its database applications, including client/server and Web-based applications. Oracle declares, "With its integrated team-working environment, Oracle Designer delivers a dramatic increase in productivity for application developers."

THE CORNUCOPIA CASE

The formal analysis of the current system concludes with the development of system models, namely the user's system diagram. At this point in the process the analysts have a good understanding of the old system and, perhaps surprisingly, a rough idea of what the new system will include. The transition from analysis of the existing system to design of the new system is gradual, as evidenced by an informal list of ideas for the new system.

The continuing interaction between the analysts and owner produces a general awareness of their approach to the project, but they have not formally presented their ideas to the enterprise. To structure their efforts and assure the owner that the analysts have the necessary project controls in place, the analysts must prepare several project management products and present a coherent plan. The ensuing analyst presentation, referred to as the Preliminary Presentation deliverable, is described in the next chapter. The concluding paragraphs of this section illustrate how the analysts developed several project management tools for the Cornucopia project. For your reference, a complete description of project management fundamentals appears in Appendix A.

System Model

The first-level DFD developed in Chapter 4 (Figure 4-13) can be transformed into the more user-friendly USD, which substitutes familiar icons for the formal symbols and eliminates some of the detail. The first challenge for the analysts is to choose appropriate icons; however, clip art and paintbrush packages make this task fairly easy. The next challenge for the analysts is to label each icon. The analysts should use verbs or verb phrases to label processes. This makes it easier to distinguish processes from external entities and data stores, which are labeled with nouns or noun phrases. Figure 6-8 shows the USD for the current reordering system.

FIGURE 6-8 / *Cornucopia Existing USD*

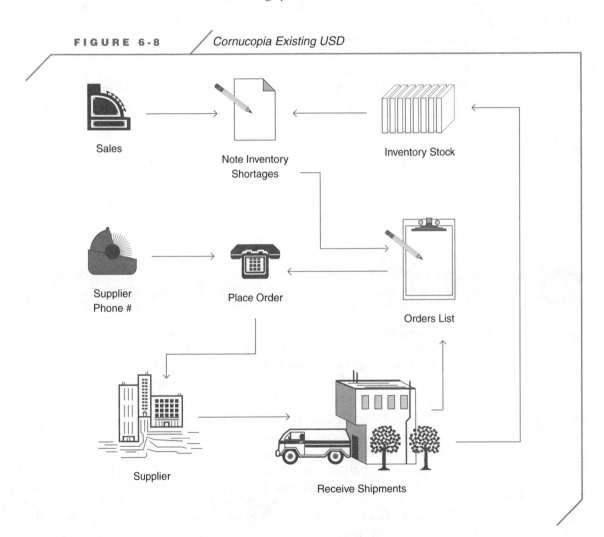

Remember that the purpose of this model is to facilitate communication between the analysts and the user. A meeting with the owner confirms that this indeed is the way the reordering system works today. She is intrigued by the artwork, but she wonders how necessary it is and how much it cost to prepare. When the analysts pull out the first-level DFD and explain the importance of understanding the current system in order to avoid costly redesigns later, she is temporarily satisfied. She shows no interest in discussing the DFD.

Because the current system is entirely manual, the analysts did not need to prepare a menu tree diagram or a system flowchart. This saves time and avoids further explanations to the owner about their methodology. Also notice that the analysts did not include customer correspondence on this USD. For now, this element of the system is totally independent of the reordering system and therefore would unnecessarily clutter the diagram.

Ideas for the New System

The analysts' efforts to date have not focused totally on the old system. The project contract specifies four areas for improvement:

1. Customer record keeping
2. Product reordering
3. Customer communications
4. Sales trend analysis

Each of the activities of the analysis phase stimulates ideas for the new system as outlined in these broad terms, but two thoughts are very clear at the moment. First, the current system modeling confirms that no provision exists for capturing sales data for future sales trend analysis. This suggests that the design needs to incorporate a sales subsystem of some kind. Second, the reordering system relies far too much on the owner's memory. The new design needs to provide some consistent way to associate supplier phone numbers and the products that Cornucopia needs to order.

The analysts should make notes of such design ideas as they occur. Not only does this provide a stimulus to the formal design process, but it also offers an important insight into the complexity of the project, which will certainly help the analysts develop a more detailed project budget and project status report.

The Build-or-Buy Decision

Before the analysts proceed much farther, however, they must consider the results of their research into vertical software commonly used in the industry. As mentioned in the feasibility analysis discussion in Chapter 2, the analysts don't want to reinvent the wheel. A complete turnkey product may be available for less money than the one the analysts propose to assemble. A little research shows that an existing point-of-sale (POS) product is available, one that can considerably reduce the work: Ozware Computer Systems markets several variations of "Phono-Scan," at an initial cost of between $1,990 and $4,990, plus $700 for installation, another $1,000 for peripheral equipment, then $1,280 in annual update and maintenance charges. The full-featured system comes to $6,690, plus $1,280 annual fees. The system does not include the customer correspondence subsystem elements outlined by the user, but the analysts could add this feature and still keep the overall cost below $10,000.

Although tempted, the analysts won't adopt this vertical product in the case study. The purpose is to provide you with a complete product design and development experience in this sample case. It is instructive, however, to see that there are real-world vertical software products designed to serve small-enterprise computing needs.

Project Management

Projects of any size require planning. Projects involving the efforts of more than one person require coordination. Projects with projected costs that exceed mere pocket change require financial resource controls. *Project management* is a term used to describe a collection of activities that include planning, supervision, and cost control. Cornucopia is a modest project with a modest set of project management tools in place. Although sophisticated project management programs are available, the analysts decide to use a spreadsheet program to implement budget and project status reports.

Project Budget

Budgets can be built from the bottom up or from the top down. Mostly they are built with some of each method. This project has some very clear constraints: don't spend more than $10,000, and complete it in 16 weeks. The three major cost components are: hardware, software, and labor. With these givens, and a little experience, the budget is constructed using the format described in Appendix A. The budget breakdown shown in Figure 6-9 reflects a slightly refined version of the $10,000 budget first presented in the feasibility analysis in Chapter 2. Figure 6-10 presents the budget as of the third week.

This budget reflects an estimate of 86 labor hours, billed at $50 per hour, to complete the project. The hardware and software purchases are estimated to occur during the sixth and ninth weeks. The budget update for Week 3 shows that the analysts have not billed as many labor hours as estimated, which is shown as underbudget amounts in the weekly and cumulative sections of the project budget in Figure 6-10.

FIGURE 6-9

FIGURE 6-9   /   *Cornucopia Cost Breakdown Estimate*

Cost Category	Estimated Dollar Cost	Percent of Total
Hardware	$4,000	40.8%
Software	$1,500	15.3%
Labor	$4,300	43.8%

FIGURE 6-10   /   *Cornucopia Project Budget—Week 3*

	A	B	C	D	E	F	G	H	I	J	K	L	M	N	O	P	Q	R
1	Date:		Cornucopia Project Budget							As of: Week		3						
2																		
3		1	2	3	4	5	6	7	8	9	10	11	12	13	14	15	16	Total
4	Estimates Hardware						2500	500	500	500								4000
5	Software						1000	250	250									1500
6	Labor	200	250	250	400	200	250	250	400	200	250	250	400	250	250	250	250	4300
7	Total	200	250	250	400	200	3750	1000	1150	700	250	250	400	250	250	250	250	9800
8	Actuals Hardware																	0
9	Software																	0
10	Labor	100	200	150														450
11	Total	100	200	150	0	0	0	0	0	0	0	0	0	0	0	0	0	450
12	Weekly +/- Hardware	0	0	0	0	0	0	0	0	0	0	0	0	0	0	0	0	0
13	Software	0	0	0	0	0	0	0	0	0	0	0	0	0	0	0	0	0
14	Labor	-100	-50	-100	0	0	0	0	0	0	0	0	0	0	0	0	0	-250
15	Total	-100	-50	-100	0	0	0	0	0	0	0	0	0	0	0	0	0	-250
16	Cum +/- Hardware	0	0	0	0	0	0	0	0	0	0	0	0	0	0	0	0	
17	Software	0	0	0	0	0	0	0	0	0	0	0	0	0	0	0	0	
18	Labor	-100	-150	-250	0	0	0	0	0	0	0	0	0	0	0	0	0	
19	Total	-100	-150	-250	0	0	0	0	0	0	0	0	0	0	0	0	0	

Budget3.xls

Budget / Sheet2 / Sheet3 /

Project Status Report

Figure 6-11 presents the labor hour estimates for each phase of the project. Figure 6-12 presents the project status report as of the third week. Notice that the 86 labor hours are distributed across the phases of the project.

FIGURE 6-11 / *Cornucopia Labor Hour Estimate*

Project Phase	Estimated Hours	Percent of Total
Analysis	18 hours	20.9%
Design	27 hours	31.4%
Development	27 hours	31.4%
Implementation	14 hours	16.3%

FIGURE 6-12 / *Cornucopia Project Status—Week 3*

	A	B	C	D	E	F	G	H	I	J	K	L	M	N	O	P	Q	R	S	T
1	Date:	Cornucopia Project Status							As of: Week 3											
2																				
3	Activity	% Comp.	Status	1	2	3	4	5	6	7	8	9	10	11	12	13	14	15	16	Total
4	Analysis - Estimate	75%		4	5	5	4													18
5	Actual	75%	++	2	4	2														8
6	Design - Estimate	0%					4	4	5	5	4									22
7	Actual	5%	ok			1														1
8	Develop - Estimate	0%									4	4	5	5	4					22
9	Actual	0%	ok																	0
10	Impl. - Estimate	0%													4	5	5	5	5	24
11	Actual	0%	ok																	0
12	Total - Estimate			4	5	5	8	4	5	5	8	4	5	5	8	5	5	5	5	86
13	Actual			2	4	3	0	0	0	0	0	0	0	0	0	0	0	0	0	9
14																				
15	Contract	100%	ok	C																
16	Prelim. Present.	90%	ok			S														
17	Design Review	0%	ok							S										
18	Prototype Review	0%	ok										S							
19	Training Session	0%	ok															S		
20	Final Report	0%	ok																S	
21																				

The status of the major SDLC activities appears in the upper half of the document. The analysts report a total of eight hours were devoted to developing the various models for the existing system, and estimates that the analysis phase is now about 75 percent complete. In addition, one hour was devoted to some preliminary sketches of the new system design, which allows the analysts to report a five percent completion of the design phase. The analysis phase completion percentage coincides with the estimated completion figure, but the analysts accomplished this in six fewer hours than estimated. Therefore, a "++" in the status column indicates that the analysts are performing better than planned. Work on the design phase actually began a week earlier than expected, which accounts for the difference in the estimated and actual completion percentage for the design phase activities.

The status of project deliverables appears in the lower portion of the document. The analysts indicate that the contract was completed as scheduled with a "C." Preparation for the preliminary presentation is nearly complete, with the actual presentation scheduled ("S") in the near future.

The status report reflects your best judgment of your progress toward project completion. It is composed of estimates about the future and historical data about the past. In practice, its preparation is a difficult, time consuming, and imprecise task. For example, hours billed to any given project are not necessarily expended in one time frame. Many unrelated interruptions and intervening activities are inevitable. Furthermore, it is common for an analyst to work on more than one project at a time. While waiting for an answer on one project, the analyst can work on another project. This demands that the analyst develop a procedure to record and report project hours accurately—and to the correct project. Figure 6-13 illustrates one solution to this data collection activity.

FIGURE 6-13   *Cornucopia Project Analyst Hours Log—Week 3*

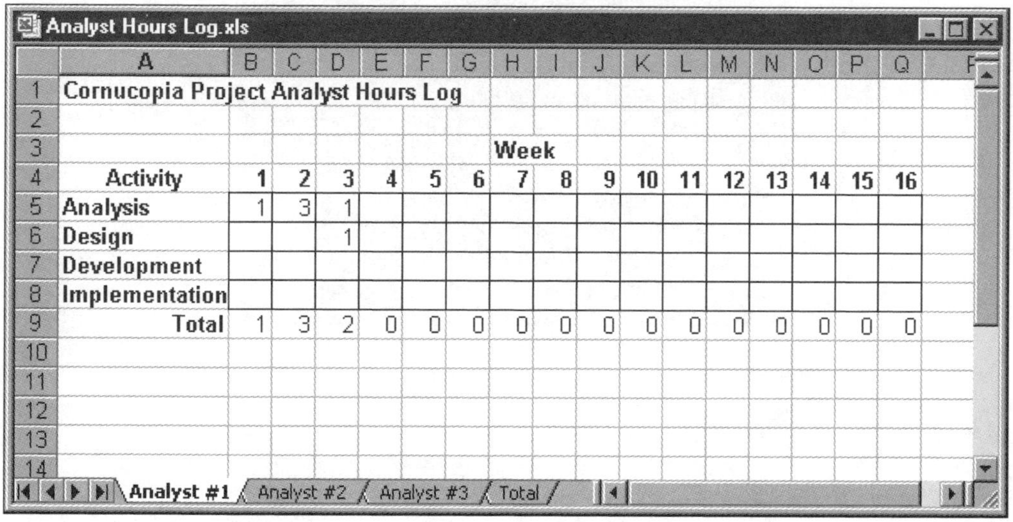

Analyst Hours Log.xls

Activity				Week													
	1	2	3	4	5	6	7	8	9	10	11	12	13	14	15	16	
Analysis	1	3	1														
Design			1														
Development																	
Implementation																	
Total	1	3	2	0	0	0	0	0	0	0	0	0	0	0	0	0	

Cornucopia Project Analyst Hours Log

Analyst #1 / Analyst #2 / Analyst #3 / Total

Detailed Task List

Figures 6-14, 6-15, and 6-16 detail the Cornucopia tasks, subtasks, and events associated with each phase of the project. They correspond to the activities presented in very broad terms by the enhanced SDLC model. Future chapters describe each of these activities in detail.

Notice that the design and development stages are combined to reflect the back-and-forth relationship between design and prototyping activities. The numbers following each task coincide with the job estimates in the status and budget reports.

FIGURE 6-14 / *Cornucopia Analysis Phase Tasks and Events*

T.1 Initial Consultation (4)
  T.1.1 Interviews (1)
  T.1.2 Feasibility Report Preparation (2)
  T.1.3 Contract Preparation (1)
E.2 Contract Completion

T.2 Full Analysis (14)
  T.2.1 Interviewing (3)
  T.2.2 Industry Research (2)
  T.2.3 Existing System Diagramming (4)
    Context Diagram
    Data Flow Diagram
    User's System Diagram
    Entity-Relationship Diagram
    Object Model
  T.2.4 Build versus Buy Analysis (1)
  T.2.5 Develop Project Budget (1)
  T.2.6 Develop Project Status Report (1)
  T.2.7 Prepare Preliminary Presentation (2)
E.3 Preliminary Presentation

FIGURE 6-15   *Cornucopia Design and Development Phase Tasks and Events*

T.3 Initial Design Sketch (7)
   T.3.1 Incorporate Preliminary
   Review into Design (1)
   T.3.2 Develop Alternative New
   System Proposals (3)
      User's System Diagram
      Data Flow Diagram
      Entity-Relationship Diagram
      Object Model
      Menu Tree
      Basic I/O Formats
   T.3.3 Develop Cost/Benefit Analysis (1)
   T.3.4 Prepare Detailed Hardware
   and Software Specs. (1)
   T.3.5 Prepare Design Proposal (1)
E.4 Design Review Session

T.4 Create Prototype (27)
   T.4.1 Incorporate Design Review into Design (1)
   T.4.2 Revise New System Models (2)
   T.4.3 Design Menu Tree Screens (1)
   T.4.4 Develop Master File I/Os (4)
   T.4.5 Develop Query I/Os (5)
   T.4.6 Develop Report I/Os (5)
   T.4.7 Develop Process Designs (5)
      Program Module Structure Charts
      System Flowcharts
   T.4.8 Unit Testing (2)
   T.4.9 Prepare Prototype Demonstration (2)
E.5 Prototype Review Session

T.5 Final Product Development (20)
   T.5.1 Incorporate Prototype Review into Design (2)
   T.5.2 4GL Programming (10)
   T.5.3 Build System Environment (4)
   T.5.4 System Testing (4)
E.6 Training Session

FIGURE 6-16   *Cornucopia Implementation Phase Tasks and Events*

T.6 Develop System Documentation (8)
   T.6.1 Assemble Project Binder (1)
   T.6.2 Prepare Reference Manual (2)
   T.6.3 Prepare Procedures Manual (5)
      System Description
      Operating Instructions
      User Interface Illustrations
      System Security Provisions
      Emergency Instructions
      Appendices
E.7 Product Delivery

T.7 Develop Training Material (4)
   T.7.1 Establish Training Schedule (1)
   T.7.2 Prepare Training Manual (3)
      System Overview
      Demonstration Outline
      Hands-On Exercises
      Quick Reference Guide
E.6 Training Session

T.8 Installation (2)
   T.8.1 Prepare Conversion Plan (1)
   T.8.2 Supervise File Conversion and Creation
      (Under our contract, this is primarily a user activity.)
   T.8.3 Project Review (1)
      User Acceptance
E.7 Product Delivery

PERT Chart

Some project tasks must be completed in a specific sequence, while others may be completed in a parallel fashion. For example, the analyst must complete the context diagram before attempting the first-level data flow diagram, but project documentation may occur while both of these tasks are in progress. Understanding this, a project manager can assign analyst resources more efficiently. In the Cornucopia case, one analyst can work on the USD, while another analyst works on the menu tree. A *PERT chart* is a diagram of the relationships between tasks and the events that mark the beginning and ending of tasks. It can also indicate the amount of time estimated to complete a task.

In order to document the relationship between tasks, the analyst develops a set of worksheets (Figure 6-17) from the detailed task list. In turn, the worksheets provide information for the analyst to develop a PERT chart (Figure 6-18). The chart shows that tasks T.6 (Develop System Documentation) and T.7 (Develop Training Material) can be worked on at the same time as several other tasks.

FIGURE 6-17   /   *Cornucopia PERT Worksheet*

Major Event	Week Due
E.1 Start	Week 1
E.2 Contract Completed	Week 1
E.3 Preliminary Presentation	Week 3
E.4 Design Review Session	Week 8
E.5 Prototype Review Session	Week 11
E.6 Training Session	Week 15
E.7 Product Delivered	Week 16

Event	Begins Task	Ends Task
E.1	T.1	
E.2	T.2	T.1
E.3	T.3	T.2
E.4	T.4, T.6	T.3
E.5	T.5, T.7	T.4
E.6	T.8	T.5, T.7
E.7		T.6, T.8

Project Phase	Major Tasks	Labor Estimate
Analysis	T.1 Initial Consultation	4 hours
	T.2 Full Analysis	14 hours
Design and Dev.	T.3 Initial Design Sketch	7 hours
	T.4 Create Prototype	27 hours
	T.5 Develop Final Product	20 hours
Implementation	T.6 Develop Sys. Doc.	8 hours
	T.7 Develop Training Mat.	4 hours
	T.8 Installation	2 hours

FIGURE 6-18   /*Cornucopia PERT Chart*

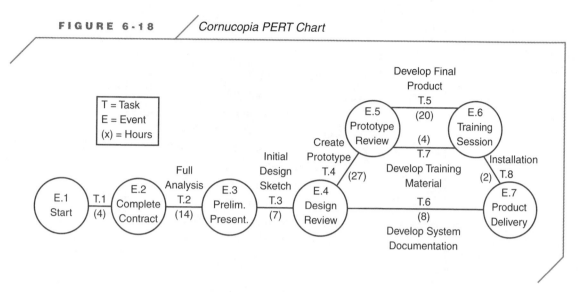

Project Dictionary

A *project dictionary* is like a catalog—providing a quick reference for the file name, storage location, size, type, creation date, and descriptions of all the project materials. After just three weeks, there are more than two dozen different files associated with the Cornucopia project. To implement the project dictionary, the analysts use a Windows Explorer directory structure (Figure 6-19) and a simple file naming convention in which an abbreviated document name is followed by its version number, such as ES-Cntx-v1 for the first version of the existing system context diagram.

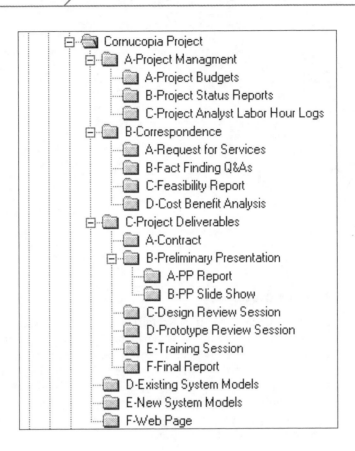

Cornucopia with Visible Analyst

Visible Analyst can analyze each model for syntax and normalization through the Analyze option in the Diagram menu. Figure 6-20 illustrates the analysis performed on the existing system's ERD. The warning messages refer to the fact that the analysts did not enter two-way labels, even though they selected two names per relationship in the Create New Project dialog box (see Figure 2-11).

FIGURE 6-20 / *Cornucopia Model Analysis with Visible Analyst*

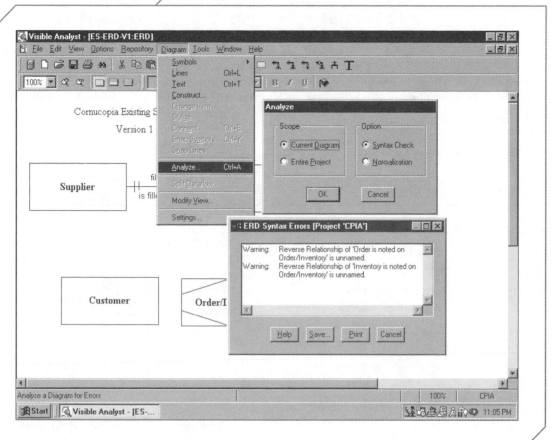

SUMMARY

This chapter completes the modeling process of the six components of a computer information system. These models work together to give the analyst two important, but distinct, views of the system: the technical view of the computer professional and the friendly view of the user. A clear understanding of both views is necessary to prepare the analyst for full-scale work on the design of the new system.

The modeling techniques presented as part of the analysis phase are part of a well-developed branch of computer science. At some point in the analyst's academic or professional career, a detailed study of each model is recommended. This chapter offers a practical introduction to the concepts and techniques required to develop three system views. First, the system flowchart is a technical, analyst-friendly model showing the relationships between files, programs, and subsystems. Second, the menu tree, with its emphasis on system functionality, is useful to both the user and the analyst. The user can see what services and information products are available, while the analyst can see what subsystems need to be developed. Third, a new adaptation of an old picture-based storytelling technique, the user's system diagram, is designed to help the user understand how subsystems interface with one another.

In the case of model building, you must be able to abstract what you see, hear about, and read about. This process sharpens your understanding of the system. Furthermore, and more importantly, building models provides you with a symbolic notation that you can easily manipulate during the remainder of the SDLC. Therefore, mastering each modeling skill is not necessarily the key point. Your ability to conceptualize the whole system without losing track of its major pieces is much more important.

TEST YOURSELF

Chapter Learning Objectives

1. In what ways does the USD facilitate communication between the analyst and user?

2. How does the menu tree organize and display information system functions and services?

3. In what ways does a system flowchart depict the physical characteristics of an information system?

4. What commonplace spreadsheet features are used to implement project budgets and status reports?

Chapter Content

1. The user's system diagram provides sufficient detail for the analyst and software engineer to identify areas that will require significant programming effort. True or False?

2. There is a well-established set of processing, file storage, and external entity icons to use in constructing the user's system diagram. True or False?

3. The USD models all six components of the information system. True or False?

4. While the menu tree model is especially useful in documenting the various functions performed by an information system, it offers little help in describing how those functions are carried out. True or False?

5. The user is a particularly good source of information about the various menu tree options of the existing information system. True or False?

6. The menu tree is of little use to the analyst. True or False?

7. The system flowchart is likely to identify the application software used to carry out the processing of the information system. True or False?

8. There is a well-established set of system flowcharting symbols the analyst uses to model the information system. True or False?

9. The rectangle is used to represent a computer process, which can only be a custom program written in the C++ language or an off-the-shelf application package such as Microsoft Office. True or False?

10. Project budgets and status reports can be prepared with a spreadsheet program such as Microsoft Excel. True or False?

11. Why might the analyst prepare the USD before the DFD?
    a. The USD is easier to construct than the DFD.
    b. It is very difficult to develop the USD after the DFD.
    c. The USD facilitates more effective communications with the user about the nature of the information system.
    d. Once the USD is constructed, the analyst may decide not to build the DFD.

12. Why should the analyst review a wide-angle system flowchart that shows how the system under study connects with other systems within the enterprise?
    a. The analyst is responsible for all of the systems, not just the one under study.
    b. These other systems may influence the performance of the system under study.
    c. The analyst should not waste time reviewing other systems within the enterprise.
    d. Such a view is almost always called for in the project contract.

13. What is the best method for estimating labor costs for a computer information systems project?
    a. The U.S. Labor Department provides an extensive catalog of software development tasks and labor hour estimates.
    b. It's all guess work, so any numbers will do to get started.
    c. Use experience based on prior work of a similar nature.
    d. Divide the client's budget in thirds, with one third dedicated to labor expense.

14. What is the value in gathering actual labor hour data for each analyst, each task, and each week?
    a. Analysts are always paid by the hour, so you need to know how many hours each analyst works.
    b. Actual data provides a valuable historical record for use in developing future estimates.
    c. Any deviation from the estimated labor hours needs to be documented for analyst performance reviews.
    d. The data needs to be reported to the U.S. Labor Department so it can maintain its labor estimates catalog.

15. How might a PERT chart help a project manager reassign analyst responsibilities?
    a. A PERT chart is no help in this regard because it doesn't show analyst skills.
    b. A PERT chart automatically assesses analyst work schedules, generating recommendations for improvements.
    c. A PERT chart can help the project manager identify tasks that can be worked on at the same time.
    d. A PERT chart matches estimated labor hours with actual labor hours, highlighting any differences for management.

ACTIVITIES

Activity 1

Review the narrative and the detailed file layouts for the existing Silhouette Sea Charter information system in Chapters 3 and 4.
  a. Create a user's system diagram.
  b. Create a system flowchart.

Activity 2

Conduct an Internet search for vertical software products designed to help manage a typical small enterprise in your community. Prepare a brief summary of your findings and compare the cost of the vertical software to the $10,000 small-enterprise project budget used in this text.

Activity 3

Develop a personal time budget for the next 24 hours. Identify specific tasks (eat dinner, sleep, read Chapter 6 again, etc.) and the estimated time you expect to devote to each task. Then, carefully monitor your activities for this period of time, recording the actual times devoted to each task. Compare the estimates to the actual times.

DISCUSSION QUESTIONS

1. Find one system (information or otherwise) that uses a variation of the USD to communicate its functions. Describe the effectiveness of the icons used in the diagram. Attach a copy of the diagram to your answer.

2. Using the product documentation, locate the menu descriptions for your favorite spreadsheet program. Describe how the menu is implemented in the computer environment (e.g., buttons, pull-down menus, command line, etc.). Develop a quick reference guide or menu tree for the six or seven commands you use most often.

3. Explain why the nonpiracy agreement and roster data stores appear on the TKSystem USD but not on the system flowchart.

4. How does the USD compare with the UML's use case diagram?

5. Explain why it is a good idea to begin updating a project dictionary from the very beginning of the project.

PortfolioProject

Team Assignment 6: System Modeling

After building process, data, and object models of the existing system, it may seem pointless to construct another diagram of a system you intend to replace. Why not simply move on to the new system design? The answer to that question is twofold. First, the USD and menu tree models provide the best opportunity for you to verify your understanding of the existing system with the user. Second, these models are the most easily altered to help you communicate your first model of the new system to the user.

In practice, the investigation, analysis, and model building of the existing system is not nearly as orderly as the sequence suggested by this and the previous three chapters. Nor is it likely that you will approach the end of the analysis phase with a complete set of existing system models. It is common for analysts to jump from building a model of the existing system into designing the new system with little concern for violating a textbook-driven methodology. The purpose is to understand the existing system well enough to eventually construct a collection of models to guide the development and implementation of the new system.

This chapter presents several project management tools in conjunction with the Cornucopia case. Team Assignment 3 required you to prepare a draft of your project budget and status report. If you are required to submit regular budget and status updates, you should carefully consider adopting the data collection and reporting methods described in the Cornucopia section and Appendix A. Additionally, the directory structure used to organize the files associated with the Cornucopia case might prove useful as a simple project dictionary for your portfolio project.

In order to complete the existing system models and prepare documents for the preliminary presentation described in the next chapter, you must complete the following tasks:

1. After reviewing the project management discussions in Appendix A and the related Cornucopia case examples in this chapter, revise the draft project budget and status report prepared in Team Assignment 3.
2. Prepare a user's system diagram for the existing system. Submit a copy of this diagram to your instructor.
3. File a copy of the existing system USD and the revised project budget and status report in the Portfolio Project Binder behind the tab labeled "Assignment 6."

Design

systemdesign

OVERVIEW

In this chapter, you officially begin the design phase of the systems development life cycle. This chapter discusses several popular design strategies, introduces design prototyping, and suggests a specific sequence of form design activities. Throughout the previous four chapters, you studied several modeling techniques, each with distinguishing characteristics and purposes. These abstractions remove much of the system detail in order to present different, but complementary, perspectives of the existing information system. During the design phase, the analyst manipulates these abstractions to produce models for the new system design.

Design work is the analyst's most elusive skill and requires an unusual mix of technical expertise and creative inspiration. Many practitioners will confide that design work is what they really are paid to do. Good design has no substitute—bad design no escape.

- Transform the project objectives into a preliminary design model that the user can follow and understand.

- Explain why the focus of design work begins with the user's perspective and ends with the hardware perspective.

- Adapt the joint application design (JAD) methodology to fit a small-enterprise systems project.

- Design a set of user-friendly screen forms.

DESIGN STRATEGIES

Pinpointing exactly when, or with what model, the design process begins is impossible. Even though analysts initially use DFDs, ERDs, USDs, and so on during the analysis phase, they also use them as design tools. Although this text presents an idealized sequential chronology for using these tools, in practice, they are used in many different sequences. Design is a circular process, with repeated evaluation and revision. Still, the chronology presented in Figure 7-1 is useful because it can help organize your thoughts and provides an activity sequence to start the process.

FIGURE 7-1   / Modeling Chronology

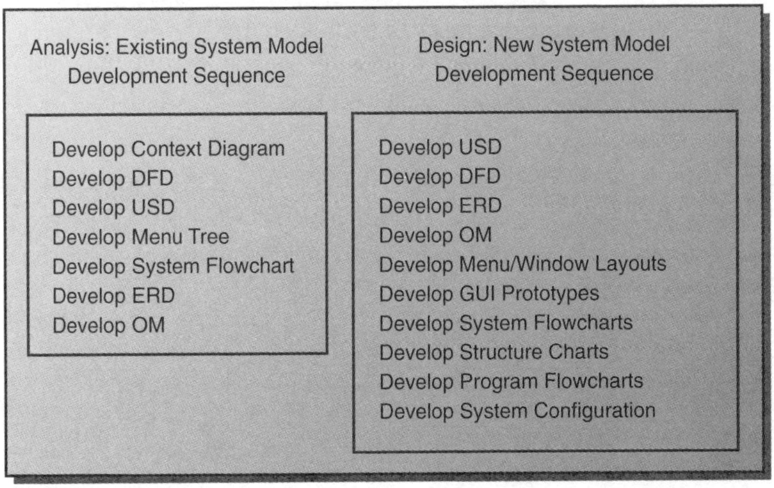

Analysis: Existing System Model Development Sequence	Design: New System Model Development Sequence
Develop Context Diagram Develop DFD Develop USD Develop Menu Tree Develop System Flowchart Develop ERD Develop OM	Develop USD Develop DFD Develop ERD Develop OM Develop Menu/Window Layouts Develop GUI Prototypes Develop System Flowcharts Develop Structure Charts Develop Program Flowcharts Develop System Configuration

Another important element of your design strategy involves the user. Figure 7-2 suggests that your strategy begin with the user, proceed to a definition of output, file, and input requirements, and conclude with processing structure. Platform and peripheral specifications naturally follow these design considerations. This is called *user-driven design*.

FIGURE 7-2   / *User-Driven Design*

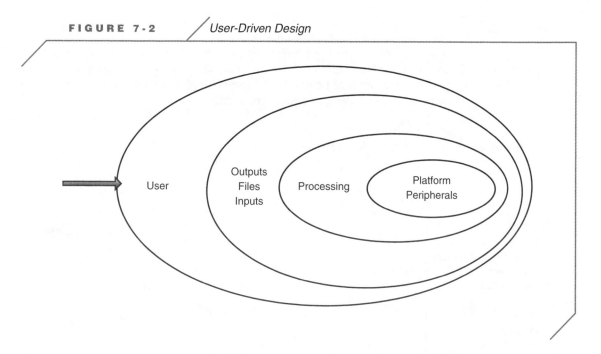

Although the analyst is responsible for the design process, ultimately, all design questions are answered in the context of what is best for the user. When the user is satisfied with the proposed user interactions and functions of the new system, then you can move on to finalize the processing design. Although this process is fundamentally circular, you can safely terminate all of the design elements, knowing that the user will be satisfied with the product. The analyst must remember that the user cares most about the system outputs, cares least about the internal process, and will tolerate only so much input inconvenience. Given the traditional academic emphasis on the technical aspects of computer information systems, putting the user first requires considerable discipline from the analyst. In theory, the user-driven design approach permits the user to decide when the design is complete.

In practice, another concern often forces design activities to a close. The project cost and due-date constraints require that design be completed with remaining budget and time sufficient to finish developing and implementing the project. In other words, although we cannot ignore the overlapping nature of the phases of the SDLC, neither can we revisit the major design issues indefinitely. The analyst should have a design completion date in mind well before the end of the design phase. If the analyst cannot meet this date, the final project completion date is compromised.

Finally, our design strategy focuses on the functions performed by a typical small-enterprise information system. Chapter 1 introduced the information system hierarchy and described the general information needs of the small enterprise. In Figure 7-3, we fashion the traditional information system categories into a series of progressively more complex information systems against three life-cycle periods of several years each.

FIGURE 7-3 / *Information System Stages*

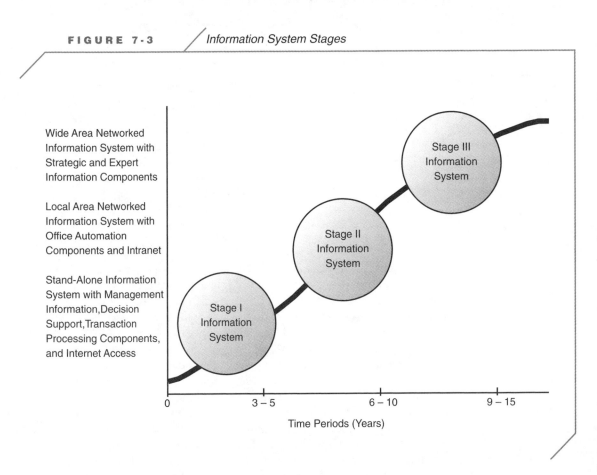

As a small enterprise reaches the end of one system's functional life, it is likely to graduate to a more ambitious system. The three successive systems are labeled Stages I, II, and III. The systems under study in this text are all in Stage I, which considerably narrows the discussion of the available design options. This *three-stage concept* is also applicable to medium- and large-enterprise information systems, with appropriately more complex systems available for the different stages.

TECHNOTE 7-1

TRADITIONAL INFORMATION SYSTEM TYPES

Many properties can characterize information systems: their cost, the audience they serve, the equipment they use, and the functions they perform. The following categories include some of each in their descriptions and are generally considered more complex as you go from one to the next.

Transaction-Processing System—a system that captures data that describes events. An online system processes the data immediately or very soon after the event occurs. These are called **online transaction-processing systems (OLTP)**. Data is treated differently in **batch systems**, which gather the captured data for processing sometime after the event. Automated teller machines, which immediately process the data describing your financial transaction, are a good example of OLTP. An employee payroll system, which records your presence or absence each day but delays payroll calculations until the end of the week, illustrates a batch transaction-processing system.

Management Information System (MIS)—a system that integrates information collected from different parts of the enterprise. For example, your course registration system is probably a part of a large management information system that includes scheduling, room utilization, faculty assignments, and so on.

Decision Support System (DSS)—a system that allows the user to develop information as the need arises. Such systems often require the user to learn only a few simple query sentences rather than a complete programming language. For example, a transportation manager, suddenly faced with a broken down vehicle, might ask the decision support system to display the available vehicles closest to the needed route.

Local Area Networked (LAN) Information System—a system that connects computers located in a small geographic area, usually the same building. A local area network allows users to share data stores and hardware peripherals, making enterprise-wide office automation possible. Electronic mail is a good example of office automation.

Wide Area Networked (WAN) Information System—a system that expands the principles of local area networking to a wider geographic area. A wide area network, which most often allows an enterprise to connect computers located in different buildings, can be expanded to connect the enterprise to regional, national, or global information networks and databases.

Expert Systems Information Network—a system that can analyze and diagnose problems. These systems offer possible solutions based on an electronic knowledge base and an inference engine designed to apply the necessary expertise and experience to generate recommended actions.

Strategic Information System—a system that reaches beyond the enterprise to focus on external data. For example, such a system might collect and process data concerning the enterprise's competition, industry trends, or both.

The input-process-output (IPO) framework introduced in Chapter 3 (Figure 3-2) provides an outline for a more detailed look at the realistic design choices available to the analyst working on a Stage I, small-enterprise information system. Remember that transaction processing is a major component of such systems. This usually places great pressure on the design for data collection and data entry. Likewise, processing design must accommodate high volumes of file access, and output design must handle heavy loads during any batch-processing activities. Finally, Internet technology must be integrated into the system. Figure 7-4 provides some broad guidelines and concerns associated with the Stage I, small-enterprise system's input, process, and output design.

FIGURE 7-4 / *Small-Enterprise Design Choices*

Input

Manually keyboarded input is time consuming and produces a lot of errors. The best alternative is to use a machine-readable format, such as the universal product code (UPC), along with a bar code reader or scanner. Some screen dialogues will, of course, require a keyboard and/or mouse input. The Internet is a possible input medium as well.

Processing

Transactions can occur at regular, predictable intervals or at random. The decision to delay processing of transactions for batch processing or to dedicate a portion of the system resources to process transactions online as they occur is critical to processing design.

Output

Hardcopy output is expensive to produce, transport, and store, yet most systems require some amount of printed output. The softcopy alternative is favored whenever possible. The Internet is a possible output medium as well.

Alternative Approaches to Design Work

The analyst applies many tools as part of an overall problem-solving strategy. For example, earlier chapters allude to the use of several modeling tools. It is impossible to prepare you for every situation presented by the six CIS components and the SDLC. Your ability to adapt these tools to new circumstances will determine your level of success. This section offers both familiar and unfamiliar problem-solving strategies. In all likelihood, you will not be able to adhere to a single strategy for any particular problem. Instead, experience will allow you to blend several approaches, creating a unique, personal, and comfortable problem-solving style of your own.

Before discussing several well-defined strategies, a short, diversionary list of alternative approaches might stimulate your thoughts about problem solving in general. As you read the ideas presented in Figure 7-5, try to think of situations in which you might use—or have already used—these techniques.

FIGURE 7-5   /*Problem-Solving Strategies*

1. Work forward . . . work backward . . . work inside-out.
2. Work independently . . . work in a group.
3. Build models (physical, logical, visual).
4. Rely on heuristics . . . rely on algorithms.
5. Think linearly . . . think extemporaneously.
6. Employ environmental inducements
   (silence, music, chatter, light, dark, etc.).
7. Use diversions (play, art, exercise, etc.).
8. Reduce the problem to a more solvable size.
9. Expand the problem to extreme proportions.
10. Solve a different but similar problem.
11. Assume a solution, and then simulate its implementation.
12. Look for solutions that already exist.

Each strategy stimulates different thought patterns, contributing to a rich collection of ideas from which solutions will appear. At this stage, it is very important to keep many solution alternatives on the table. As the process continues, new alternatives may unfold, as old alternatives are combined or transformed.

Finally, the proliferation of application software increases the chances that some, if not all, of the solution already exists. The old saying, "Don't reinvent the wheel," is certainly applicable. Chapter 2 suggests several ways to research existing products. Your own experience may suggest solution adaptations as well. Nevertheless, at times you may think you have stumbled onto a problem that has never been solved before. When this occurs, try to abstract the problem to a higher level; that is, reduce the problem statement detail. This refocuses your attention to the essence of the problem, perhaps sparking the insight necessary to find similarities where none seemed apparent before.

Structured Design

After the preceding advice about free thinking, it may seem somewhat ironic that the discussion now turns to *structured design*. However, you must remember the lessons learned during the past 40 years. Undisciplined design leads to implementations that are difficult to maintain. Some very straightforward principles define modern design work. Essentially, structured design directs the analyst to create modular solutions to complex problems, thereby dividing the problem into smaller, more manageable parts. However, analysts cannot use this technique indiscriminately; they must also consider the extent to which these modules are cohesive and coupled.

Today's 4GL products naturally encourage the public to view information systems as a collection of functional modules. The typical system presents the user with a menu or window of choices. This fits nicely with the long-held view that the programs that make up information systems should be designed in a modular fashion. Today, the analyst can substitute modular 4GL products for the custom-designed programs of old. Thus, a correspondence subsystem might use a word processor, a budgeting subsystem might use a spreadsheet, and an inventory subsystem might use a database. Each of these subsystems can be considered a separate module, but together they function as an integrated information system, accessible under a single menuing umbrella.

Figure 7-6 illustrates a window with several application choices represented as shortcuts to common Microsoft Office products. Users quickly understand that selecting an icon launches the program. In fact, users often come to regard such GUIs as the only way they can access programs on their system. The analyst, on the other hand, might view these icon collections differently, especially if an information system uses each of these products as interrelated subsystems. In this case, the analyst should consider each application to be a module, thus applying *modular design*, a very successful programming practice, to system design.

FIGURE 7-6 / *Modularity and 4GL Products*

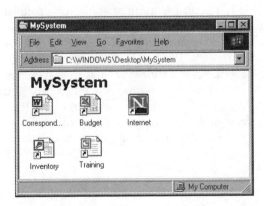

Coupling and cohesion (Figure 7-7) are terms often associated with structured programming. As such, they describe the conflicting consequences of modular program design. Modules that are completely independent of one another are highly *cohesive*. Such modules contain a well-focused set of tasks and exhibit little or no data sharing with other modules. Modules that are closely tied together, via the data they pass from one to another, are called highly *coupled*. The perfect program exhibits maximum cohesion and minimum coupling. This creates solutions that are easy to understand and change, which lengthens the effective life of the program.

FIGURE 7-7   *Modular Design: Coupling and Cohesion*

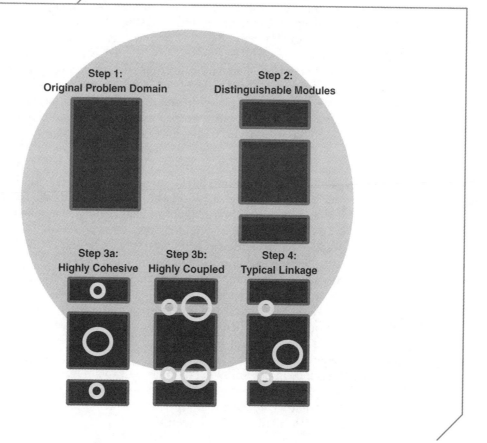

Step 1:
Original Problem Domain

Step 2:
Distinguishable Modules

Step 3a:
Highly Cohesive

Step 3b:
Highly Coupled

Step 4:
Typical Linkage

Extending this prescription to small-enterprise computing raises the question: how can you assemble completely independent (cohesive) software packages into a collection of interrelated (coupled) subsystems? Although early word processors, databases, and spreadsheets were designed to perform identifiable and distinct functions independent of one another, today, traces of each product's functionality are found in all three, as well as built-in, data-sharing capabilities through file import, export, and object linking and embedding (OLE). You will learn more about these topics in later chapters. To the extent that data is shared between and among application products, those products are considered to be coupled. Understand, coupling is not necessarily a design flaw—it merely requires special consideration because it does compromise the simplicity of your system.

Object-Oriented Design

Object-oriented design (OOD) calls for further refinement of the object model. Recalling the OM created during object-oriented analysis, we have already identified objects that support the existing system. We have specified the data attributes and methods associated with each object, and we have studied the relationships between the objects. Formal object-oriented methodology provides very little distinction between analysis and design activities. To support the new system, you most likely will need to design some new user interface objects and you may need to define some new attributes and methods for existing objects. Nevertheless, the process is essentially the same, which makes for a smooth transition from analysis to design and design to development.

To explain, let's review the model building sequence for a moment. You begin with the data flow diagram (a process model), move on to the entity-relationship diagram (a data model), and conclude with the object model (an encapsulated data and process model). System diagrams (user's system diagrams, menu trees, and system flowcharts) provide a holistic view for users and analysts. A fully object-oriented approach would skip the DFD and ERD modeling processes and move directly to the OM process. The approach in this text is to develop each model in succession, not only to better understand the existing system and design a new system, but also to gently introduce an emerging object-oriented methodology.

From a practical point of view, our approach adds slightly to the overall amount of work performed by the analyst. However, we are rewarded for the extra effort when it comes time to implement our design. Almost all of the system-building software we use employs object-oriented techniques. For example, in the following chapters we explain how to design interface objects, which you will recognize as nothing more than custom screen forms. These new objects fit neatly into the object model, while at the same time, they retain their familiarity to Windows users.

Joint Application Design

Throughout the analysis activities, the user-analyst relationship is characterized by repeated question-and-answer sessions. These sessions open communications, improve mutual understanding, and establish an atmosphere of trust. As mentioned, few users have great interest in the analyst's process and

TECHNOTE 7-2

FORMAL JOINT APPLICATION DESIGN

Joint application design (JAD) provides a substitute for the traditional analysis and design cycle in which the analyst does the problem solving, the user offers a reaction to the analyst's work, the analyst then makes modifications as required, and the cycle repeats itself until all parties are satisfied with the product. With JAD, both the analyst and the user engage in the problem-solving process through intense workshops in which all participants are expected to contribute to the solution. JAD is designed to produce the same set of traditional analysis and design documents, but in much less time and at considerably less expense. The JAD process may also bring the user into the SDLC with ownership rights. That is, users who have invested their own resources into the solution are more likely to feel committed to the success of the project.

The following is a generic description of the formal JAD method:

Participants:
- Users
- Analysts
- Observers—technical advisors
- Scribe—a person to take notes
- Facilitator—someone to direct the process

Setting:
- The meeting place is away from the normal enterprise activities so that interruptions and distractions are minimized.
- Users and analysts sit at tables arranged in a semicircle, with the facilitator in the front and the scribe off to one side.
- Audiovisual aids such as blackboards, flip charts, and computer projection equipment are positioned behind the facilitator so that each participant can see them.

Agenda:
- Evaluate the existing system models and documentation.
- Agree on the new system goals.
- Develop alternative designs.
- Select the best design based on previously determined criteria (see Figure 7-8).

Write-Up:
- Background summary
- I/O interface designs and definition
- Operational menu design and definition
- Processing rules
- Operating procedures

Notice that the JAD process does not generate detailed data, processing, or system models. Nor does it generate detailed system resource requirement specifications or detailed programming designs.

data modeling work. The USD, on the other hand, transforms these models into something the user can relate to as you move into the design phase, when even closer ties develop between the analyst and user.

Joint application design (JAD) is an analysis design method developed during the 1980s. It brings the analyst and user together in a short, but intense, formal design workshop session. A variation of this method is used here to describe a prolonged user-analyst design partnership. The enhanced SDLC, originally illustrated in Figure 1-11, shows that the design partnership begins during the analysis phase and extends well into the design phase. This is different from the formal JAD concept, but it serves our purposes very well. While acknowledging the need for the user's undivided attention and active participation in the analysis and design effort, this design partnership avoids an overly formal setting that might not be appropriate for small-enterprise work. Rather than one concentrated design session, our partnership calls for several, less formal meetings. The project's preliminary presentation initiates this transition from analysis to design, at which point the USD provides the common ground for the initial analyst-user design work and the adaptation of JAD.

Evaluating Alternative Designs

As with any complex problem-solving process, several alternative designs will emerge. Because each of these designs satisfies the project requirements, some criteria must be applied to select the "best" approach. This always involves a compromise because of the competing interests of users and developers. Figure 7-8 lists several criteria for judging alternative designs.

FIGURE 7-8 / *Design Evaluation Criteria*

1. End-user orientation
2. Understandability
3. Expandability
4. Security
5. Execution speed
6. Expense to develop
7. Time to develop
8. Accessibility

Thinking Critically

Design by Committee

You may have heard of products designed "by committee." Typically, this results in a product that has been compromised until it is very dysfunctional. Knowing that there is always some truth in such humor, how do you reconcile this story with the current trend toward JAD?

The analyst may choose to optimize a system, as measured by one or two of these criteria, but carrying this over to all criteria is impossible. For example, the very user-friendly system may be easy to understand but also difficult to expand. Likewise, an inverse relationship often exists between execution speed and the cost to develop the system.

As discussed earlier, although inexpensive to purchase ($250), the CIS Lab timekeeper system was not well documented. Yet, this shortcoming was acceptable because the product satisfied almost all the constraints (budget, time frame) and the user believed that computer personnel could live with the existing documentation until something better could be developed. Thus, selecting the best approach is often a matter of concentrating on what is most important to the user.

FOR EXAMPLE—SUNRISE SYSTEMS

Sunrise Systems is a small investment management company that can be used to illustrate the design activities of the SDLC. Serving about 250 clients with financial planning, investment analysis, and brokerage services, the firm deals almost exclusively with stocks, bonds, and mutual funds, but can refer the client to real estate and venture capital investment professionals if necessary.

Sunrise already has a Stage II, small-enterprise information system with which it is happy. In this case, you will work backward from a fully operational system to demonstrate file and form, report and query, and processing design concepts. This is called *reverse engineering*, in that an existing system is taken apart to learn how it was put together.

Figure 7-9 presents the first-level DFD for Sunrise Systems, and Figure 7-10 illustrates the corresponding user's system diagram. A small local area network is at the heart of the Sunrise system. The investment analysts' desktop computers are connected to a server to provide shared access to data storage and peripherals. The network provides electronic communications with investment information services and trading exchanges. Sunrise interacts with clients in one of three ways: during regular, semiannual, face-to-face consultations; quarterly portfolio statements; and on-demand requests.

FIGURE 7-9 / *Sunrise Systems DFD*

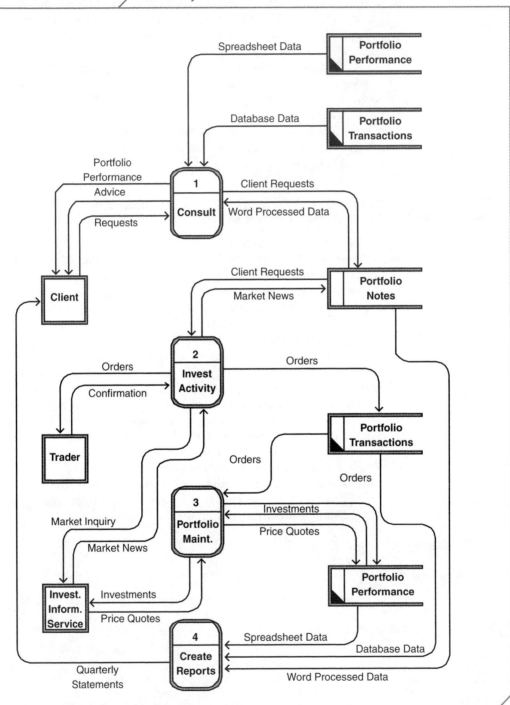

FIGURE 7-10 / *Sunrise Systems USD*

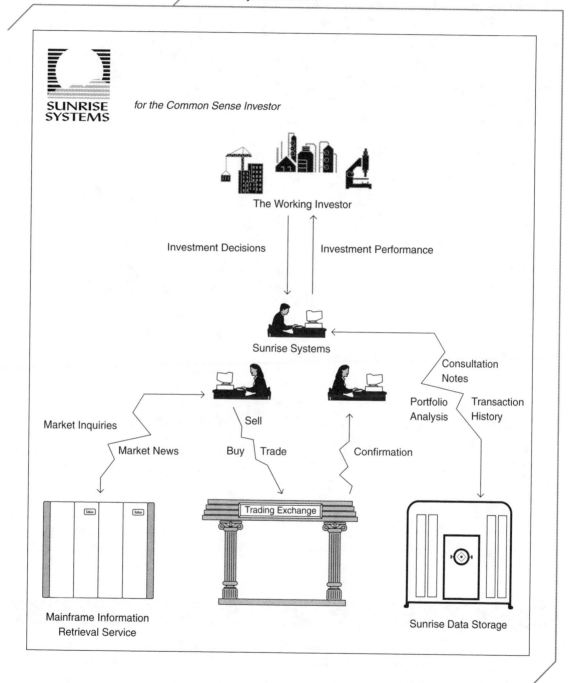

TECHNOTE 7-3

INTERNET TECHNOLOGIES

The Internet should be no stranger to systems analysis students. Indeed, even the most casual computer user is familiar with the Internet. However, to integrate this tool effectively into an information system, the analyst must possess an intermediate to advanced understanding of Internet technology. The following presents a collection of important Internet and telecommunications fundamentals, organized into four skill levels. All of these terms appear in the glossary and many of them are discussed in detail in future chapters.

1.   Basic Internet Skills:
     A typical Internet user with basic Internet skills is able to use the following features:
         E-mail
         World Wide Web (WWW)
         Browsers
         Search engines

2.   Intermediate Internet Skills:
     Someone with intermediate skills most likely subscribes to the Internet via an Internet service provider (ISP) and designs, creates, and maintains his or her own Web page. Someone with intermediate skills is able to use the following features:
         Client-server computing
         Web page design and construction

         File transfer protocol (FTP)
         HTML, XML
         Scripting languages (VB Script, Java Script)
         Animation, audio, and interactivity (Flash)

3.   Advanced Internet Skills:
     Those with advanced skills most likely maintain an Internet host computer. A host computer is a network server that is also connected to an ISP, thus providing Internet access to all of the host's clients. Someone with advanced skills is comfortable using the following features:
         Firewalls
         Web site management
         Communications protocols (TCP/IP)
         Transmission mediums
         E-commerce
         Intranets
         CGI programming
         Java programming

4.   Network Specialist:
     Network specialists design, install, and maintain networks. They are able to perform the following functions:
         Local area network (LAN) specification and installation
         Network management

DESIGN SPECIFICATIONS

This chapter began with the description of the circular nature of design work and a caution against entering the development phase without an agreed upon design. The design specifications, in presenting that design, serve two major purposes: first, they provide the user with the first detailed look at the system; second, they provide the analyst with the blueprint from which the system will be built. We return to this topic in Chapter 11 when we discuss the design report and review session that occurs near the completion of the design phase. At this point in the SDLC, however, the analyst is ready to develop a preliminary overview of the new system design.

Developing the New System USD

As suggested at the end of the last chapter, ideas for the new system design begin to accumulate during the earliest activities of the project. Some of these ideas require careful design work before they can be included in the new system. For example, during the early analysis of the Sunrise Systems project, a need emerged for the new system to integrate documents generated by three different horizontal applications: a word processor, a spreadsheet, and a database. However, the detailed design of this process was not obvious. Rather than try to rush this work, the analyst simply included a design note describing the expectation that this need must be addressed by the new system.

The major components of the new system were easy to identify directly from the project contract. The excerpts from the Sunrise project in Figure 7-11 contract illustrate this point.

FIGURE 7-11 / *Sunrise Systems Project Contract Excerpts*

Problem Summary:
The system must...
1. Record client-analyst notes.
2. Preserve portfolio history.
3. Calculate portfolio performance.
4. Allow for timely market transactions.
5. Provide quick access to market information.
6. Permit several analysts access to the system.

Sunrise Systems USD is based on several items from the preceding summary, the integrated document idea mentioned earlier, and some insight into the investment business. Notice that the diagram (see Figure 7-10) is not specific about implementing this design, but it does use familiar icons to suggest the major features of the new system. The multiple computer icons identify a network solution to permit several analysts access to the system. The jagged lines connote a fast, timely, and secure telecommunications system that can give analysts the ability to make investment transactions based on the latest information. The three external entities to the system are featured with well-labeled descriptions. The data stores are combined into an image designed to promote a secure feeling about how such sensitive financial data will be handled.

Design Prototyping

A *design prototype* is very much like a movie set—all show. It serves as a cheap, expendable stand-in for the real thing. During the design phase, prototypes of screen forms, source documents, and reports can be created to illustrate much of the client interaction possibilities. This lends a certain amount of realism to the design specifications and is especially valuable during the design review with the user. These prototypes can also serve as a springboard to development phase prototyping. The next three chapters heavily illustrate the design prototypes for the Sunrise Systems example. Chapter 11 is devoted entirely to prototyping.

PROJECT DELIVERABLE: THE PRELIMINARY PRESENTATION

The term *preliminary presentation* is used here to mean a formal report, both oral and written, that describes the project's major objectives, general design, budget, and present status toward completion.

Because this is the first opportunity you have to address the user in a formal setting, you should be prepared to introduce yourself, your general background, and your understanding of the project itself. Figure 7-12 provides general guidelines for the report and oral presentation, the details of which may vary greatly depending upon the project and the relationship between analyst and client. To assist you in preparing this report, Appendix C offers discussions on technical writing and presentations.

FIGURE 7-12 / *Contents of the Preliminary Presentation Report*

Written Material:
1. Summary of Project Requirements
2. Overview of General Design Concepts
3. Proposed Timetable (Project Status Report)
4. Proposed Project Budget

Oral Presentation:
1. Introductions
2. Visuals for Items 1-4 Above
3. Question and Answer Period
4. Preview of the Next Step

Previewing the Design Report and Review Session

The preliminary presentation is a transition event, marking the end of the analysis activities and the start of many detailed design tasks. When the formal design phase is complete, the analyst prepares a design report for the next project deliverable—the design review session. This session is extremely important because it provides an opportunity for the user to react to specific design features, such as user interfaces, report layouts, and Web site pages. The use of the user-driven design model (see Figure 7-2), along with design prototyping, can reduce the oral explanations to a brief summary. However, the report and visuals require careful supporting documentation. Therefore, it is important to continually update your project dictionary to make it easier to locate project documents.

FORM DESIGN FUNDAMENTALS

After the broad, USD treatment of the new system design, the analyst must create more specific and detailed design elements. Form design, given its crucial role in developing a user-friendly interface, should begin very early in this process. Even before the new system models are completed, you can be sure that the data stores of the DFD and the entities of the ERD will require numerous interface forms; that is to say, files require procedures and processes that provide easy user access. In the past, this involved the batch update, in which separate programs processed add, change, and delete transactions periodically. Today's systems, however, are more likely to feature an online option that activates a screen form. This approach is sometimes referred to as *event-driven computing*, which describes an information system that is always available to respond to a button-pushing user. In such an environment, the design of the interfacing screen form becomes an important part of input design. The previous joint application design discussion identifies the user as an important partner in the interface design process. This is reflected in the step-by-step form design activities presented at the end of this chapter.

Form design fundamentals also extend to Web pages on the Internet. However, the very public nature of a Web page requires some consideration for the diverse audience it may attract. Enterprise personnel must use internal information system forms even if they are poorly designed, tedious to use, or just plain boring. Further, enterprise personnel can get by with confusing or contradictory forms because of their familiarity with the terms and functions of internal forms. Workers possess a great deal of contextual knowledge about forms simply because they are involved with the enterprise on an everyday basis. The Internet audience, on the other hand, may not be nearly so well informed or motivated to wade through a not-so-friendly Web page to get to the intended message. In such an environment, the quality of the design can determine the success or failure of a Web site.

TECHNOTE 7-4

WEB PAGE DESIGN

Each day millions of people use the Internet to visit one or more of the millions of host computers that make up the World Wide Web. As these numbers continue to grow, so does the need for the enterprise to present a well-designed electronic storefront to the world. In addition to the considerations given to in-house form design, the Internet presents special design challenges, a few of which are itemized below.

Transmission Speed

The allure of transmission speeds in the upper hundreds of thousands of bits per second encourages designers to stretch the speed boundaries set by today's connection options. Pages with sophisticated graphic images, full-motion video, audio, and animation require transmission speeds in excess of those available to the typical Internet user.

Client-Side Computing

Because of the wide range of capabilities of computers used to access the Internet, many experts recommend that pages be designed for computers most commonly used by your target audience. You may need to make significant compromises in your use of some of the more exotic features described previously.

Construction Tool Sets

Web page construction software is among the most volatile in the software industry. To maintain your skills you must stay in tune with the industry newsweeklies, magazines, trade shows, conferences, and Web sites.

Competition

The competition for visitor attention continues to increase as the potential rewards for visitors increases. You must employ proven marketing and business strategies to define the purpose and target audience for your Web site.

Content Currency

Web page content may be the single most important factor in attracting and keeping Internet visitors. This means that your page content must meet many of the same criteria established for successful traditional information systems. It must be relevant, accurate, timely, usable, and accessible. Design your Web pages so they can be easily maintained. Some estimates put the annual cost of Web page maintenance at 50 percent of the original startup cost.

Navigational Strategy

Be very careful to develop a consistent navigational system so that the user can quickly learn to move from page to page. If you decide to include links to sites other than those you maintain, be sure to check those sites regularly for continuity of service and relevancy.

Source Documents

Form design for a 4GL environment is complicated by the fact that the electronic form has become the principal source document for many information-processing procedures. A *source document* contains the original data-capture entries. The credit card application you fill out is an example of a paper source document, which is later used as a reference when the data is keyed into the system. However, where is the source document when you order a product over the telephone or when you use your ATM card? Paperless information systems are the continuing trend. What does this say about the need for effective interfacing screen forms? Without a person available to help the user, such stand-alone systems require designs that anticipate the difficulties and questions that arise during the data collection and entry activity so as to neither frustrate users nor degrade system performance.

Traditionally, source documents have been considered essential to the audit trail necessary to reconcile data inconsistencies and correct errors. When the form displayed on the screen is intended also to function as a source document, the analyst must address these audit needs as well. Requesting immediate user verification and error correction can reduce audit work, but will never eliminate it. Therefore, screen-based transactions should be written to history files for later reference.

Graphical User Interface Dialogs (GUIDs)

A *graphical user interface dialog (GUID)* is a sequence of screen forms that not only instructs the user on how to proceed with system operations but also accepts user selections and specifications. A menu sequence is an example of a GUID, as is a *screen form* that allows the user to identify a particular customer address or account activity to be displayed on the screen.

As a source document, screen forms must be designed to allow for updating master files as well as collecting and processing transactions (both online and batch). Furthermore, the screen forms must be designed to allow users to select from the various subsystem options available to them. Remember, in this respect, small-enterprise information systems are fundamentally different from the old corporate approach, in which the user had little control over the operation of the information system. At this point, the analyst should schedule another JAD working session with the user to solicit ideas about how the system will actually look when it is operational.

Screen designs should have a consistent format, regardless of their specific function. The existing source documents should provide you with a starting point. After consulting with users about the conveniences and deficiencies associated with the current forms, try to preserve something of the look with which users are familiar as you fashion the new designs. Incorporating the enterprise logo on screen forms is particularly appreciated, even if it has no real processing value. The point is that users will be more receptive to change and more easily trained if the new interfaces are not quite so unfamiliar. Figure 7-13 illustrates an interface template for Sunrise Systems.

FIGURE 7-13   / *Sunrise Systems Interface Template*

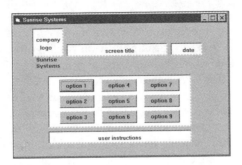

A menu form such as is shown in the figure might present one menu option for each subsystem, with some options indicating that another sub-menu will appear upon selection. An alternative screen format might include pull-down menus as well as button options. Many menuing software packages are available to make the construction of these screens a simple operation.

Screen forms are also used to provide access to a database file. These screens are referred to as *front ends*, meaning that they function as the entry point to update the database files. Commercial products are available to automate their construction. For example, Microsoft Visual Basic can be used to create custom access screens to most products.

In this text, these screen sequences are collectively referred to as a graphical user interface dialog (GUID). When designed properly, they provide a highly effective user-system feedback cycle. Users who want something from the system can request it, and when the system requires further input, it can prompt for it. Users are immediately informed of their errors, making error detection and correction a routine part of data collection and entry.

Automated Input

The input bottleneck has always frustrated users and analysts. Scanners, bar code readers, and other direct input devices considerably increase the input speed and reduce errors. Early in this text, such technology was suggested as a remedy for this common performance defect. The design question remains, however: how far do you remove the human operator from this process? Notice that the supermarket scanner presents a screen echo for the checker and the customer, as well as an audio cue when human intervention is required. If you use these labor-saving devices, you need to consider which kinds of system monitoring options are prudent and cost effective.

Editing for Errors

No matter which input system you settle on, errors will eventually occur. Obviously, it is better to find and correct an error sooner than later. Effective data editing is now possible through the popular database front-end procedures described earlier. Everything from capitalization (e.g., all customer names should be in uppercase) to range checks (e.g., the year must be less than or equal to the current year) can be easily incorporated into the screen forms that service your files. The time you devote to building these edits into your design will be well rewarded through improved system performance and integrity.

THE CORNUCOPIA CASE

As the analysts embark on the design activities of this project, their first task is to develop a preliminary view of the new system design. The project contract and the analysis phase activities present a fairly clear picture of what the new system must provide. Figure 7-14 summarizes these requirements.

FIGURE 7-14 / *Cornucopia New System Requirements*

Areas for Improvement:
1. Customer record keeping
2. Product reordering
3. Customer communications
4. Sales trend analysis

Ideas for the New Design:
1. Incorporate a sales subsystem
2. Associate suppliers with products
3. Associate customers with sales transactions
4. Provide Internet access

New System USD

Remember that the existing system USD (Figure 6-8) shows only the reordering system. The analysts' first thoughts about design for the new system should also center on the reordering system. The owner has already shown what is important here. She wants a series of sales trend reports to better manage her inventory dollars and improve sales. This can be done only if the analysts capture the sales transaction data in a Sales subsystem and incorporate a Sales Trends subsystem. CD and Supplier subsystems seem reasonable, so that the analysts can create reports that identify specific product trends and orders and provide supplier telephone numbers. In a similar fashion, the request for customer record keeping and communications leads to penciling in Customer and Correspondence subsystems. Counting the obvious Reordering and Internet subsystems, there are eight processes to work with, as listed in Figure 7-15.

FIGURE 7-15   / *Cornucopia New Subsystems*

1. Sales (sales transaction capture)
2. Sales trends (sales summarization)
3. Reordering (CD orders based on sales)
4. Customer (Customer master file)
5. CD (CD master file)
6. Supplier (Supplier master file)
7. Correspondence
8. Internet

Understand that this is a preliminary look at the new system. But, it does give the analysts the chance to develop a new USD (Figure 7-16) that they can share with the owner. Notice that this diagram builds on that of the existing system. This allows the owner to place the new system elements in a familiar frame of reference. The owner's first reaction is, "I thought we were talking about one computer, not eight!" Of course, only one computer will be used, but it will feature several menu selections that correspond to each subsystem. The computer icon is used to alert the owner that this is a machine operation.

FIGURE 7-16 / *Cornucopia New System USD*

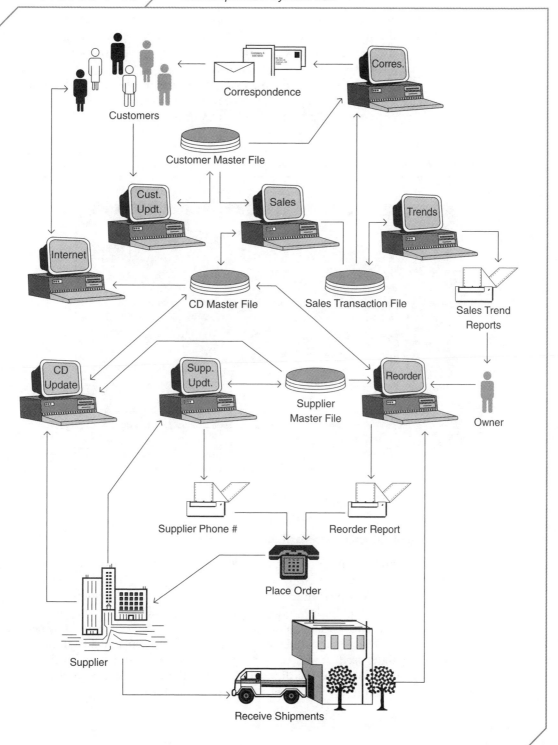

Customers

Correspondence

Corres.

Customer Master File

Cust. Updt.

Sales

Trends

Internet

CD Master File

Sales Transaction File

Sales Trend Reports

CD Update

Supp. Updt.

Supplier Master File

Reorder

Owner

Supplier Phone #

Reorder Report

Place Order

Supplier

Receive Shipments

New System Form Design

First impressions are important to the overall acceptance of any product. With an information system, the user's first impression is greatly influenced by the system's screen form design. With that in mind, the analysts developed a general form design layout and color scheme, as pictured in Figure 7-17. The form includes a custom Cornucopia logo, an informational hot pad, form title, and custom tool pad in the upper portion, leaving the lower portion for data. The analysts expect to use this design for most of the forms.

FIGURE 7-17   / *Cornucopia Form Design*

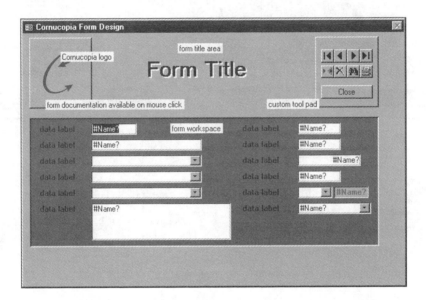

Preliminary Presentation

The preliminary presentation includes a hardcopy report and an oral presentation, supplemented by a PowerPoint slide show. The slide show consists of seven slide topics: title, staff introduction, project requirements, form design, timetable and budget, question and answer, and next meeting. In developing the slide show design and selecting content for each slide, the analysts' intent was to find the right balance between detail and generality. Figure 7-18 shows the slide used to describe the project timetable and cost breakdown. The images chosen for the slide are similar to the project budget and status report illustrated in the last chapter (Figures 6-9 and 6-10).

FIGURE 7-18   /*Cornucopia Preliminary Presentation Slide Show*

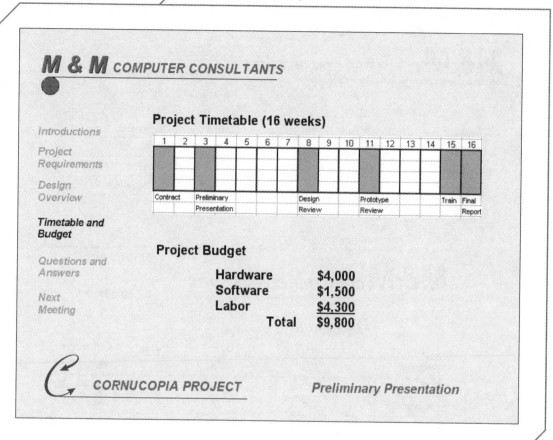

Time and Money

During the fourth week of the project, three hours were spent completing the analysis phase and three hours were spent on the design phase, as reflected on the Billable Hours forms that appear in Figure 7-19. The analysis phase is virtually 100 percent complete, but because the design work that lies ahead may require the analysts to reopen some part of the analysis, they record this as only 95 percent complete. With the first attempt at the new USD completed, the analysts show the design work as 20 percent complete. Illustrations of the updated project status report and budget are omitted in this chapter.

Cornucopia Billable Hours Form—Week 4

M&M *COMPUTER CONSULTANTS*

Billable Hours Form

Analyst: Jamie

Project	Activity/Task	Date	Hours	Comment
corncpia	T.2.3	2/22	.5	Existing USD
corncpia	T.2.7	2/24	1.0	Finish Prelim. Pres.
corncpia	T.3.1	2/24	1.0	Prelim. Pres. Review

M&M *COMPUTER CONSULTANTS*

Billable Hours Form

Analyst: Scott

Project	Activity/Task	Date	Hours	Comment
corncpia	T.2.2	2/23	1.5	Industry Res.
silhouet	T.1.2	2/24	1.5	Feasibility Rpt.
corncpia	T.3.2	2/25	2.0	New USD

Cornucopia with Web Authoring Software

Dreamweaver, from Macromedia, is a good example of a sophisticated Web site design and construction package. While this software provides a user-friendly GUI tool to help the analysts build and maintain a complex site, it is not the only product needed to create attractive, informative, and easy-to-use Web pages. TechNotes 7-3 and 7-4 are devoted to detailing many important Web technologies and design features. Dreamweaver can help assemble the outputs from these products. Custom imaging, audio and video streaming, and interactive programming are examples of the products Dreamweaver can help add to Web pages. Figure 7-20 displays Cornucopia's new home page, with several overlaying floating windows to illustrate the feature-rich nature of Dreamweaver.

FIGURE 7-20 / *Cornucopia with Dreamweaver*

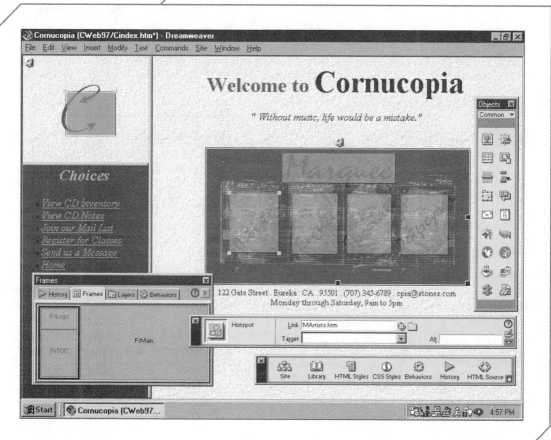

SUMMARY

Most projects of any consequence require planning to coordinate the resources that are brought together for such endeavors. Projects of the size and nature addressed in this text require more than planning. They require an overriding strategy that provides a direction and sequence to problem solving. Such a strategy is referred to as a methodology. This chapter presents some guidelines for the design portion of the eclectic methodology introduced in the first chapter.

The discussion bridges the analysis and design material, beginning with a summary of the modeling tools employed in both activities. The emphasis on design as a circular, user-centered process is crucial to the sections on developing alternative designs and prototypes.

Some structured programming terms (modularity, coupling, and cohesion) introduce the notion that modern 4GL information systems require careful assembly to ensure that the subsystems are well coordinated, easily understood, and easily maintained. A brief introduction to some of the design choices available to small-enterprise systems is offered to get the analyst started. Of course, such a list must be continually updated as the industry changes.

The chapter outlines the preliminary presentation and provides a preview of the design report required at the end of the process.

Finally, this chapter introduces the fundamentals of form design, emphasizing the graphical nature of the project interfaces. The term graphical user interface dialog (GUID) describes the interactive sequence between the user and the information system.

TEST YOURSELF

Chapter Learning Objects

1. How does the USD transform the project objectives into a preliminary design model that the user can follow and understand?

2. Why does design work begin with the user's perspective and end with the hardware perspective?

3. How would you adapt the joint application design (JAD) methodology to fit a small-enterprise systems project?

4. What makes a GUID user-friendly?

Chapter Content

1. Information system design is a carefully sequenced series of modeling activities, requiring that each model be complete before moving on to the next. True or False?

2. Joint application design is too formal and burdened with bureaucratic procedure to be of any use to the analyst working on a small-enterprise information system project. True or False?

3. Select the appropriate staging label (Stage I, II, or III) for the following information system description: local area networked information system with office automation components and intranet.
a. Stage I
b. Stage II
c. Stage III

4. Select the appropriate staging label (Stage I, II, or III) for the following information system description: stand-alone information system with management information, decision support, and transaction processing components, as well as Internet access.
a. Stage I
b. Stage II
c. Stage III

5. Select the appropriate staging label (Stage I, II, or III) for the following information system description: wide area networked information system with strategic and expert system components.
a. Stage I
b. Stage II
c. Stage III

6. Assume a structured design specifies a software suite, such as Microsoft Office, to support a small-enterprise information system. Select the sentence that best describes the degree of coupling and cohesion displayed by the various pieces of application software in the suite.
   a. The application software is tightly coupled because of the common graphical user interface.
   b. The application software is completely cohesive, with no coupling whatsoever, because the various pieces of application software do not communicate directly with one another.
   c. The application software is both loosely coupled, because data sharing is well managed, and highly cohesive, because each application provides self-contained functionality.
   d. The application software meets the perfect requirements of modular design because it is completely cohesive, with absolutely no coupling.

7. Select the statement that best describes how object-oriented design fits into the SDLC methodology articulated in this text.
   a. Object-oriented design has no relationship to the models or activities associated with the SDLC methodology discussed in this text.
   b. Object-oriented design is simply the latest application development fad, providing new names to familiar model components.
   c. Object-oriented design provides the missing link between the traditional SDLC models.
   d. Object-oriented design is a natural extension of the data and process models, providing an intuitive structure called an object.

8. Which of the following documents would *not* be helpful to the analyst attempting to construct the user's system diagram for the new system design?
   a. project contract
   b. existing system context diagram
   c. existing system USD
   d. new system DFD

9. During the project preliminary presentation, the new system design can best be described by use of the:
   a. project contract
   b. new user's system diagram
   c. project status report
   d. new system data flow diagram

10. What is a GUID?
    a. a set of program specifications
    b. a graphical user interface dialog
    c. a screen-painting utility
    d. another data model

11. The fundamentals of screen form design are best learned by:
    a. developing one good form design and sticking with it
    b. reading this chapter three more times
    c. learning more about graphic design from a trade book or a course in graphic design
    d. hiring an artist to design your forms

12. Which of the following events cause the design process to end?
    a. The user begins to lose interest in the design process.
    b. The design is perfect.
    c. The projected design completion date may be compromised if the discussions continue.
    d. The user is satisfied with the design.

13. A design prototype is best described as a:
    a. movie script
    b. collection of images representing the system interfaces
    c. black box
    d. sample computer program

ACTIVITIES

Activity 1

Review the descriptions and illustrations that introduce Sunrise Systems' Stage II computer information system (Figures 7-9 and 7-10).

a. Prepare a list of the processes, external entities, and data stores that appear in the data flow diagram.

b. Associate the information needs summarized from the project contract with the processes on the above list.

c. Associate the icons used in the user's system diagram with the items on the above list.

d. Explain any discrepancies or omissions you find between the list above and the needs or icons.

Activity 2

For the next several days, keep a brief journal describing any logistical or computational problems you encounter and the strategies used to solve the problems. Prepare a summary of the strategies—noting similarities, differences, patterns, and so on. Indicate which strategies, if any, you might apply to your project design activities.

Activity 3

Critically evaluate the GUID associated with an information system you use on a regular basis, such as an ATM, a library catalog system, or one of your school's online services. Evaluate the user-friendliness of the screen forms. Describe the form design elements (icons, layout, colors, type face, etc.). Suggest improvements.

DISCUSSION QUESTIONS

1. To what degree would you expect that the user's level of computer expertise hinders or enhances the analyst's design work?

2. If we agree that design is a circular process, what clues might tell you when to stop the process?

3. From the problem-solving strategies presented in the chapter, select one that you have used and, for a specific case, describe your success (or failure) in applying the strategy.

4. Is it possible to eliminate subsystem coupling in the small-enterprise information system? If so, is this desirable?

5. How are GUIDs similar to the UML state-chart diagram discussed in Chapter 5?

6. Define the word "prototype." Provide two examples of prototypes you have seen, read about, worked with, or designed.

PortfolioProject

Team Assignment 7: Preliminary System Design

Given the increased emphasis on user participation throughout the SDLC, it is very likely that you and the user have already discussed your progress during the analysis phase of the project. Regardless of the number and frequency of such informal conversations, the preliminary presentation provides an opportunity for you to document formally your transition from analysis to design. The descriptions and visuals of your preliminary system design will stimulate further dialog between you and the user and set the foundation for your detailed design activities.

The specific content and sequence of the presentation can vary considerably, depending on the complexity of the project and the user's prior experience with the SDLC. For this assignment, and the ensuing preliminary presentation, you are required to develop and deliver a standardized, streamlined collection of project documents. Appendix C offers more detailed instruction on technical writing and presentations, but remember that you must make every effort to connect with the user if you expect the user to be an effective participant in the process.

In order to develop the new system's preliminary design, you must complete the following tasks:

1. After reviewing the build or buy discussions in Chapters 2 and 6, research the availability of turnkey or vertical software solutions to your project. Prepare a brief summary of your findings. Submit a copy of your summary to your instructor.
2. Prepare a user's system diagram for the new system. Submit a copy of this diagram to your instructor.
3. Develop a screen form design for the new system GUIDs. Submit a copy of this design to your instructor.
4. File a copy of the research summary, new USD, and screen form design in the Portfolio Project Binder behind the tab labeled "Assignment 7."

Project Deliverable: Preliminary Presentation

1. Submit a preliminary presentation report to your instructor containing the following items:
    a. Cover letter
    b. Summary of the project requirements
    c. Overview of the proposed new system (USD)
    d. Preview of the screen form design (GUID)
    e. Project timetable
    f. Project cost breakdown
    g. Oral presentation slide show handouts (three slides per page)
2. Prepare a slide show and the appropriate handouts to support a 15–20 minute oral presentation covering items b–f.
3. File a copy of your report in the Portfolio Project Binder behind the tab labeled "Preliminary Presentation."

databasedesign

When you complete this chapter, you will be able to:

OVERVIEW

At this point in the systems development life cycle (SDLC), the analyst must develop the first complete set of models for the new system. This task may be partially complete, depending on how radically the new system differs from the old. In this chapter, you learn how to make the transition from the preliminary user's system diagram (USD) to the new data flow diagram (DFD), entity-relationship diagram (ERD), object model (OM), menu trees, and implementation overviews.

You will see that input/output design and file specifications go hand in hand. The information presented on a query screen or report originates in a file, which itself is dependent upon input data forms. As the centerpiece of this relationship, this chapter addresses file design fundamentals, followed by some additional coverage of form design.

- Develop the new system process model (DFD) from the preliminary USD.

- Develop the new system data model (ERD) and object model (OM).

- Develop estimates of the resource requirements associated with file-processing design.

- Design graphical user interface dialogues (GUIDs) and screen forms to access system files and processes.

CREATING NEW SYSTEM MODELS

To some degree, the new information system design begins the moment the SDLC process begins. Certainly, by the time the analyst has completed most of the analysis phase, a skeleton design exists within the new USD. Now we can reverse the process, working from the new USD back to the other models.

The combination of the existing system models, the user requirements, and the analyst's understanding of the different stages of small-enterprise information systems produces a set of ideas about how the new system might look. The analyst and user should openly discuss these ideas. Unfortunately, no single path can lead you to "the answer," no single tool can help you distill the eventual new design from this collection of ideas. Your only comfort may come from knowing that the designs you develop at this time do not have to be perfect—you can revise these designs many times before you have to commit to the final version.

New System Modeling: Step by Step

Figure 8-1 shows a segment of the enhanced SDLC, along with a suggested pattern for the development of the new system models. Remember that the shaded areas represent the user's active participation in the process. In this case, the modified joint application design method is featured to emphasize the importance of user interaction. The preliminary USD sketch provides the basis for the first JAD activity. Even if this diagram was part of the preliminary presentation, the analyst should schedule a separate JAD working session to analyze the document. This sketch of the new system design is user-friendly, and its informal structure invites the user to suggest changes. In other words, it is a working draft. The critique should produce a revised USD that the analyst can then use to develop the preliminary process and data models.

The first step in creating a complete set of new system models is to sketch the new system DFD. We have already seen that the USD is nothing more than a simplification of the data flow diagram; therefore, developing the new DFD should be as simple as reversing that process. Each of the processes on the USD is also a process on the data flow diagram, so this is a good place to start. Remember that each process is connected to a data flow. Ask yourself where the data comes from or where it goes—another process, a data store, or an external entity? Also, consider whether the data is part of a larger data structure or whether it is independent. Do not be afraid to rely on the old system data definitions and structures to get you started on the new system DFD.

/ *Design Activities*

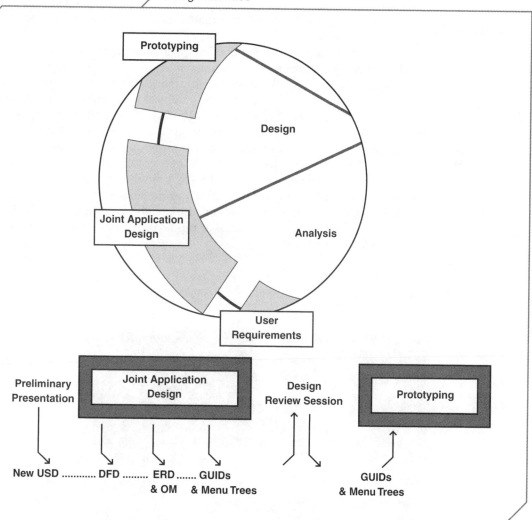

The second step is to sketch the new system ERD. The sketch of the preliminary new system entity-relationship diagram can also follow directly from the USD, but doing so is easier if you also have the new data flow diagram to use as a reference as well. After all, the DFD identifies the data stores and many of the data elements that are part of the new system design. Start by drawing a rectangle for each data store and then noting the file structure immediately below the rectangle. Connect the rectangles for those files that have common data elements. Now you must complete the normalization (see Chapter 4)

and file design process. This has very practical consequences. Through normalization, you can identify the files that need to be designed and, in turn, the forms and graphical user interface dialogues (GUIDs) that need to be designed to service the files.

The third step is to sketch the OM. As discussed previously, the OM can be developed independently from the data model. To simplify your introduction to object modeling, we will simply transform the data model into an object model.

The fourth step is to design the graphical user interface dialogs and the menu tree. Figure 8-2 presents the beginning of the GUID design for Sunrise Systems. The design reflects a common 4GL strategy in which the user is repeatedly prompted for more specific information, which reduces screen clutter and helps the user focus on the desired system function. The Main Menu command button captions, taken directly from the project contract introduced in Chapter 7 (Figure 7-11), are refined as the analyst proceeds further in the system design process. In a similar fashion, the menu tree, which presently consists of a handful of choices that mimic the command button captions, is expanded and refined later.

FIGURE 8-2   /   *Sunrise Systems GUID Design*

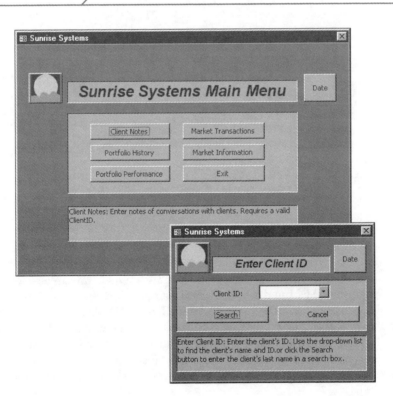

Figure 8-2 also includes a preview of how design prototyping relates to joint application design. As first described in Chapter 7, design prototyping allows the user to see the design in a computer environment rather than a paper representation. This is a powerful design tool because it eliminates one level of abstraction, which makes the system design more real to the user. After the design review session is completed, the analyst creates the first prototypes, and thus begins yet another user-analyst partnership, this time with prototyping as the facilitating medium.

FILE DESIGN FUNDAMENTALS

User-centered design always demands that the analyst return to the question, "What information does the user want from the system?" A broad answer to this question is embedded in the project contract and the preliminary presentation, where several output products are described. The new system USD icons and DFD data flows identify system outputs as well. All this suggests that we begin our detailed design work with output specifications rather than file specifications, which would require the analyst immediately to resume JAD working sessions to design detailed paper-based output layouts. However, because our intention instead is to create actual sample output designs on the computer, we must first develop the files upon which the outputs depend. In practice, you may find yourself switching among file, output, and form design as the design activities progress.

Before delving into the step-by-step activities for designing files and the forms that service the files, this next section reviews *file design fundamentals*, which should complement the database and file-processing fundamentals presented in Chapter 4.

Traditional File Types

Files serve many purposes. Using the commonly accepted definition that a file is simply a collection of like items, then an appointment calendar, a payroll timecard, and a phone book can all be considered files. These files have one thing in common: they all have an underlying structure. Each item within these files can be distinguished by its particular attribute values. For example, if the appointment calendar requires location, client, and topic entries for each day and hour, the 1/10/2004 @ 10:00 entry could be "Room 102, Ms. Jones, job interview," whereas the entry for the 11:00 hour could be "cafeteria, Lisa, coffee break." You probably recognize this structure as the familiar record layout for a file. What you may not be so familiar with is how these files differ in the way they are implemented in a computer setting.

The appointment calendar must be available at all times for reference and to be updated. The payroll timecard need only be updated once a day (to show exceptions to the norm), accessed once each pay period, and then saved for year-end processing. The phone book rarely changes. These differences lead to important consequences in terms of file design. Figure 8-3 summarizes the *traditional file types*. Can you associate one or more of these file types with the preceding examples?

Traditional File Types

Master File	Contains data that is seldom changed but is often used referentially by processes within the system
Transaction File	Contains data that describes an event that pertains to some part of the information system; one kind of transaction is associated with batch processing, the other with online transaction processing (OLTP)
Batch File	Transactions that are collected together for processing sometime after the event they describe
Online	Transactions that are individually processed as soon as possible and then usually placed in a history file
History File	Contains data that describes what happened in the past
Backup File	Contains duplicate data from another file
Temporary File	Contains data that need not be retained beyond the current processing
Table File	Contains data that is used referentially and that can change periodically

4GL File Types

4GL software produces 4GL files, commonly distinguished by their file extension. A *4GL file type* is any file known by its file extension. Common examples of 4GL file types are .doc, .xls, .mdb, and .ppt, which correspond to Microsoft Word, Excel, Access, and PowerPoint. Students familiar with 4GL products may wonder where these file types fit into the traditional file types presented in Figure 8-3. These programs promise to satisfy almost any user's information needs, seemingly with no required thought about file-processing fundamentals. Spreadsheet users rarely concern themselves about the distinctions between master and transaction files. And the only file types that most users have heard about are backup and temporary files. These questions arise: should the analyst be concerned about 4GL file types? Are the newly invented classifications really new? The answer is both yes and no.

This text assumes that 4GL products play a central role in small-enterprise information systems. It follows, therefore, that the analyst must understand how 4GL files are composed, stored, accessed, and cross-referenced. For example, a spreadsheet file is an interesting combination of master and transaction files (you will learn more about this later in the chapter). Thus, we are looking at a new crossbreed of files, not an entirely new set of file types.

Although we can associate these files with the traditional classifications, something *is* different about them. The project budget and project status report introduced in Chapter 6 are implemented with a spreadsheet program. They function as master files in that they retain project information that helps to more fully describe the project as well as information that changes very seldom (historical cost estimates, activity descriptions). They function as transaction files because they record information about very specific events (actual costs, activity completion dates, and status). The analyst has immediate, continuous access to these files through the spreadsheet interface. Prior to 1980, budgets and status reports were output products of third-generation programs (usually COBOL) that transformed files of data (master and transaction). Users were restricted to the regimen established by the Data-Processing Department, with little latitude for quick adjustments. Thus, it is not the file types that have changed so much as the file access methods that have improved.

Figure 8-4 illustrates a few situations where 4GL products can be employed in the same fashion as the traditional file types. You should notice that the database files form the heart of our data model, the entity-relationship diagram. The added description of these files in terms of master, transaction, or historical file is important because of the form, report, query, and processing design implications of each type, as discussed in this and the following two chapters.

Indeed, current practices produce an ever increasing multitude of file types, distinguished more by their file extensions than by their uses. These include ASCII files (.txt), graphics files (.bmp, .pic), programming files (.bat, .pas), executable files (.bin, .exe), as well as others too numerous to list here. Technically, the file extension is meant to describe the way the data is formatted, that is, the data's physical organization, the special characters used to delimit data, and so on. This proliferation of file distinctions is not likely to subside. Most of these files can and should be incorporated into the analyst's thinking as possibly serving in a traditional way, even though they require nontraditional software sponsors.

File Organization

Access to file data is greatly affected by the way in which the file is physically and logically organized. *File organization* describes the way records are physically stored on disk and the way in which those records are accessed. Remember that a disk is a direct access device, allowing any data to be retrieved in roughly the same amount of time, regardless of its physical position in the file. The obvious advantage of a direct access device such as a floppy disk is that you can collect and process data through the input interfaces the instant the data is available, without fear of degrading overall system performance.

FIGURE 8-4   *4GL Products and File Types*

4GL Product	File Type	Description
Database	Master	Personnel records
		Product records
		Customer records
	Transaction	Payroll exceptions
		Sales activities
		Satisfaction surveys
	Historical	Cumulative wages
		Summarized sales
		Survey summaries
Spreadsheet	Master	Budget categories
		Activity descriptions
	Transaction	Actual expenditures
		Completion dates
	Historical	Past-year expenditures
Word Processor	Master	Personnel résumés
		Product use instructions
		Customer psychological profile
	Transaction	Performance review notes
		Product development notes
		Customer contact notes
	Historical	Personnel dossiers
		Product patent correspondence
		Customer correspondence

The analyst, anticipating the need for direct access, builds indices into the file design. An *indexed file* permits logically ordered access to physically unordered data. To access data in logical order, the computer uses the index the same way you might use the index to this text. Thus, even though we know that product sales are random throughout the day, we can generate a daily sales summary for each product quite easily if we have provided for an index on product code. Likewise, a daily salesperson summary can be derived from the same file simply by activating the salesperson code index.

However, the *sequential file* still has its uses. Because it requires no indexing algorithm, sequential file processing has very little system overhead and therefore can be much faster than indexed access. If a significant majority (say, more than 80 percent) of the records on a file must be accessed, sequential processing is probably a better choice than indexed processing. For example, paycheck-producing subsystems are good candidates for sequential processing: we assume almost every record on the payroll master file needs to be accessed to generate a check. By definition, the payroll exception transaction file affects only a small percentage of the master file records, so why should we pay an index-processing premium to visit every record on the master file? Such are the considerations the analyst must entertain when file organization is factored into file design.

File Structure

After you address questions of file type and organization, you must be concerned with exactly which data elements are relevant to the anticipated file-processing requirements. For example, if the user wants a report on sales by the hour, you should include the time of sale as part of the sales transaction file. Incidentally, this seemingly innocent request for time-sensitive information has profound effects on data collection hardware and software requirements. Check your next few sales receipts to see which retailing systems provide time-of-sale data. What does this suggest to you about input design?

Traditionally, we describe *file structure* in the context of the broad *data structure hierarchy* shown in Figure 8-5. Each of these structures, with the possible exception of the knowledge base, should be familiar to you. In fact, programming students will also want to add some of their favorite structures, such as arrays, stacks, queues, and trees. However, file and database processing dominate small-enterprise information systems and are the focus of our continued discussion of file structures.

FIGURE 8-5   *Data Structure Hierarchy*

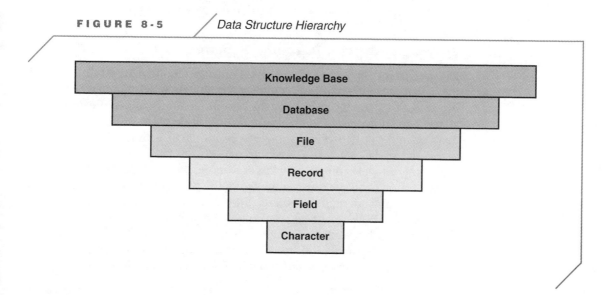

File structure is defined by the individual fields that make up the records. Each field must have a unique name (within the file), a type (numeric, character, logical, date, etc.), a length, and an indication about whether the field is to be used as an index. Of course, all these definitions must be entered into the data dictionary.

File Security and Controls

Given the proprietary nature of file contents, every enterprise is concerned about safeguarding files against inadvertent damage or intentional misuse. Virus protection, regular backup procedures, and basic password protection are only a few of the file security measures that should be built into your file design as a matter of course.

The physical security of files can often be overlooked in the small enterprise, where there probably are no off-premises storage facilities for archiving historical documents. But storing backups in a drawer across the room from the computer cannot prevent data loss in the event of fire or flood. File copies must be stored in a separate facility.

Data integrity is a qualitative term describing the degree of accuracy of the data in a file. An *internal control* is a procedure designed to help maintain a high degree of data integrity. Maintaining consistent internal controls on data files is an important part of the overall process to ensure data integrity. The data integrity process begins with data collection and verification, which is discussed further in the next section. Even after the data safely resides in a file (master, transaction, historical, etc.) it can still be corrupted through processing errors, misdirected updates, or both—not to mention the famous system anomaly (i.e., "We don't know what happened"). One successful internal control measure is to regularly check for reasonableness. For example, a simple query to display all accounts with extremely high or low balances or activity might catch duplicate or missing data. Another method is randomly to select a small percentage of records (say, five percent) to be thoroughly audited for all activity during a period. The point is that the analyst should not assume that the system runs error free.

Estimating System Resource Requirements

File designs influence the software and hardware requirements of the system. If the design calls for a database, you most certainly need a database management program and a fast computer on which to run it. Furthermore, file design also affects data storage requirements. Estimating disk space needs used to be a quasi-scientific task involving some simple arithmetic with file sizes, record volumes, and update estimates. Now the rule is to buy the fastest and biggest of everything, while keeping a keen eye on your budget constraints. Databases, for example, can consume vast amounts of disk storage when they incorporate audio, video, and graphic objects into their structure. In addition, some resources (CD-ROM, scanners, speakers, etc.) may be required to support certain file designs. Such resource requirements should be cataloged at this point in the SDLC to help prepare for the future tasks of developing a detailed resource needs list, a Request for Proposal, and cost data for the cost/benefit analysis.

File Design: Step by Step

Figure 8-6 suggests a sequence for the analyst to follow when working through the file design for the new system. Several of the file design steps refer to files that may be part of a larger database, where a file is called a "table" or a "relation." It is very likely that all of the small-enterprise projects you encounter will have a database at the very heart of their file design. For this reason, the next section revisits relational databases and offers a sample implementation.

FIGURE 8-6   *File Design Steps*

1. Inspect the preliminary new system design models (USD, DFD) to locate the files (data stores) required.

2. Identify the file types (master, transaction, history, etc.), beginning with the master files if possible.

3. Determine if some of these "files" are really databases, which themselves might consist of several related files. Develop a sketch of the preliminary ERD.

4. Decide how to organize these files (indexed, sequential) and how they are accessed (batch processed or OLTP).

5. Define the data elements required to support the processing and output requirements associated with these files.

6. Develop specific data element definitions, including key fields for indexing as appropriate (see Step 4 above).

7. For the database files, identify the common data elements and define the file relationships (one-to-one, one-to-many, many-to-many). Normalize the preliminary ERD to eliminate repeating groups and many-to-many relationships. Sketch the OM.

8. Build the 4GL file data structures and enter a small sample of data (ten database records, one spreadsheet page, one word-processed paragraph).

9. Estimate resource requirements necessary to support the files.

10. Update the data dictionary to include the files and data elements.

RELATIONAL DATABASES REVISITED

The modeling process identifies the data files required to support the information system without specifying exactly how these files are to be implemented. Previous discussions on data modeling and file design fundamentals suggest that data files can be implemented as independent, sequential files or indexed files. Typically, this approach requires a considerable programming effort. The relational database approach offers a convenient alternative. The database implementation is usually preferred because it requires only minimal programming skill and it provides a wide range of user-friendly, graphical tools.

Chapter 4 presented a theoretical perspective on relational databases and normalization. Here, we turn to a more practical view. Microsoft Access is a popular relational database management system (RDBMS) tailored to the small-enterprise project. Access offers a user-friendly graphical interface, provides the developer with a wide array of convenient and sophisticated application building tools, incorporates many object-oriented features, and is very affordable. In short, Access is well suited for the single-user database application stored on a single user's desktop computer.

Database Implementation with Microsoft Access

With Access, the analyst can move directly from the normalized entity-relationship diagram to database design, with each entity represented as a table in the database. Figures 8-7 and 8-8 show images of the Northwind sample database that is packaged with Access. These figures illustrate the graphical manner in which the developer can implement the database design. There are obvious similarities between the table relationship image in Figure 8-8 and the entity-relationship diagram, notwithstanding the nonstandard cardinality notation offered by Access.

ThinkingCritically

RelationalDatabase**Expertise**

It is assumed that the analyst has a passing acquaintance with most of the technology employed in the SDLC process, but lacks expert knowledge in any area. Most of the time, project teams are assembled to tackle the problem. Each team member is responsible for one or more aspects of the work. This chapter emphasizes the importance of database design. How much time should the analyst, who does not aspire to become a database specialist, invest in learning about relational database software?

FIGURE 8-7 / *Database Table Definition with MS Access*

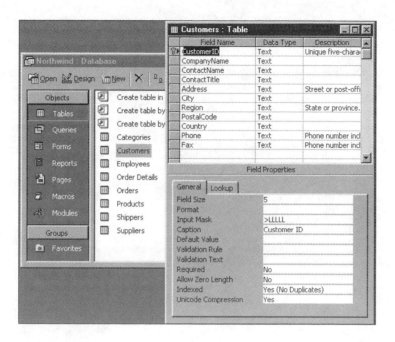

FIGURE 8-8 / *Database Table Relationships with MS Access*

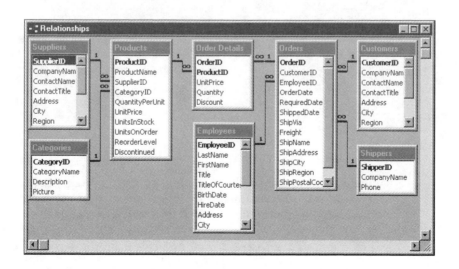

TECHNOTE 8-1

OBJECT-ORIENTED DATABASES

A database management system solves many of the problems associated with complex file processing. For example, there is no need to design, develop, test, and maintain dozens upon dozens of computer programs to perform such mundane tasks as data entry and validation, file update, and data summarization and reporting. The database engine that provides these services has undergone some dramatic changes over the last forty years.

The first-generation database of the 1960s and 1970s was built on a hierarchical and network data model. The second-generation database of the 1980s and 1990s was built on the relational model. Some analysts believe that the third generation, object-oriented database, will overtake the relational model to become the new standard. This probably won't happen any time soon, because only three percent of all database applications employ truly object-oriented database products. Nevertheless, there are sound reasons for the estimated 50 percent annual growth in the number of object-oriented applications.

First, the rise in computing power has fueled an expansion of computer applications. For example, computer assisted design, multimedia communications, and virtual reality are three application areas experiencing impressive growth. Each expansion creates new, more complex data types, most of which are difficult, if not impossible, to incorporate into the relational database. The object-oriented database provides for such complex objects.

Second, a growing number of satisfied computer professionals who have enjoyed the productivity gains associated with object-oriented programming and objectlike applications are eager to duplicate their experience with database applications. Why not extend the benefits of encapsulation, inheritance, and polymorphism to all enterprise objects? In response to this challenge, two approaches emerged: the **object-oriented database management system (OODBMS)** and the **object-relational database management system (ORDBMS)**.

The OODBMS provides all of the services required to create, maintain, manipulate, and report on objects stored in a persistent and shareable repository. There is no agreement on exactly how to implement such a system. Everything from extending an existing object-oriented programming language with database capabilities to developing a completely new data model and data manipulation language has been suggested.

The ORDBMS is really an extension of the traditional relational model. As before, there is no single extended relational model. Individual relational database vendors tailor object-oriented capabilities or extensions to exploit a particular feature or accommodate a particular market segment.

As the boundaries of the information system grow, it is likely that the underlying database product will need to grow as well. Larger applications, involving multiple, simultaneous users, spread over an office complex or extended geographical area, require a client/server database solution. In such cases, the database resides on the database server, not on the client station. Client requests are passed across the network to the server, where database access and processing are completed before the server sends the response back to the client. Oracle is one of the leading providers of such client/server database products.

FILE DESIGN FOR SUNRISE SYSTEMS

The data flow diagram in Chapter 7 (Figure 7-9) identifies three data stores (Notes, Transactions, and Performance), all of which appear to be transaction files. Figure 8-9 and Figure 8-10 illustrate the preliminary ERD and the normalized ERD, along with field definitions and some sample data. Notice that the newly identified Investor master file is required to normalize the ERD. The OM, which is not illustrated, follows directly from the normalized ERD.

FIGURE 8-9   *Sunrise Systems New System Preliminary ERD*

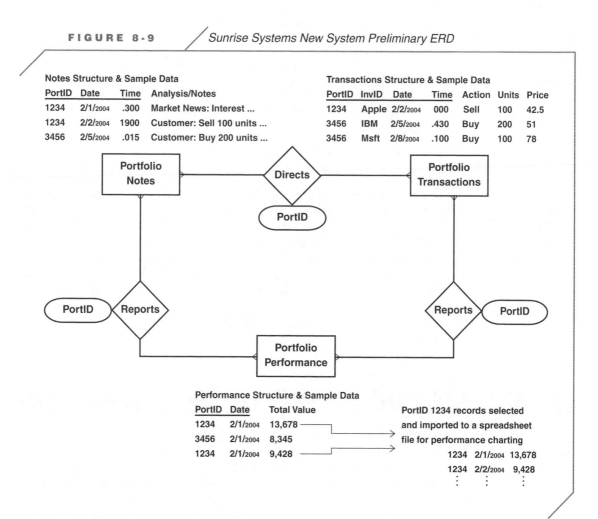

Notes Structure & Sample Data

PortID	Date	Time	Analysis/Notes
1234	2/1/2004	.300	Market News: Interest ...
1234	2/2/2004	1900	Customer: Sell 100 units ...
3456	2/5/2004	.015	Customer: Buy 200 units ...

Transactions Structure & Sample Data

PortID	InvID	Date	Time	Action	Units	Price
1234	Apple	2/2/2004	000	Sell	100	42.5
3456	IBM	2/5/2004	.430	Buy	200	51
3456	Msft	2/8/2004	.100	Buy	100	78

Performance Structure & Sample Data

PortID	Date	Total Value
1234	2/1/2004	13,678
3456	2/1/2004	8,345
1234	2/1/2004	9,428

PortID 1234 records selected and imported to a spreadsheet file for performance charting

1234	2/1/2004	13,678
1234	2/2/2004	9,428

FIGURE 8·10 / *Sunrise Systems New System Normalized ERD*

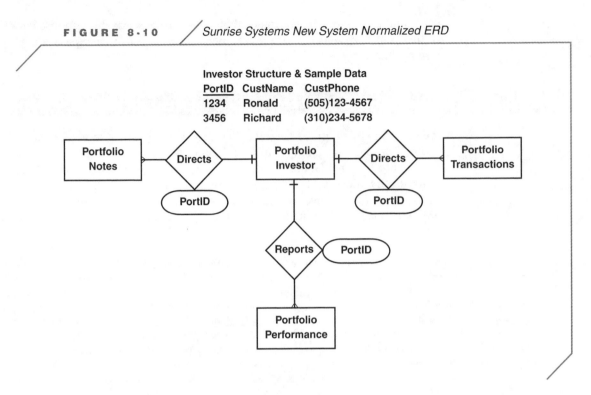

Investor Structure & Sample Data

PortID	CustName	CustPhone
1234	Ronald	(505)123-4567
3456	Richard	(310)234-5678

Each file presents some interesting features. The Notes file combines some elements of a word processor with the database. The analysis and notes attribute is called a memo field, but it can be manipulated like a word processor text file. The Transactions file requires the combination of four fields to define a key value. The Performance file contains a record with the total value of each portfolio for each day. Records for a particular portfolio can then be selected and charted. The Investor file structure would probably include more attributes, such as the investor's address. For simplicity, we assume that each investor has only one portfolio and that each portfolio belongs to only one investor (i.e., a one-to-one relationship between the investor and his or her portfolio).

FORM DESIGN REVISITED

The form design fundamentals discussed in Chapter 7 are clearly applicable to the forms required to service the tables in a database. Database master files are subject to maintenance activity that includes the addition of new records and changes to existing records. Database transaction files change as transaction records are added or updated on a regular basis. Figure 8-11 suggests a sequence for the analyst to follow when working through the form design for the new system's database.

FIGURE 8-11 / *Form Design Steps*

1. Draft the basic screen design templates (menu, database and transaction). A paper sketch is sufficient at this point. These basic designs will be revised as specific screen samples are developed below. Schedule a JAD working session to work through Steps 2–10 below.

2. Determine what the user must supply to initiate updates for one of the master files. This usually includes an action (add, change, delete) plus a key field value.

3. Determine which file data the user needs in order to verify the update.

4. Determine how to inform the user about invalid or incorrect data.

5. Develop a set of sample "dialog sequences" (GUIDs) that illustrate the series of screen inputs and responses defined by Steps 2, 3, and 4. This set can be used as a reference for each of the master files rather than repeating these steps over and over.

6. Perform Steps 2–5 for one of the transaction file data entry operations. This work will serve as a guide for designing the other transaction file "servicing" screen forms. It should be duplicated only if very different or highly specialized features are associated with the other files.

7. Develop a menu tree diagram sketch showing how all the master file and transaction file screen forms interact.

8. Build one master file update screen form using a database screen painting utility or a front-end product. (This assumes that you have already built the sample file.)

9. Repeat Step 8 for one of the transaction files.

10. Steps 8 and 9 produce the first design prototype product. Work with the user to incorporate revisions agreed to through repeated use of this process.

11. Estimate resource requirements necessary to support the forms.

12. Update the data dictionary to include the forms that you and the user have designed.

FILE AND FORM DESIGN WITH CASE TOOLS

By itself, the CASE tool cannot design the system. As you can see from the material in this and the previous chapter, design work requires creative problem-solving skills. However, CASE tool products can help you implement your design. For example, Excelerator has an option called Screens & Reports that enables the user to construct file-based screen forms. Because of the built-in, CASE data-element controls, you are guaranteed that your forms are tied to the files you have previously defined through Excelerator's DFD and ERD options.

Most database products offer similar form-building, or screen-painting, features. Of course, these products might not automatically update the data dictionary, so when using these alternatives to CASE, the analyst must be extremely careful about choosing data elements that are consistent across the entire system design.

THE CORNUCOPIA CASE

You should note that as the discussion moves more into the user-centered activities of the SDLC, the owner's role as a principal user of the system is emphasized by referring to her as the user. The new USD introduced in Chapter 7 (Figure 7-16) provides the springboard for the first serious discussions with the user about the new system design. The analysts' first JAD working session focuses on the USD. The user's initial concern about the multiple computer icons is satisfied after they explain that they represent menu options, not separate computers. The analysts also discuss the consequences of the Customer Maintenance and Correspondence subsystems now being connected to the rest of the information system. The analysts explain that the new Internet subsystem provides another means of communication with customers. Finally, the analysts point out the changes to the present Sales subsystem, which includes new CD inventory maintenance functions. The user agrees that the new system offers some great improvements over the old system. She is especially interested in the Internet connection, asking about what her page will look like and wondering if visitors will be able to "mess up" her computer in any way. The analysts assure her that the design calls for limited Internet access, which does not endanger the integrity of her data files.

New System DFD

The analysts return to model building, comfortable with the knowledge that the user is in basic agreement with the new system design. First, the process model (DFD) is constructed from the USD, and then the data model (ERD) is constructed from the previous two models. Finally, the object model (OM) is fashioned. This exercise serves the same purpose as it did during the analysis of the existing system. It helps the analysts better understand the relationships among the external entities, processes, data flows, and data stores in the new system. In addition, on a very practical level, it helps to identify the file types (master, transaction, history) and file-servicing forms that must be designed and prototyped during the next JAD working session.

The new data flow diagram (Figure 8-12) differs only slightly from the new USD, but there are some important design features to note. First, the Sales subsystem is connected to Correspondence through the CD Transactions data store. This permits the user to fashion letters in a more personal way, including references to past customer purchases, and so on. Second, the one-way Internet connection to the CD Master file implies that users are able to look at the file without changing the file. Third, the Sales subsystem updates the CD Master file in order to maintain an electronic inventory system. Fourth, the CD Orders subsystem is fully automated, connecting to three data stores (CD Master, Supplier Master, and Order Detail).

The data flow diagram affords the analysts the opportunity to inspect the design in even more detail. Notice the Auto Order description above the flow line connecting the CD Orders subsystem to the CD Master data store. The analysts intend to allow the user to have CDs ordered automatically when the quantity on hand falls below the reorder level. Furthermore, note that the owner is able also to enter custom orders.

FIGURE 8-12 / *Cornucopia New System First-Level DFD (version 1.0)*

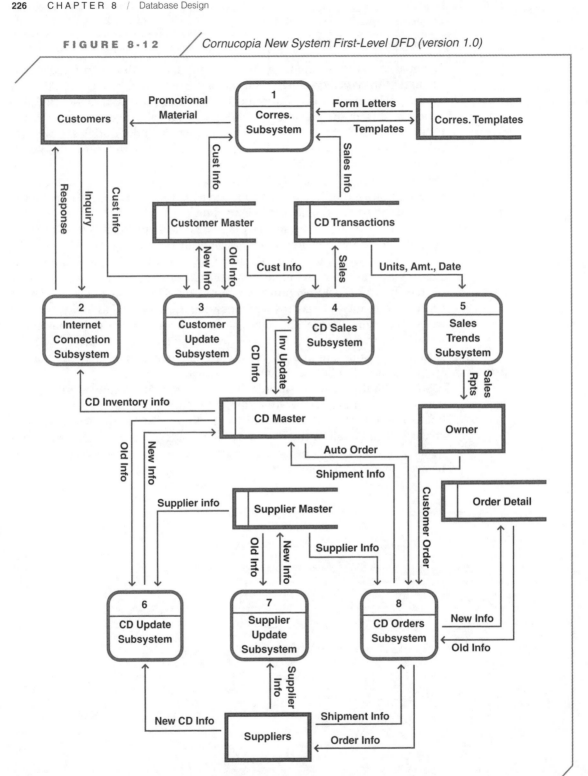

New System ERD

The data model, shown as the ERD in Figure 8-13, is derived from the data flow diagram. The data stores on the DFD become the data entities on the entity-relationship diagram. Notice that the relationship file CD Transaction Detail has been added to eliminate the potential many-to-many relationship between CD Master and CD Transaction, which occurs when a single transaction involves more than one CD. This illustrates an important advantage of data modeling: what appears as a single data store on the DFD may in fact become several files in a normalized database schema.

FIGURE 8-13   *Cornucopia New System ERD (version 1.0)*

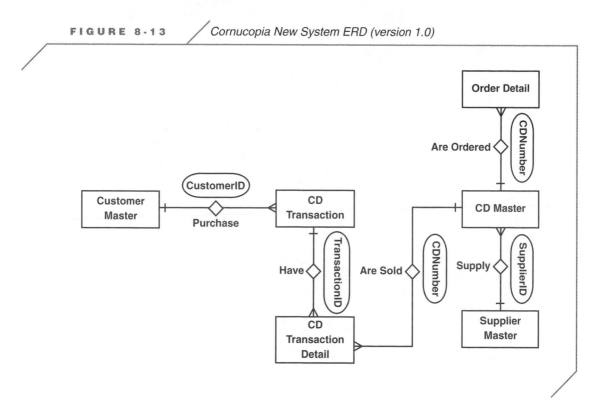

New System OM

The object model displayed in Figure 8-14 is a virtual copy of the ERD, except that the OM includes class behaviors. As noted previously, object modeling can be an alternative to data and process modeling. In this situation, the analysts use the OM to detail the data attributes of the class objects. Because they intend to use an objectlike relational database product (Microsoft Access) to implement their design, many elements of the object-oriented methodology are relevant to their work.

These model-building activities do not involve the user in the same way as do the building and verification of the user's system diagram. The user is brought back into the design process during the form design and prototyping work that is soon to follow.

New System File Design

Now that the analysts have identified some of the file requirements for the new system, they can develop file structures that include specific field attributes. Nevertheless, this does not signal that the file design is complete. The report, query, and process design activities described in the next two chapters almost certainly require some modifications to the new system models. This explains why these models were assigned version numbers.

FIGURE 8-14 / *Cornucopia New System OM (version 1.0)*

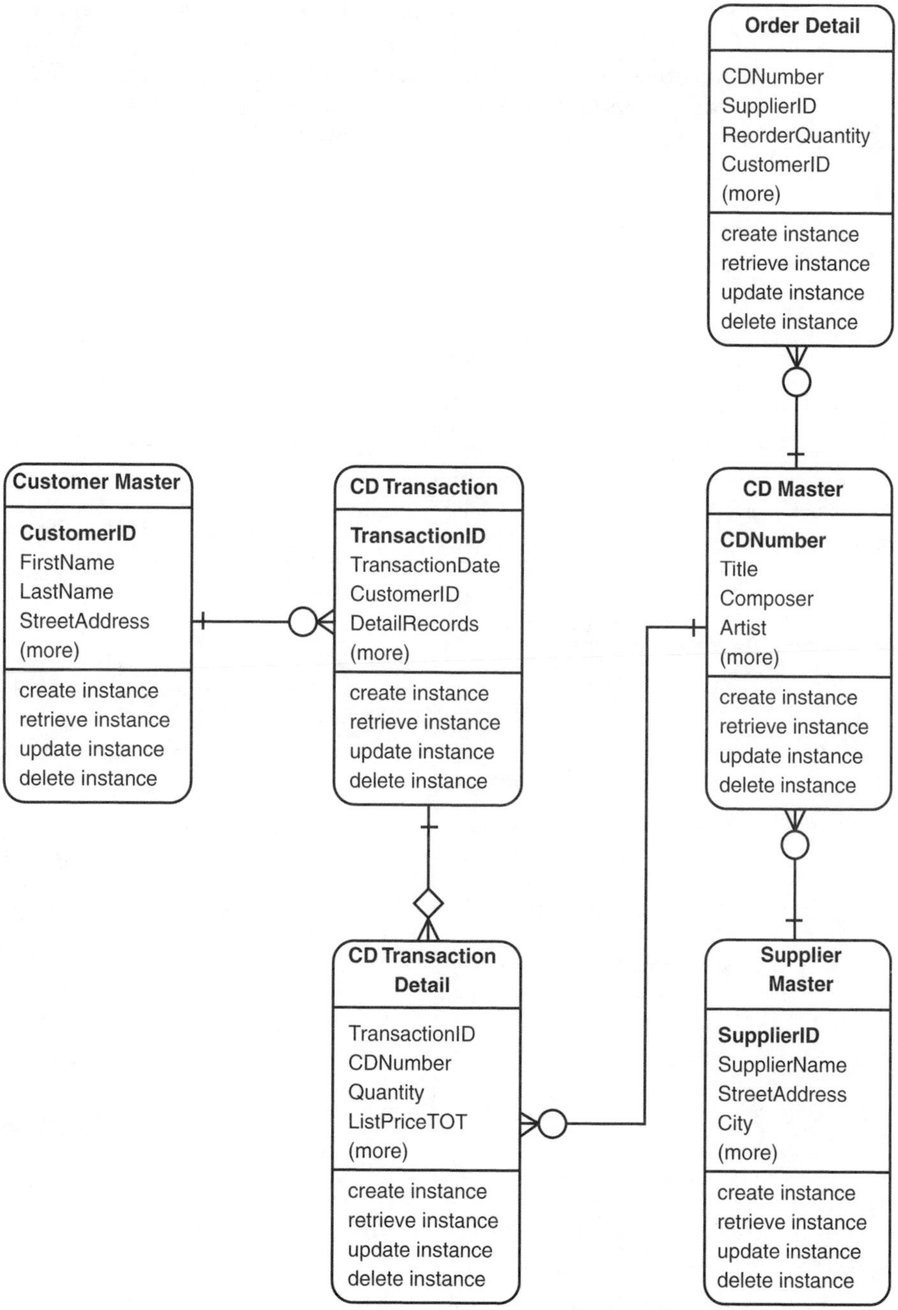

New System Menu Tree

In keeping with the desire to inform and involve the user in the design process, a very simple menu tree (Figure 8-15) provides some perspective to the series of choices that the user will have when the system is operational. Notice that in version 1.0 of this model, the Sales Trends subsystem is left blank. Details of this portion of the system await future JAD working sessions that occur during the report and query design activities.

FIGURE 8-15 / *Cornucopia New System Menu Tree*

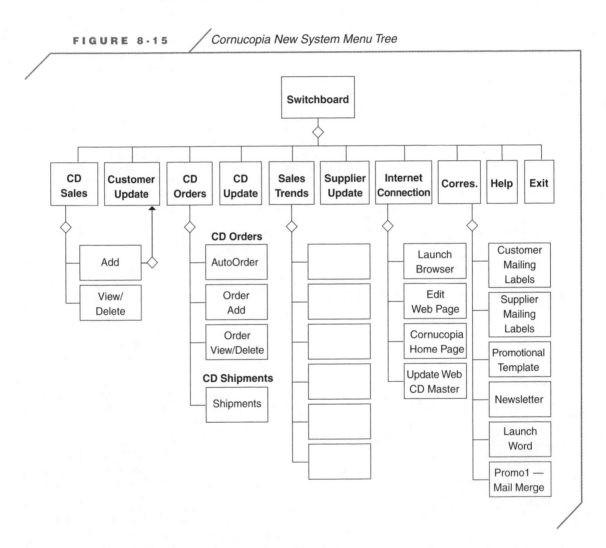

Detailed Form Design

With the completion of the first versions of the new system models, the analyst and the user hold another JAD working session to develop the forms that will service the new system files. In preparation for this session, the analyst should create the database file structures and enter a small amount of sample data. This allows the analyst and the user to create form prototypes using the screen-painting features common to most relational database packages. The following forms are designed to support the three major processing files in the system (Customer, CD, and Sales).

Figure 8-16 illustrates the CD Update Form activated by the CD Update option on the switchboard menu. This form was created with Microsoft Access. Most relational database management system products have similar screen-building tools. As a future upgrade, the CD file maintenance could be greatly enhanced by downloading new product updates through the Internet. An even less complicated updating procedure might be a simple floppy disk-driven system, in which the suppliers would provide the retailer with a new disk each month.

FIGURE 8-16 / *Cornucopia CD Update Form*

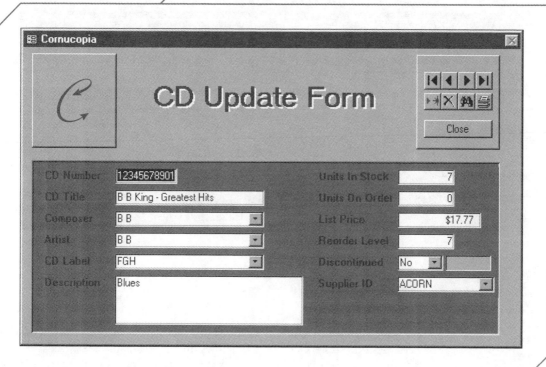

Figure 8-17 illustrates the CD Sales Form. This form is much more complex than the Master File Update form, because it interacts with both the CD and Customer tables in order to create entries in the CD Transaction and CD Transaction Detail tables. To be sure, such complexity presents a challenge to the analysts when it comes time to implement the design. Fortunately, several tools are available to help with the construction tasks that lie ahead. You will address these issues in future chapters. Note that in order to be consistent with Microsoft Access terminology, the term "table" rather than "file" is used.

Figure 8-18 presents the CD Order Form.

FIGURE 8-17 / *Cornucopia CD Sales Form*

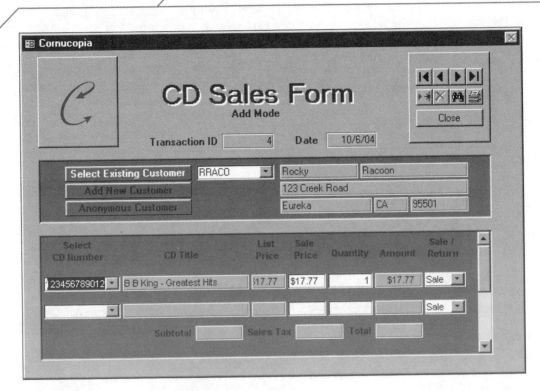

Time and Money

The analysts report eight hours against this project during the week, all of which were spent on design. Figure 8-19 shows the project status as of Week 5. Notice that the actual hours reported to the design activities total more than 50 percent of the estimated total, but the percentage complete is only 30 percent. This illustrates the difficulty of estimating systems work. Overall, the project is still within its budget for labor because the actual hours required to do the analysis are comfortably less than the estimates.

FIGURE 8-18 / Cornucopia CD Order Form

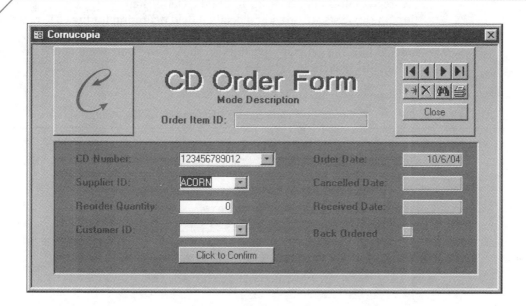

FIGURE 8-19 / Cornucopia Project Status—Week 5

Status5.xls

Date:	Cornucopia Project Status							As of: Week 5												
Activity	**% Comp.**	**Status**	**1**	**2**	**3**	**4**	**5**	**6**	**7**	**8**	**9**	**10**	**11**	**12**	**13**	**14**	**15**	**16**	**Total**
Analysis - Estimate	100%		4	5	5	4													18
Actual	95%	ok	2	4	2	3													11
Design - Estimate	30%					4	4	5	5	4									22
Actual	30%	ok			1	3	8												12
Develop - Estimate	0%								4	4	5	5	4						22
Actual	0%	ok																	0
Impl. - Estimate	0%												4	5	5	5	5		24
Actual	0%	ok																	0
Total - Estimate			4	5	5	8	4	5	5	8	4	5	5	8	5	5	5	5	86
Actual			2	4	3	6	8	0	0	0	0	0	0	0	0	0	0	0	23
Contract	100%	ok	C																
Prelim. Present.	100%	ok			C														
Design Review	0%	ok							S										
Prototype Review	0%	ok										S							
Training Session	0%	ok													S				
Final Report	0%	ok														S			

Status / Sheet2 / Sheet3 /

Cornucopia with Visible Analyst

Figure 8-20 illustrates one way to use the Define Item dialog box. In this situation, the analysts are providing detailed field definition for the CD Transaction file. While these definitions are specifically entered for this file, they are also entered into the general repository so that at any time the analysts can easily call up this reference.

FIGURE 8-20 / Cornucopia Data Store Definition with Visible Analyst

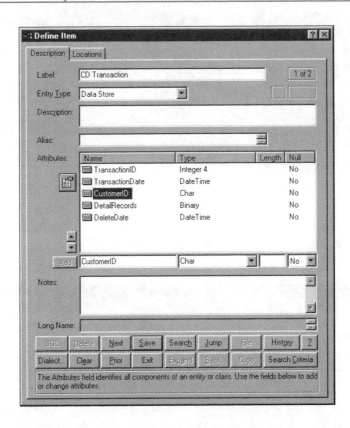

SUMMARY

With new system model building, database design, and detailed form design activities, substantial progress is made toward designing the new system. You have clearly expanded your efforts to include more tools, such as GUIDs and screen painters, and the beginnings of other, complementary methodologies, such as JAD and design prototyping.

Remember, these methodologies, and the tools that go with them, are nothing more than methods or strategies for working through the project phases, which are described collectively as the SDLC, or the enhanced SDLC. Furthermore, each of these phases must address the six components of a computer information system. Finally, remember that design is a process that requires constant evaluation and revision of its products.

TEST YOURSELF

Chapter Learning Objectives

1. What steps are necessary to develop the new system process model (DFD) from the preliminary USD?

2. What steps are necessary to develop the new system data model (ERD) and object model (OM)?

3. What process would you use to estimate the resource requirements associated with file-processing design?

4. What are the basic graphical user interface dialogs (GUIDs) and screen forms that are required to access system files and processes?

Chapter Content

1. Select one of the following statements that does *not* reflect the basic step-by-step process of creating screen forms that service an information system.
   a. Begin with a basic screen design template.
   b. Develop master file-servicing forms before transaction file-servicing forms.
   c. Use the JAD process to refine the forms.
   d. Ask the user to supply a detailed sketch of the form.

2. To normalize an ERD is to:
   a. remove all invalid data from the tables
   b. add a great deal of descriptive comment to the model
   c. change the entity names to reflect a person, place, thing, or event
   d. eliminate data redundancy and repeating groups from the database

3. What is the proper sequence in which the new system models should be developed?
   a. DFD, ERD, OM, USD
   b. ERD, DFD, OM, USD
   c. USD, ERD, OM, DFD
   d. USD, DFD, ERD, OM

4. The DFD is the best model to use when communicating with the user about the new system design. True or False?

5. Select the best file type description for a file of student petitions to add and drop courses.
   a. master file
   b. transaction file

6. Select the best file type description for a file of scholarship descriptions.
   a. master file
   b. transaction file

7. Select the best file type description for a spreadsheet file containing project status information.
   a. master file
   b. transaction file

8. Select the best file type description for a spreadsheet file containing one row of grade data for each student in a course.
   a. master file
   b. transaction file

9. Sequential file processing is outdated. True or False?

10. Database processing has replaced traditional file-processing methods such as sequential and indexed access. True or False?

ACTIVITIES

Activity 1

Refer to Sunrise Systems' ERD (Figure 8-10) and database interface template (Figure 7-13) to answer these questions.
   a. How is the normalized ERD affected if it is possible for a single investor to have more than one investment portfolio? Develop a new normalized ERD to address the problem.
   b. Develop a GUID sequence to handle the Investor master file maintenance subsystem.
   c. How does the addition of the new Investor master file to the ERD affect the DFD presented in Figure 7-9?

Activity 2

Design the necessary master and transaction files to keep track of your college transcript. Prepare file definitions that include file type, organization, and structure.

Activity 3

Design the maintenance screen forms required to service the master and transaction files you designed to keep track of your college transcript (see Activity 2).

DISCUSSION QUESTIONS

1. What would you do to encourage a reluctant user to participate in the SDLC process?

2. In what ways do appointment calendars, payroll timecards, and phone books resemble master files and/or transaction files?

3. In what ways does the need for the time of sale of a transaction affect the design of a transaction file and the design of the input form?

4. Under what circumstances should the analyst specify direct file access and sequential file access in the new system design?

5. What are the advantages of designing the new system database before designing the new system detailed form design?

6. In what ways does the Microsoft Windows Save As dialog box function like a GUID?

PortfolioProject

Team Assignment 8: New System Modeling and Database Design

Building new system models leads directly to the design of the underpinning element of your small-enterprise information system. Database design, a challenging task under any circumstance, is easier if you first develop a high-quality ERD. Do not be surprised if your ERD requires several revisions before you can declare it normalized.

This is a lengthy assignment. It requires every team member to take on substantial responsibilities and effectively communicate with every other teammate. Everyone should make the effort to become familiar with the database design, regardless of his or her past experience or present project responsibilities.

In order to develop the new system file and form designs you must complete the following tasks:

1. Develop the new system context diagram and first-level data flow diagram. Submit a copy of your diagrams to your instructor.
2. Develop the new system entity-relationship diagram. Submit a copy of this diagram to your instructor.
3. Develop the new system database table definitions, including field names and types. Submit a copy of these definitions to your instructor.
4. Develop the new system object model. Submit a copy of this diagram to your instructor.
5. Develop the new system menu tree. Submit a copy of this diagram to your instructor.
6. Develop a detailed form design for updating one new system master file.
7. File a copy of the new system diagrams, database definitions, and detailed form designs in the Portfolio Binder behind the tab labeled "Assignment 8."

reportand**query**design

OVERVIEW

In this chapter, the focus shifts to output, an element of such primary concern to users that it is often closely associated with the information needs analysis and project contract that began the systems project. Therefore, analysts and users often begin the design process with output design. It seems very natural to follow the question, "What are the information needs of the enterprise?" with "In what specific format should the system provide this information?"

The discussion of design activities began with file and form design for two reasons. First, the data stores (files) are prominently identified on several of the system models, which give the inexperienced analyst a convenient and rather obvious place to start. Second, these files provide the necessary foundation for the input form and output report prototypes. Actually, the input, file, and output design activities are all underway in the same time frame. The joint application design (JAD) working sessions are not necessarily restricted to one or the other topic.

This chapter discusses several important characteristics of output content and presentation frequency and then addresses the question concerning the specific format of the system's output.

- Match different users with the appropriate output content and presentation frequency.

- Design multi-sourced outputs that rely on sophisticated file-sharing capabilities of 4GL application software.

- Distinguish between periodic and on-demand reports.

- Provide a way for users to design and create their own extemporaneous output.

- Develop resource requirement estimates associated with output design.

OUTPUT CONTENT

We have already addressed the broad information-processing requirements of relevancy, accuracy, timeliness, usability, affordability, accessibility, and adaptability (Chapter 2). Both user and analyst join in making decisions regarding exactly what information to include in a particular output.

Tailoring Reports to Accommodate the Audience

In Chapter 1, Figure 1-1 showed the different levels of personnel within an organization, along with their matching information needs. The manager needs to see the summarized picture, whereas the everyday worker needs to see the detailed picture. This view of enterprise information strata serves medium-size and large organizations very well. Small enterprises, on the other hand, present different challenges to the analyst.

In the small enterprise, the same person may need access to strategic, enterprise-wide reports in the morning but operational, job-specific screen images in the afternoon. The design must allow users to select what they want, when they want it, rather than requiring the user to live with a preset schedule of output product deliveries. The menu and icon-based graphical user interface dialogs (GUIDs) discussed in previous chapters provide an effective design solution to this challenge.

Figures 9-1 and 9-2 present a collage of illustrations to demonstrate how a menu hierarchy can assist the small-enterprise user to access the appropriate output. Notice that the same basic information is presented in different ways.

FIGURE 9-1 / *Sunrise Systems Report Hierarchy*

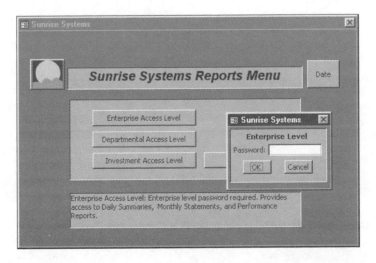

The investment analyst needs detailed information about the transactions for a particular portfolio, whereas the manager needs summarized information about all the transactions.

FIGURE 9-2 / Sunrise Systems Report Design

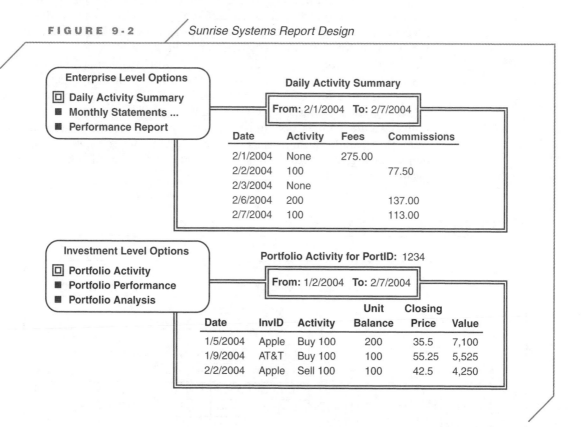

Combining Information from Several Files

In the early days of electronic data processing (EDP), output was tied to a singular computer activity. Source documents were packed off to the data-entry station. Punched cards, magnetic tape, or floppy disks were fed into computer programs that sorted, combined, and summarized the data into clearly recognizable outputs. Such straight-line systems—with no confusing uplinks, downlinks, subscriptions, or broadcasts of data native to some other computer or information subsystem—were easy to portray in the input-process-output (IPO) chart. In short, it was easy to identify where outputs came from.

In contrast, this textbook is composed of information from many different files: the word-processor files, one for each chapter; the graphics files, which come from many sources, such as paint files, screen files, and text files;

and the spreadsheet and database files captured into many figures of the text. Today, electronic textbooks containing audio and video information are available on CD-ROM and on the Internet. *Document processing* is the collective term applied to output in the 4GL age, where a single document is actually an assemblage of products created by different applications, each with its own specialized function. Although document processing may be confusing at first, it actually increases productivity when applied correctly.

The illustration of the Sunrise Systems Report Hierarchy (Figure 9-1) suggests an output strategy similar to that of the early days of computing. The portfolio transaction file, combined with stock exchange pricing information, fee schedules, and commission formulas, provides sufficient input to create both reports. Figure 9-3 presents a much different approach. In this case, the outputs from three application packages are combined into a single *multisourced output*, giving the investment analyst greater access to all the information regarding a portfolio.

FIGURE 9-3 / *Sunrise Systems Multisourced Output*

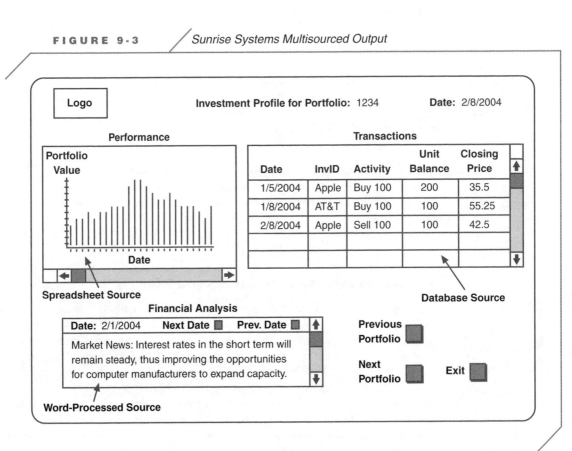

The analyst should recognize the file-sharing potential afforded by 4GL products. Most modern application software is designed to work with objects, allowing the user to combine objects from different sources into one document. This permits the analyst and user to consider how the system outputs are actually used, rather than simply what kind of outputs are required. For example, the investment profile for Sunrise Systems was designed after a user-analyst JAD working session in which the user described the inconvenience of referring to three separate outputs during investment advising sessions. After the user sketched an ideal output screen, the analyst agreed to investigate the possibility of using an object-oriented relational database product or a computer macro language to implement the design.

One of the fastest growing segments of information technology is *multi-media communications*. This describes an approach to combining multiple data formats into one product. By adding sound, video, and animation to well-established text and graphics formats, multimedia specialists can create computer-based information system products that mimic motion pictures and television. The addition of interactive user interfacing rounds out the sextet of multimedia elements.

Output to a Data File

As mentioned, subsystems sometimes create outputs intended for other systems or subsystems. This is the case with the investment activity subsystem for Sunrise Systems discussed in Chapter 7 (Figure 7-9), in which the transaction data store is considered an output of the activity process and an input to the consult process. Specifically, one possible implementation design for the investment profile output calls for a spreadsheet program to use the transaction data to create the performance chart. Sometimes, the analyst must redesign such dual-purpose files to accommodate the subsystems. Notice that the transaction window in the multisourced output image includes a field for unit balance, which does not appear on the entity-relationship diagram (Figure 8-10). This new field was added to allow the spreadsheet to calculate the portfolio's total value.

The history, backup, and temporary file types introduced in Chapter 8 are also an important part of the output design process. Users never see such outputs, so they are usually the sole concern of the analyst. The *history file* is of special concern in systems that use screen-prompted input and softcopy output. The history file is a file of data records that reflect historical information, usually at the transaction level of detail. In the absence of a physical record of the information system inputs and outputs, the analyst must create a way to provide an electronic audit trail of the data or information. In some ways, the analysis window provides such a history for Sunrise Systems. The conversations, advice, and instructions that concern a portfolio can help the investment advisor and the investor reconstruct past events. Retaining history and backup files for some period—the length determined by the nature of the enterprise's business—is well advised. Moreover, in certain businesses, the analyst should carefully investigate any legal requirements for document retention.

TECHNOTE 9-1

MULTIMEDIA COMMUNICATIONS

Multimedia communications combines text, graphics, audio, video, and animation into one interactive product. As such, it represents the ultimate multisourced document. Spontaneous interactivity distinguishes the medium from traditional information system products. The event-driven, graphical interfaces of 4GL products, such as Microsoft Office, pale in comparison to the experiential environment of multimedia entertainment, advertising, and education.

Multimedia is relevant to the systems analyst for several reasons. First, the analyst can use it to enhance Web-based communications. Second, it can be used to develop training materials associated with information system operations and functions. Third, the analyst can use multimedia to create effective presentations for his or her clients.

Several specialized hardware and software products are available with which the analyst should be familiar in order to understand and appreciate the work of the multimedia expert.

Hardware
  Visual-digitizing devices (scanners, cameras, camcorders)
  Audio-digitizing devices (MIDI synthesizers, digital recorders, microphones)
  Mass storage devices (CDs, DVDs, Jaz drives)

Software
  Image-manipulation programs (Photoshop, Morph)
  Audio-manipulation programs (Sound Forge)
  Text-formatting programs (PageMaker)
  Authoring programs (Director, Authorware)

In addition, the analyst should be aware of several technical issues that influence the performance of multimedia outputs. Multimedia products must be designed for distribution and playback on client stations. Recall that Internet transmission speeds and playback quality are governed by several client station attributes (modem speed, processor speed, memory, monitor resolution, and so forth). Consequently, the multimedia developer must search for a compromise between what is possible and what is practical. That compromise often revolves around the following:

Image Quality

Image **resolution** is directly proportional to the number of pixels per inch, and **bit depth** determines the number of colors available to reproduce the image. Higher resolutions and bit depths result in images that are more accurate. Because both of these attributes require memory, we can say that image quality is directly related to memory requirements.

Full-Motion Video

Smooth running video requires many images. Because there is little point in using full-motion video with poor image quality, you can see that from the previous discussion video requires a lot of memory.

Sound Synchronization

The ear is very unforgiving; it can recognize the slightest interruption or distortion immediately. Thus, quality digital sound requires high sampling rates, which translates into a lot of memory. Furthermore, the playback processor must be capable of handling large audio and video files without perceptible pause.

Data Compression

Many, but not all, of the technical issues previously raised can be made less of a problem by good compression software because smaller files can be transmitted to client stations more quickly.

Output Security

Controlling access to information has always been a concern to system designers. Obviously, some information is so sensitive or private that only specific people within the enterprise should be able to see it or change it. How do you provide for easy, universal access to the open area of the system while maintaining the appropriate level of information security? Passwords, hidden directories, locked files, masked fields, and encryption schemes are some of the techniques that make inadvertent access unlikely and unsavory access difficult. In the Sunrise Systems information system, passwords can be applied to the menuing system to prevent unauthorized access to the various reports. In situations requiring some individuals to have multiple levels of access authority, a *hierarchical password system* should be used, which allows users to supply a series of passwords to access all levels of information equal to or below their own.

Another security concern involves the physical storage of the data and information. Backup files are essentially duplicates of the original files, with one important exception—the file name. Some files are backed up daily, some less frequently. For example, if you intend to keep the three most recent versions of master files, you need to develop a file-naming procedure to distinguish the files from one another. Sunrise Systems backs up the investor master file each month, using the year and month as part of the backup file name. Thus, IMF072004 identifies the investor master file backup for July 2004. Given this revolving, three-month *file retention period*, this particular file is retained until October 2004.

REPORTING FREQUENCIES AND PRESENTATION

The what-you-want, when-you-want-it expectations of today's computer users contrast with the rigid computer schedules of the past. Programmers refer to this as event-driven computing, meaning that the programs they write must execute in response to a screen button or menu selection from the user. Sunrise Systems' Monday morning Sales Summary report is now available any day, any time, and anyone with the proper access codes can produce it. The analyst need not fully succumb to this trend, but must be ready to defend the design that calls for the traditional end-of-the-period output.

Regularly Scheduled Periodic Reports

Every enterprise has an information rhythm that dictates how and when it must generate reports. Some outputs are required once a month, others are required daily, whereas still others are required immediately after the event they describe. Reports that are produced in such a regular fashion are often referred to as *periodic reports*. Although it's tempting to design a system in which everything is available at the click of a mouse, such an approach is costly. First, processing and operating system capacity, as well as memory and disk space, are finite resources. Second, adding unnecessary access to output not only slows the performance of the system but also shortens the system's functional life by leaving little room for future upgrades. Therefore, the analyst must be discriminating in designing output schedules to fit the differing information needs and rhythms of the enterprise.

Figure 9-4 details some of the common output reports that the system can produce on a regular or periodic schedule. Of course, a reasonable assumption is that these report choices will appear in the familiar GUID sequences already discussed. Remember, the small enterprise, in particular, does not have an Information Technology (IT) Department to watch the calendar, run the programs, and deliver the reports. Furthermore, it is safe to assume that many of these periodic reports will be in softcopy, rather than hardcopy, form.

FIGURE 9-4   / *Scheduled Reports*

Description	Periodicity
Payroll	Weekly, Monthly, Annual
Drafts	
Withholdings	
W2s	
Financial	Monthly, Quarterly, Annual
Income Statement	
Balance Sheet	
Accounts	Monthly
Payable	
Receivable	
Master Files	Monthly
Employee	
Customer	
Product	
Vendors	
Management	Daily, Weekly, Monthly
Work Schedules	
Productivity	
Inventories	

On-Demand Reports

As the name suggests, *on-demand reports* are produced on an irregular basis, at the specific direction of the user. The user may see little difference between on-demand and periodic reports, given that they both are initiated via menu commands. Small-enterprise users, having a menu of reports to choose from, will probably not care to distinguish between the regular Daily Work Schedule Report they call for at 8:00 a.m. and the on-demand Daily Sales Summary they call for at 10:00 a.m. and then again at 11:00 a.m. To the user, a report is a report is a report! The question is, does the analyst care, and if so, why?

The analyst does indeed care about the distinction because it significantly affects several of the components of the information system, as well as the user's perception of the overall responsiveness of the system. Let's follow the impact that on-demand reports might have on a system. To begin, the processing design (software) must include provisions for the user to interrupt, and later resume, whatever else is running on the system. Then, the hardware must be fast enough to make a quick transition from one program environment to another, which is called *context switching*. Finally, the data files required to create the response must be available to the system in an online fashion. That is, you should not expect the user to go through a complicated disk-swapping procedure to bring the files into the system. All these concerns complicate the analyst's job and therefore require careful planning and design.

A small enterprise usually offers personalized, customer-oriented service. Chapter 1 discusses several information needs of the small enterprise, most of which require the analyst to design a system that is responsive to the user and sensitive to the context in which the information is to be used. In other words, the highly regimented set of operating procedures necessary to serve a complex corporation would be inappropriate in the small enterprise where more latitude for flexibility and personal initiative is to be expected. On-demand reports provide the user with a tool that supports this kind of behavior, thus enhancing the image of the small enterprise. Figure 9-5 provides a sample list of on-demand reports that commonly appear in small-enterprise information systems.

FIGURE 9-5   *On-Demand Reports*

Productivity Summaries
    (daily sales as of this moment)
Inventory Summaries
    (number of "widgets" on hand)
Master File Activity
    (adds, changes, deletes)

User Inquiry

An enterprise of any size can benefit from the *user inquiry*, which provides a way for users to respond to information needs that were not anticipated when the system was designed. Well-trained users should be able to query their system for such information without involving the analyst. This is especially true in the small-enterprise environment, where the analyst may not be available when unexpected information requests occur. For example, six months after the computer analyst delivered Sunrise Systems' information system, the manager wanted to contact all investors who had purchased a particular stock. This was a special, one-time need that neither the analyst nor the user anticipated during their JAD working sessions. Relating this request back to the earlier discussion on output security and backup files, what file retention period would you recommend to Sunrise Systems?

Thus, user inquiries differ from periodic and on-demand reports in one significant way: the analyst does not design inquiries—they are conceived, planned, and executed by users. This requires some knowledge of the data files in the system and the syntax rules of the accessing software. Although voice recognition software is common, systems in which a user can simply issue verbal commands to the computer are still beyond the means of the typical small enterprise. Therefore, the analyst must provide clear instructions, operating procedures, and training on the user-inquiry capabilities of the system.

OUTPUT DESIGN FUNDAMENTALS

Chapter 8 emphasized the importance of file processing and design of the forms that service files. Output design is certainly no less important. It, too, requires faithful attention to the user's fundamental information requirements and preferences about computer interfaces. As already mentioned in this chapter, significant differences exist between content and presentation of system output. With increasing user participation through JAD working sessions, users now assume considerable ownership of such system products. The analyst can expect that the user will have clear preferences about the specific format of system outputs.

ThinkingCritically

ImageCopyrights

ICONs is a small advertising firm located in St. Louis, Missouri. It specializes in creating logos, advertising slogans, and jingles for businesses that advertise in hometown newspapers and radio stations. The owner often hires students from surrounding colleges to assist the two full-time staff members with office computing and correspondence. Occasionally, ICONs hires marketing and fine arts students to work out preliminary designs and compositions.

As the firm's contract service computer consultant, you are responsible for maintaining the hardware and software for ICONs. During your regular monthly visit, you discover a series of files containing clip art images in one of the students' work directories. The student admits to loading these files and confides that sometimes, unmodified images appear in the final versions of advertising products delivered to clients.

What action do you recommend? Would your recommendation differ if ICONs owned the clip art software?

Hardcopy versus Softcopy

Today, printed materials still dominate information system output, but visual formats (color, graphics, animation, and video), powerful operating systems, and networking and wireless computing have expanded the reach of computer applications. Slowly but surely, the general-purpose user is gaining confidence in softcopy output. *Softcopy* is any output displayed on a screen that includes all of the information products discussed later in this section. E-mail is a good example of how quickly people will abandon the paper document when given a reliable alternative. E-mail users can race through messages, discarding the unimportant, refiling some for later use, and printing only those few that require human handling. Because the computer is now a general-purpose communication device, the percentage of softcopy output will dramatically increase. Soon, hardcopy output could be the exception rather than the rule, its use driven by the user's traditional level of comfort rather than actual need.

As noted by our enhanced SDLC, the analyst and user work with electronic prototypes of input forms and output products. Thus, it seems reasonable to assume that the small-enterprise system user will increasingly choose softcopy whenever possible. This presents some special design considerations for the analyst. First, some softcopy output must be written to disk for the reasons stated earlier in the discussion regarding output directed to a data file. Second, the analyst may need to design split screens to accommodate user softcopy output and information system operating menus, messages, and instructions. Finally, the system resource requirements must include fast video hardware and software. Moreover, softcopy permits several new and exciting design possibilities, as detailed in Figure 9-6.

FIGURE 9-6 / *Softcopy Design Options*

Characteristic	Examples
Animation	Product design
	Entertainment
Video	Database enhancement
	Computer-assisted instruction
Navigational aids	Buttons
	Scroll bars
Audio	Database enhancement
	Computer-assisted instruction
	Entertainment

Hardcopy output will probably never be eliminated, so every analyst should be aware of the basic characteristics of this format. Although the hardcopy design options embrace a great variety of output, as detailed in Figure 9-7, some features are particularly relevant to the small enterprise. To the extent that the hardcopy product is distributed outside the enterprise, the quality of the output image creates a long-lasting impression about the quality of the enterprise itself. Therefore, the analyst should seriously consider recommending a laser-quality printer, a variety of print fonts, and some good graphics software when the time comes to update the system resource requirements. In addition, you need to incorporate a print queue into the design if the printer is used frequently as a part of routine processing (receipts, packing slips, delinquency notices, and so forth). A *print queue* is a software routine that temporarily stores print files until the printer is free.

One of the most convenient output design features is the Print or View option that allows users to choose the output medium during processing. This may encourage users to rely on softcopy because they know they can always generate a printed copy if necessary. To reinforce the concept that they are of equivalent validity—as well as to reduce confusion—hardcopy and softcopy outputs should follow the same general design pattern whenever possible.

FIGURE 9-7   / *Hardcopy Design Options*

Characteristic	Examples
Preprinted form	Paychecks
	Financial instruments
	W2s
Boilerplate form	Form letters
	Legal forms
	Style sheets
Free form	Internal reports
	General correspondence
	Mailing labels
	Customer bills
Color	Advertising
	Product design
Graphics	Financial analysis
	Presentation transparencies
Camera ready	Publishing
	Promotional literature

Reports versus Queries

The earlier discussion on output presentation emphasizes the importance of designing flexibility into the information system output options. Both periodic and on-demand output can take the form of a report or a query, with the query providing much more flexibility. A query differs from a report in terms of scope. For example, the Sunrise Systems' report on all stock transactions for the day has a scope encompassing all records. The query report of transactions for a particular portfolio has a scope of "for PortID = 1234." The *scope* of a query defines a condition that is applied to each record to select only those entries that are of immediate interest to the user. Designing query forms requires a great deal of insight into the way the enterprise uses information to solve common problems that occur irregularly. This is another example of how a JAD working session is useful.

User inquiries, on the other hand, are neither reports nor queries. Don't confuse a query with a user inquiry! Remember, user inquiries employ a special language (e.g., structured query language, or SQL) by which users identify, design, and execute their own information requests without the analyst's help.

QBE and SQL Access

One of the important advantages of database management products is their ability to offer built-in output flexibility for user inquiries. Users can construct commands through *query-by-example (QBE)* or SQL access methods (or both) to interrogate databases without the assistance of the computer professional. SQL is a text version of the instructions required to satisfy the information request. QBE is a visual version of SQL, in which users are presented with GUI cues that lead them through building specific queries. Of course, the necessary skills to build these queries must be developed through instruction and experience. The analyst should include information about this access method in the system procedures documentation.

One further note regarding semantics is necessary. Many relational database products refer to both QBE and SQL. These are well-established, but distinctly different, techniques designed to give the user maximum flexibility in generating output. Some products (such as Microsoft Access) refer to query files, which serve as ongoing file filters for reports, forms, and labels. In such cases, the query specifications are automatically translated into SQL code, which becomes a permanent part of the information system, available for periodic or on-demand access. SQL code can also be created outside of the query process, giving the user the ability to query a database extemporaneously.

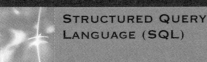

TECHNOTE 9-2

STRUCTURED QUERY LANGUAGE (SQL)

IBM developed **structured query language (SQL)** as a tool for users to interrogate a relational database extemporaneously. To do so, the user needs a detailed layout of the database file structures and file relationships. Such a document is called the **database schema**. A schema is very much like a detailed entity-relationship diagram. The user must also know the **syntax rules** of SQL, which are the rules that govern the formation of computer commands. For example, in the following SQL commands the capitalized words have special meaning and the lowercase words refer to data element names. The special characters retain their common functionality.

```
SELECT ot_amount
  FROM payroll
  WHERE empID = 12345
    AND date = {05/10/2004};
```

This statement directs the database program to look through a specific file (payroll) and display the overtime pay (ot_amount) for a specific individual (empID = 12345) who earned overtime pay on a specific date (date = {05/10/2004}). To construct this statement the user needs to know the field names in the file as well as the SQL commands.

Following is a more complex command, a statement that expands the above command to add information (name, address) extracted from a second file (employee). Notice that the two files relate to each other through a common field (empID).

```
SELECT name, address, ot_amount
  FROM payroll, employee
  WHERE payroll.empID = employee.empID
    AND empID = 12345
    AND date = {05/10/2004};
```

About 30 SQL commands are available to choose from, each requiring its own syntax and providing a specific function. Some are used to create the database, some allow the user to modify the database structure, and some, as illustrated, are used to query and update the database.

Recognizing that SQL might not be user-friendly, several popular database packages offer a visual equivalent, called query-by-example (QBE), as an alternative.

Figure 9-8 illustrates the different output options available to the Sunrise Systems user: reports, queries (periodic or on-demand), and user inquiries, all in hardcopy or softcopy format. Notice that the SQL option does *not* include the interface features familiar to the user. This deserves special consideration in the training guides and system instructions. Users need to learn from the analyst how to escape from and return to their normal information system environment. Figure 9-8 shows that the SQL inquiry (softcopy) produces the same information as does the query form (softcopy) to illustrate the parallels between the two techniques.

FIGURE 9-8 / *Sunrise Systems Output Options*

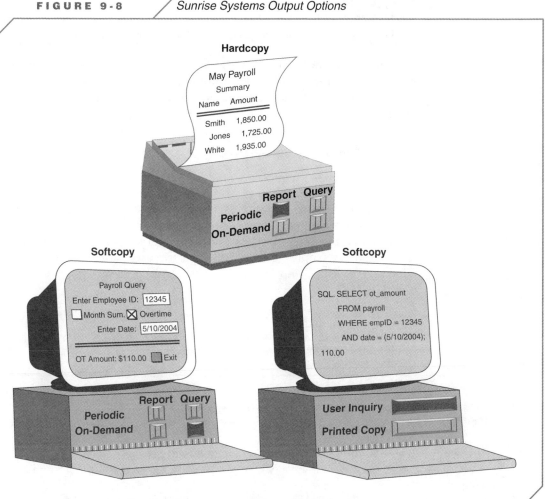

Estimating System Resource Requirements

As with file and form designs, report and query designs influence the system resource requirements. In the preceding discussions, we emphasize the need to offer the user output options that are based on careful consideration of content, presentation, and format. Each option must be analyzed to determine exactly which output peripheral is required to implement the design. For example, hardcopy periodic reports require an appropriate printing device; softcopy on-demand queries require an adequate monitor.

It is also true that these design decisions depend on the input and output peripherals that are available and affordable. The most common output devices are printers and monitors. However, choosing the specific product is not always a straightforward decision because many modestly priced variations exist. For example, laser-quality color printers and large-screen monitors can produce spectacular output at a price that is well within small-enterprise hardware budgets. Consequently, it can be very tempting to overbuy in this area of system resources.

The analyst must ask the question, "What kind of output hardware does the system require to deliver the information designed into the system?" To deliver hardcopy and softcopy output in the appropriate form, the analyst must balance cost and performance.

Output Design: Step-by-Step

Figure 9-9 suggests a sequence for the analyst to follow when working through the output design for the new system. Briefly, these steps are very similar to those identified for form design (see Figure 8-11). This should not be surprising because both input and output refer to the same data stores and processes identified on the design models. Furthermore, the same user-centered design methodology is employed. This calls for active user participation in the evaluation and design of outputs using prototypes and JAD working sessions whenever possible.

FIGURE 9-9   / *Output Design Steps*

1. List the system outputs identified on the USD and DFD.
2. Add to this list one potential report for each master file identified on the ERD.
3. Add to this list one potential activity report for each transaction file identified on the ERD.
4. Review the user requirements identified by the contract, preliminary presentation, and JAD activities to date. Add to the above list any outputs that do not already appear on the list.
5. Evaluate each output on the list in terms of its periodicity (scheduled, on-demand, query) and special characteristics (hardcopy, softcopy).
6. Develop a general output design template for hardcopy and softcopy outputs, patterned as closely as possible to the general input form design.
7. Review the list and the design template with the user.
8. Eliminate outputs that are duplicated by multipurpose input screen forms or that are unnecessary based on the above review.
9. Using a screen-painting utility or a "reports" utility, create sample outputs for each report and several potential queries.
10. Step 9 will add to your collection of system prototype products, which should be reviewed with the user as part of the normal JAD activities.
11. Estimate resource requirements necessary to support the outputs.
12. Update the data dictionary to include the outputs that you have designed.

OUTPUT DESIGN WITH CASE TOOLS

As mentioned in the last chapter, CASE products include report prototyping features that can help the analyst with the output design. Two CASE tool features are particularly helpful in keeping track of all the new data elements, file names, and output products that are added as the project design advances. First, the CASE tool checks every entry to make sure that the new element is consistent with existing elements in the data dictionary. Second, the CASE tool automatically updates the data dictionary with every new element.

Even without the data dictionary feature, many of the popular database products offer an attractive alternative technology to CASE. For example, Microsoft Access includes powerful object-oriented, import-export features (object linking and embedding), and Windows screen-manipulation tools and techniques. With such a product, the analyst and user can design and manipulate an output prototype directly from a master or transaction file. Even after the design is agreed upon and development has begun, the analyst can always return to the prototype and introduce modifications quickly and efficiently.

THE CORNUCOPIA CASE

The JAD working sessions devoted to output design begin with a series of standard outputs that are based on the data stores identified in the new system DFD discussed in Chapter 8 (Figure 8-12). The project contract and preliminary presentation are also used to determine other basic output requirements. Not surprisingly, most of the serious design work focuses on the sales and ordering subsytems. Figure 9-10 presents the list of outputs for the Cornucopia project. Notice that the list is organized into content groupings, within which are the different presentation and format descriptions. The list also helps the analysts revise and complete the system menu tree (Figure 9-11), which is identified as Version 1.1. The analysts number model revisions so that they can retrace their footsteps to earlier versions when the need arises.

FIGURE 9-10 / *Cornucopia Output List*

Report/ Output Name	Scheduled/ View/Qry	Hardcopy/ On-Demand	See Softcopy	Figure
Master Files				
1. Customer Update Form	view	on-demand	softcopy	
2. CD Update Form	view	on-demand	softcopy	8-15
3. Supplier Update Form	view	on-demand	softcopy	
Sales Transaction Files				
4. CD Sales Form	view	on-demand	softcopy	8-16
5. CD Transaction Sales Receipt	report	on-demand	hardcopy	9-12
Order Transaction Files				
6. CD Order Form	view	on-demand	softcopy	8-17
7. CD AutoOrder Form	view	on-demand	softcopy	
8. CD Purchase Order	report	on-demand	hardcopy	9-13
9. CD Shipment Form	view	on-demand	softcopy	
Sales Trends				
10. Sales Summary by Day	query	on-demand	hardcopy	9-14
11. Sales Summary by Month	query	on-demand	hardcopy	
12. Sales Range Summary by Day	query	on-demand	hardcopy	9-15
13. Sales Range Summary by Month	query	on-demand	hardcopy	
14. Sales/Profit Spreadsheet	view	on-demand	softcopy	9-16
Correspondence				
15. Letters (Promo1)	n/a	on-demand	hardcopy	
16. Newsletter	n/a	on-demand	hardcopy	
17. Mailing Labels	n/a	on-demand	hardcopy	
Internet Web Pages				
18. Home Page	n/a	on-demand	softcopy	9-17
19. Cornucopia Logo	n/a	on-demand	softcopy	
20. CD Notes	n/a	on-demand	softcopy	
21. Mailing List Form	n/a	on-demand	softcopy	9-18
22. Class Registration Form	n/a	on-demand	softcopy	9-19
23. Message Form	n/a	on-demand	softcopy	
24. View CD Inventory	n/a	on-demand	softcopy	9-20
25. Marquee	n/a	on-demand	softcopy	
26. Artists	n/a	on-demand	softcopy	
27. Events	n/a	on-demand	softcopy	
28. Specials	n/a	on-demand	softcopy	
29. Potpourri	n/a	on-demand	softcopy	9-21

FIGURE 9-11   / *Cornucopia New System Menu Tree (Version 1.1)*

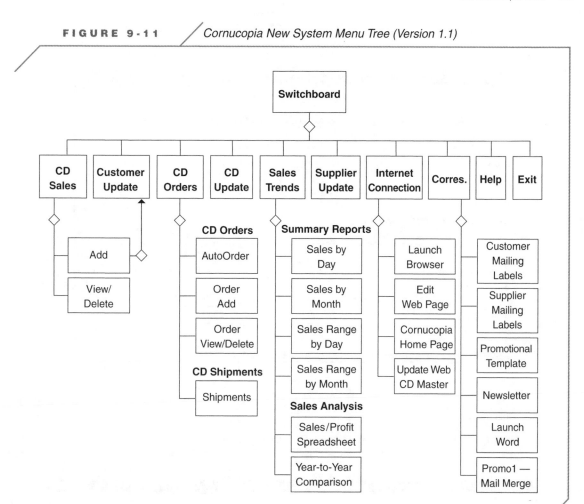

New System Output Requirements

The list of outputs includes screen views for the master file updates (Customer, CD, and Supplier) and transaction file activities (Sales and Order). Only the Sales Receipt and Purchase Order are designed for hardcopy output. All of the other outputs are softcopy, although the option to create a hardcopy of individual records is always available. Chapter 8 illustrates several of these forms. The CD Update Form (Figure 8-16), CD Sales Form (Figure 8-17), and CD Order Form (Figure 8-18) illustrate one advantage of relational database processing. In addition to providing data-entry interface for new records, each form allows access to historical records via the navigation, or Find, buttons on the toolbar. A sample of the CD Transaction Sales Receipt (Figure 9-12) and CD Purchase Order (Figure 9-13) are included. Notice that Order Item ID is composed of the supplier ID, date, and time. This allows the owner to generate a purchase order at any time.

CD Transaction Sales Receipt

Cornucopia CD Purchase Order

All of the Sales Summary queries are designed as hardcopy, on-demand output. These outputs were first described in general terms in the project contract. Because of the JAD working sessions, they now appear in much more detail. To increase their usefulness, each summary screen prompts the user to specify the number of days or months to summarize. Samples of the Sales Summary by Day (Figure 9-14) and Sales Range Summary by Day (Figure 9-15) are included. The owner agrees that year-to-year comparisons, although important, should be developed later.

The Sales/Profit Spreadsheet (Figure 9-16) is viewable as softcopy. Notice that this is really an Excel spreadsheet, with corresponding Excel editing features. In this case, the Sales Trends subsystem selects and summarizes the sales transaction data before passing it to a spreadsheet template. Once the spreadsheet is open, the owner can manipulate the profit margin values and then observe the effect on profitability for the different price ranges.

At this time, the correspondence outputs are not specified in detail. The ability to access past sales transactions will greatly enhance the mail merge operations. For example, the owner will be able to specify criteria for merging various promotional documents with customers who have purchased given amounts over given time periods.

The Web page outputs are of special interest to the owner. She has some definite preferences for the overall look of the Web site. Several examples are included in the next section.

FIGURE 9-14 / *Cornucopia Sales Summary by Day*

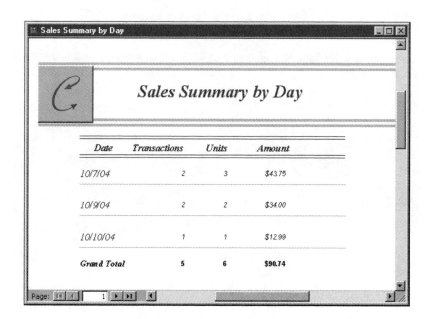

FIGURE 9-15 / *Cornucopia Sales Range Summary by Day*

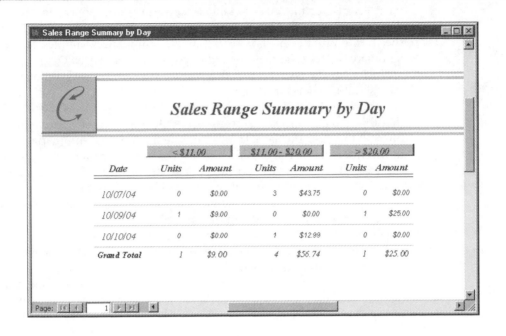

FIGURE 9-16 / *Cornucopia Sales/Profit Spreadsheet*

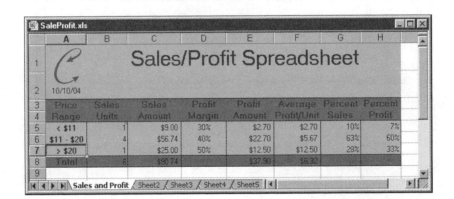

Web Page Design

Above all else, the owner wants the Web site to reflect the attitude of the retail store. She feels that the images should be visually pleasing, easy to navigate, informative, and interesting, without any assaulting fanfare. In time, she would like visitors to be able to sample audio clips and order items online. She wants the site to be stable in terms of its basic page layouts, but accepts the notion that regular content changes are required. At present, she does not want to reference other sites from the Cornucopia site. Figure 9-17 presents Cornucopia's home page, with three frames: one for the identifying logo, one to display a scrollable choice menu, and one for content.

FIGURE 9-17 / *Cornucopia Web Home Page*

The logo frame links to an information page (not pictured) that describes Cornucopia in a way consistent with the owner's wishes.

The menu choices link to pages appropriate to the menu title. Figure 9-18 illustrates a form used to add someone to Cornucopia's mailing list. Figure 9-19 shows a form used to register for one of Cornucopia's free classes. Figure 9-20 presents a form with access to the current CD inventory.

Each of the four panels within the Marquee link to special pages that the owner plans to change on a regular basis. Figure 9-21 presents the Potpourri page, which contains a short musical quiz.

FIGURE 9-18 / Cornucopia Web Mailing List Form

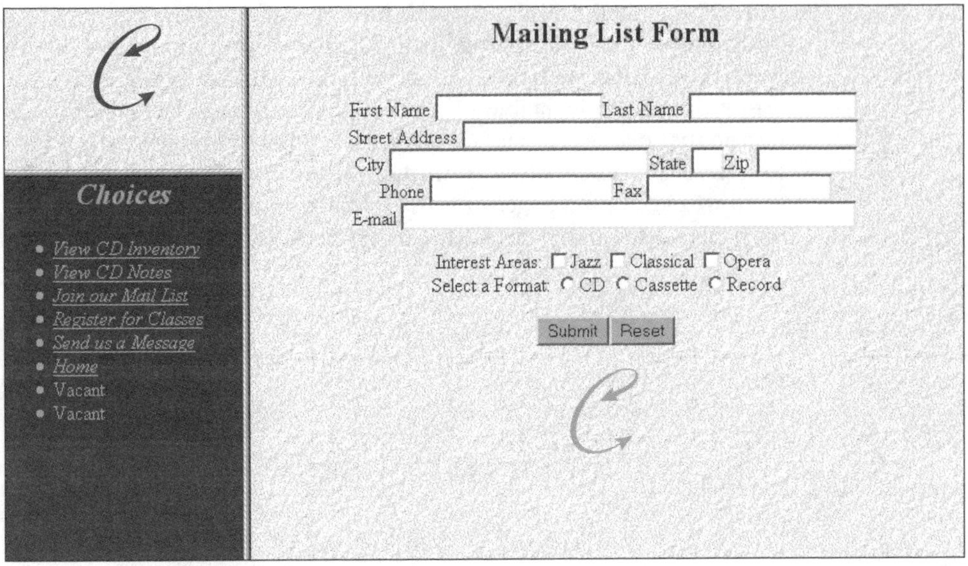

FIGURE 9-19 / Cornucopia Web Class Registration Form

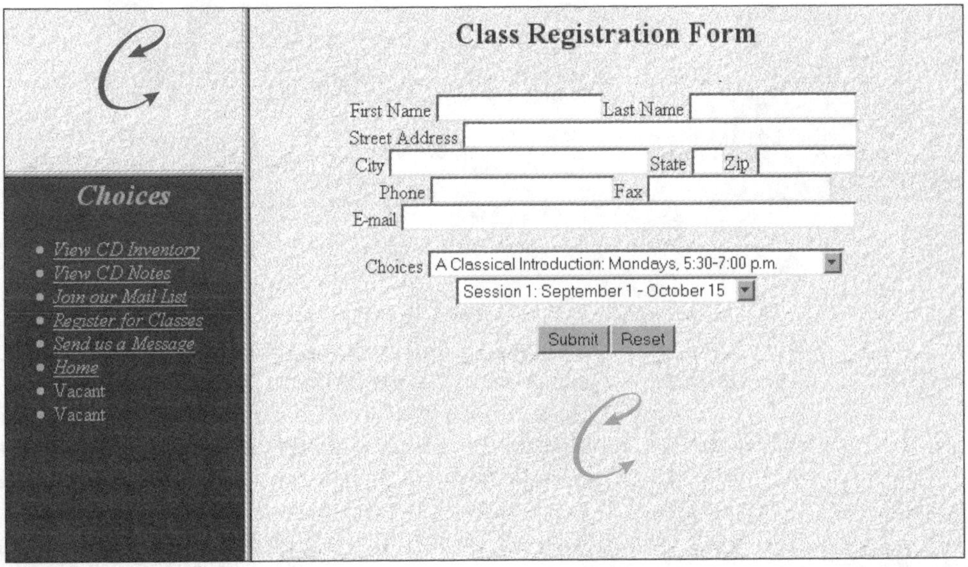

FIGURE 9-20 / *Cornucopia Web View CD Inventory*

CD Inventory Table

CD Web Query

CD Number	CD Title	Composer	Artist	CD Label	Description	List Price
123456789012	B B King - Greatest Hits	B B	B B	FGH	Blues	$17.77
234567890123	CD #2345	Bill	Susan	DEF	Blues	$12.99
345678901234	CD #3456	Dave	John	ABC	Jazz	$17.77

Choices

- *View CD Inventory*
- *View CD Notes*
- *Join our Mail List*
- *Register for Classes*
- *Send us a Message*
- *Home*
- Vacant
- Vacant

FIGURE 9-21 / *Cornucopia Web Potpourri Page*

Choices

- *View CD Inventory*
- *View CD Notes*
- *Join our Mail List*
- *Register for Classes*
- *Send us a Message*
- *Home*
- Vacant
- Vacant

Marquee Potpourri is always a surprise, even to us. The little musical quiz that follows was inspired by Harold Schonberg's *The Lives of the Great Composers* (Norton, 1981), which makes a great gift item.

Musical Quiz

Click to display a new question. [NEW QUESTION]

The question:

[   ]

Enter your answer:

[   ]

Click to check your answer. [CHECK ANSWER]

Revised New System Menu Tree

The output design activities developed the specific details for the content, presentation, and format for the sales trends reports seen earlier in Figure 9-10. This presents a practical question: how can the analysts be so far along in the new system design without having developed such detail long ago? The answer lies in the very nature of problem solving and systems work: the enhanced SDLC model is circular because information systems continue to evolve even after the initial project definition work is completed. Furthermore, the individual activity phases of the model tend to blend; it is sometimes difficult to pinpoint the current phase. Finally, the analysts often revisit specific activities within an earlier phase because of work accomplished in a later phase. Working like this resembles the way an artist works by painting the background and blocking the composition before adding the detail. In systems work this approach is inevitable. Because it is impossible to visualize the entire project at once, the analysts split the problem into more manageable parts and then solve them one at a time, with the understanding that they may need to go back to fill in some of the details from time to time.

I/O System Resource Requirements

The input and output resource requirements cannot be finalized until the process design activities are well underway. Nevertheless, identifying the general nature of these requirements as the design phase progresses is very important. The input and output design decisions are often made with specific hardware and software in mind, which can, in turn, affect the process design itself. Figure 9-22 proposes several generic hardware and software items that are required to support the input and output designs to date.

FIGURE 9-22 / *Cornucopia I/O System Resource Requirement Notes*

HARDWARE
Processing Platform:

The intended use of a database-processing package for transaction processing and master file maintenance requires a minimum processor speed of 1 GHz, with 256 MB of RAM, and at least 20 GB of hard disk storage.

Peripherals:
1. The sales transaction-processing subsystem should be supported with a bar code scanner to minimize input time and error rate.
2. The modest amount of hardcopy output (reports and correspondence) suggests a single laser-quality, black-and-white printer.
3. Efficient Internet access requires a fast modem, sound card, and speakers.
4. Placement of the unit at the checkout counter dictates a low profile (moderate size) monitor and a tower case for installation below the counter.
5. A fast CD-ROM drive is required to install software and play multimedia products, including music CDs.
6. An uninterruptable power supply (UPS) is recommended, especially with the frequent power outages during the stormy season.
7. A tape backup device, 3.5" floppy drive, and Zip drive are recommended.

SOFTWARE
Relational database, spreadsheet, word-processing, and Internet browser software is required. Web page authoring tools, virus protection, and tape backup software is also recommended.

Time and Money

A total of five hours is reported against the design activities for this project during the past week. The analysts estimate that the design is now 50 percent complete. Figures 9-23 and 9-24 illustrate the project budget and status as of Week 6. Notice that the hardware and software purchases are initiated during this period, to allow sufficient time for delivery, installation, and testing before the project implementation activities commence in Week 12. These purchases generally follow the resource requirement notes shown in Figure 9-22. A detailed itemization appears in Chapter 10.

FIGURE 9-23 *Cornucopia Project Budget – Week 6*

Budget6.xls

	A	B	C	D	E	F	G	H	I	J	K	L	M	N	O	P	Q	R
1	Date:		Cornucopia Project Budget								As of: Week			6				
2																		
3		1	2	3	4	5	6	7	8	9	10	11	12	13	14	15	16	Total
4	Estimates Hardware						2500	500	500	500								4000
5	Software						1000	250	250									1500
6	Labor	200	250	250	400	200	250	250	400	200	250	250	400	250	250	250	250	4300
7	Total	200	250	250	400	200	3750	1000	1150	700	250	250	400	250	250	250	250	9800
8	Actuals Hardware						4250											4250
9	Software						225											225
10	Labor	100	200	150	300	400	250											1400
11	Total	100	200	150	300	400	4725	0	0	0	0	0	0	0	0	0	0	5875
12	Weekly +/- Hardware	0	0	0	0	0	1750	0	0	0	0	0	0	0	0	0	0	1750
13	Software	0	0	0	0	0	-775	0	0	0	0	0	0	0	0	0	0	-775
14	Labor	-100	-50	-100	-100	200	0	0	0	0	0	0	0	0	0	0	0	-150
15	Total	-100	-50	-100	-100	200	975	0	0	0	0	0	0	0	0	0	0	825
16	Cum +/- Hardware	0	0	0	0	0	1750	0	0	0	0	0	0	0	0	0	0	
17	Software	0	0	0	0	0	-775	0	0	0	0	0	0	0	0	0	0	
18	Labor	-100	-150	-250	-350	-150	-150	0	0	0	0	0	0	0	0	0	0	
19	Total	-100	-150	-250	-350	-150	825	0	0	0	0	0	0	0	0	0	0	

Budget / Sheet2 / Sheet3

FIGURE 9-24 *Cornucopia Project Status – Week 6*

Status6.xls

	A	B	C	D	E	F	G	H	I	J	K	L	M	N	O	P	Q	R	S	T
1	Date:	Cornucopia Project Status							As of: Week 6											
2																				
3	Activity	% Comp.	Status	1	2	3	4	5	6	7	8	9	10	11	12	13	14	15	16	Total
4	Analysis - Estimate	100%		4	5	5	4													18
5	Actual	95%	ok	2	4	2	3													11
6	Design - Estimate	60%					4	4	5	5	4									22
7	Actual	50%	ok			1	3	8	5											17
8	Develop - Estimate	0%									4	4	5	5	4					22
9	Actual	0%	ok																	0
10	Impl. - Estimate	0%													4	5	5	5	5	24
11	Actual	0%	ok																	0
12	Total - Estimate			4	5	5	8	4	5	5	8	4	5	5	8	5	5	5	5	86
13	Actual			2	4	3	6	8	5	0	0	0	0	0	0	0	0	0	0	28
14																				
15	Contract	100%	ok	C																
16	Prelim. Present.	100%	ok			C														
17	Design Review	0%	ok								S									
18	Prototype Review	0%	ok											S						
19	Training Session	0%	ok															S		
20	Final Report	0%	ok																S	
21																				

Status / Sheet2 / Sheet3

Cornucopia Query and Report Design with RDBMS

Figures 9-25 and 9-26 illustrate the underlying query and report designs associated with the Sales Range Summary by Day report (see Figure 9-15). The details of the SalesTrans Temp table and query SQL code are addressed in later chapters. At this point, it is sufficient to note that relational database software, Microsoft Access in this case, provides powerful output design and implementation tools.

FIGURE 9-25   *Cornucopia Query Design with RDBMS*

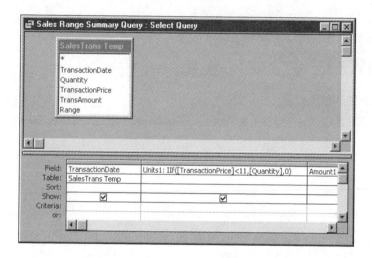

FIGURE 9-26   /   *Cornucopia Report Design with RDBMS*

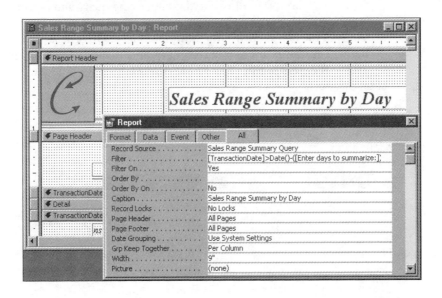

SUMMARY

Event-driven, user-directed, visually oriented computing significantly affects the way the analyst must treat information system outputs. First, the screen forms designed to support file processing are often the same for both input and output. Such visual interfacing at both ends of the system, although increasing the demands on processing, may relieve the long-standing I/O bottlenecks that have plagued traditional systems.

As users assume more ownership of the system, input error rates decline, output demands become more specific, and processing options require more flexibility. Overall, the partnership between the user and analyst strengthens through increased JAD working sessions and prototyping activities.

One final, important word of caution about output design: regardless of the advances in technology, outputs still are inextricably tied to processing design—you cannot generate reports or queries without a facilitating computer program. Each output must be designed with some sense of the accompanying processing requirements. The next chapter offers specific instruction on processing design.

TEST YOURSELF

Chapter Learning Objectives

1. In what ways do executive, supervisory, and production employees dictate output information form, content, and presentation?

2. What is a multisourced document, and how can 4GL software help you implement its design?

3. What are the fundamental design differences between periodic and on-demand reports?

4. What is an extemporaneous output and how can you make such output possible?

5. Why is it necessary to wait until the output design task is well underway before developing resource requirements associated with output design?

Chapter Content

1. In order to accommodate the differing information needs within an organization, output design should allow users to access anything at any time. True or False?

2. A multisourced document is composed by combining outputs from different software packages. True or False?

3. Not all system outputs are intended for user consumption. True or False?

4. Softcopy output precludes the creation of a verifiable audit trail. True or False?

5. Hardcopy and softcopy outputs should follow the same general design whenever possible. True or False?

6. The very first step in output design is to create output prototypes. True or False?

7. Structured query language (SQL) is easy for novice computer users to understand and apply. True or False?

8. What type of output frequency would you recommend for the output of employee paychecks?
   a. regularly scheduled
   b. on-demand
   c. user inquiry

9. What type of output frequency would you recommend for the output of sales-to-date totals?
   a. regularly scheduled
   b. on-demand
   c. user inquiry

10. What type of output frequency would you recommend for the output of the last ten sales transactions?
    a. regularly scheduled
    b. on-demand
    c. user inquiry

11. How can the analyst easily combine and update data from several sources into one output document?
    a. The only option is to cut and paste the source data into the destination document.
    b. The best option is always to embed the source data into the destination document.
    c. Object-linking and embedding (OLE) is available for Microsoft Office users.
    d. The safest way is to reenter the source data into the destination document.

12. What type of output frequency and presentation would you recommend for your college transcript?
    a. regularly scheduled, hardcopy
    b. regularly scheduled, softcopy
    c. on-demand, hardcopy
    d. on-demand, softcopy

ACTIVITIES

Activity 1

a. After reviewing the Sunrise Systems Report Hierarchy (Figure 9-1), develop a departmental level submenu with three output report options. Sketch the design for one of these reports.

b. Considering that in the Sunrise Systems Multisourced Output (Figure 9-3) the performance section information comes from a spreadsheet, speculate on where the spreadsheet data comes from and how it might be processed.

c. After reviewing the Sunrise Systems Output Options (Figure 9-8), develop two additional user query outputs—one periodic and one on-demand.

Activity 2

Using Sunrise Systems Report Hierarchy (Figure 9-1) as a reference, develop a multisourced, softcopy, on-demand report design to present Student Profile information for a college counselor.

Activity 3

Refer to the Sunrise Systems Output Options (Figure 9-8). Although the SQL syntax is correct, the query does not retrieve sufficient information to answer a manager's question about how much overtime was earned on May 10. Explain why and suggest a modification to fix the potential problem.

Activity 4

Based on the Sunrise Systems DFD (Figure 7-9), develop a minimal output requirements list for the Sunrise Systems information system.

DISCUSSION QUESTIONS

1. Describe three college information system outputs you need to plan your class schedule for next term. How would you classify these outputs: reports or queries, hardcopy or softcopy, regularly scheduled or on-demand? How well do these outputs serve the intended student audience?

2. Given the fact that user inquiries are not usually permanent parts of the information system, suggest a redesign to satisfy the following user complaint: "We have to key the same general statement many times each day to answer customers' questions about their account balances."

3. Consider the potential impact of users assuming more ownership of their information systems. How does this affect the relationship between the analyst and the user, especially in small-enterprise settings?

4. How might a user's view of hardcopy versus softcopy output be influenced by their use of the computer as a basic communications device?

5. Discuss your level of comfort with hardcopy and softcopy of the following information products: your diploma, your college transcript, the syllabus for this course, your bank statement, your paycheck.

6. To what extent do multimedia enhanced products, such as Web sites, video games, and information systems, improve or detract from the usefulness of the product?

7. After reviewing the Cornucopia Output List (Figure 9-10), explain why only two of the 29 outputs are classified as reports.

8. Use your favorite Internet search tool to investigate multimedia fair use policy. Describe how you would apply such a policy to your work as a systems analyst.

PortfolioProject

Team Assignment 9: Report and Query Design

The list of new system outputs grows with each encounter between the analyst and user. It is common for the user to focus almost entirely on this particular subject. Understanding that information system outputs are made possible by effective input capture and database design, the analyst should always be mindful of the design implications of each item on the output list. The system requirements listed initially on the project request for services and later on the project contract are typically very general. Over the course of the analysis phase and early design phase activities, the analyst refines, expands, and adds detail to these general requirements. The result is the new system output list, from which it is possible to define the specific processing functions of the information system.

In order to develop a comprehensive output design and prepare for the processing design activities ahead, you must complete the following tasks:

1. Develop a list of the new system output list (see Figure 9-10). Submit a copy of this list to your instructor.
2. Design a new general system hardcopy report layout. Submit a copy of this layout to your instructor.
3. Design a new general system softcopy screen design. Submit a copy of this design to your instructor.
4. Design two queries for the new system. Submit a copy of the QBE design or SQL code for the queries to your instructor.
5. Develop an I/O system resource requirements note (see Figure 9-22) for the new system. Submit a copy of this note to your instructor.
6. File a copy of the new system output list, hardcopy report layout, softcopy screen design, query designs, and I/O system requirement note in the Portfolio Binder behind the tab labeled "Assignment 9."

processdesign

When you complete this chapter, you will be able to:

OVERVIEW

Your input and output design work sets the stage for the process design. Most computer professionals envision the use of various general processing procedures from the very beginning of the project. In this chapter, you will learn how to develop a design that relies on three major classifications of prewritten software—vertical, horizontal, and integrated. Although these processing options relieve the analyst of considerable programming responsibilities, they introduce a new dilemma in terms of file sharing. The chapter then focuses on file sharing with object linking and embedding. Finally, you will learn how to apply programming skills to further customize popular software packages to the user's particular needs.

- Determine which prewritten software products are appropriate to incorporate into the process design.

- Construct a system menu tree and system flowchart using the data flow diagram as a starting point.

- Incorporate a variety of file-sharing techniques into the process design.

- Read and understand subsystem structure charts and program flowcharts.

- Evaluate the amount of programming expertise required to complete a 4GL-based, small-enterprise information system.

4GL OPTIONS

Systems analysts commonly use the term *4GL* to describe the products popularized by the PC. Characteristics of these products are inviting user interfaces, powerful data manipulation commands, and open-ended application. Indeed, the PC itself owes a great deal of its acceptance to one of the first of these packages, a forerunner to today's spreadsheet programs called VisiCalc. This product introduced countless computer illiterates to the potentials of microcomputing. With a minimal amount of instruction, people could generate useful information products without involving their company's IT Department.

Technically, 4GL means "fourth-generation language," a term intended to remind you that computer-programming practices have evolved to a higher, more user-oriented level. However, as you shall see, the use of the word "language" implies that users must follow a set of rules as they apply 4GL products. Furthermore, as with any true language, these rules may be used in endless ways to create new combinations of commands to exploit the full power of the product.

Vertical Software

As mentioned in the discussion about build versus buy strategies in Chapter 2, vertical software is designed to solve problems within a narrow range. The owner of a printing shop, for example, can purchase a program that helps develop price quotes for printing jobs. Such a program would be of no use to the head of the homeowners' association who wants to develop a maintenance schedule. Vertical solutions are available for many information problems of the small enterprise. The analyst must explore these solutions to determine whether they can satisfy some or all of the needs identified during the analysis stage of the systems development life cycle (SDLC).

The first step in your exploration is to identify potentially useful products that already exist. Ask users to list the professional organizations to which they belong, the conferences they attend, and the magazines and journals to which they subscribe. Request any literature the users may have that relates to their professional connections outside their own enterprise. A university or good public library subscribes to many publications that contain advertisements, summaries, and reviews of vertical software. Inquiries of this nature should lead to a list of possible contacts. A phone call, short letter, e-mail, or visit to a software vendor's Web site should produce a quick response with sufficient information or demonstration software to allow you to begin your evaluation. For example, the Time Keeper System (TKSystem), first discussed in Chapter 2, was discovered at a regional computer educator conference.

The next step is to evaluate the software. Figure 10-1 lists several ways to research vertical software. The degree to which vertical software becomes part of your process design varies from project to project. It is unlikely, however, that you can rely on vertical software alone to satisfy all the user's needs.

FIGURE 10-1   /Evaluating Vertical Software

1. Firsthand evaluations from existing users
2. Hands-on demonstrations and workshops
3. Magazine reviews
4. Product advertising literature
5. Sales personnel

Horizontal Software

The three major categories of process-specific software designed to have broad application are word processors, spreadsheets, and databases. Remember, this text focuses on the use of horizontal software in the small-enterprise environment with the assumption that you have a working knowledge of these products. However, if this is not the case, several excellent trade books are available that describe these products.

A recent trend in the evolution of horizontal software is the blurring of each product's distinctive characteristics as software publishers attempt to widen their market share by adding new features. For example, a spreadsheet package is likely to be touted for its database features, whereas the reverse claim is made for database management packages. Even though these packages are addressed separately, keep in mind that such product crossover is likely to continue.

Integrated Software

Some products combine all three processes to form an *integrated software* package with a common user interface. Significant advantages accompany this approach. The cost of an integrated package is usually less than the cost of the three separate packages, and the common user interface reduces training time. The principal disadvantage is that the integrated package is not as full-featured as are the separate packages. For the purposes of the text, this 4GL option is covered in depth because you are going to be heavily involved with file sharing, which the integrated packages do quite easily.

In addition, a noteworthy advantage of the integrated package is that it is likely to include graphics and data communication modules. These features are valuable additions to the primary file-processing products.

The decision to use integrated software is influenced by the way you expect the user to interact with the information system. If you intend to build a very protective user environment, the standardized user interface serves no real purpose. On the other hand, if your users plan to work independently with the individual subsystems you assemble, a standard interface can trim the user learning curve considerably. Fortunately, your design decisions are not restricted to stand-alone products or integrated products. The next section presents several file-sharing techniques, such as import-export, which offer a compromise solution.

Software Suites

A popular software packaging category, known as the *software suite*, assembles several horizontal software products that possess highly developed file-sharing capabilities and uniform user interfacing. Figure 10-2 identifies two such packages. Note that both suites include widely used stand-alone word processor, spreadsheet, and database management applications. Each package promises to make it easier to combine the output from these specialized products into a finished product that can include text, graphics, charts, and slide show presentations. Furthermore, each product is equipped with tools to help the user design and construct custom applications.

A software suite is particularly attractive to the small enterprise because it reduces the number of software suppliers involved in the project. Considering that the average software product is updated every 18 months, this reduces the confusion that surrounds such upgrades and reduces the possibility that some products may not remain fully compatible.

FIGURE 10-2 / *Software Suites*

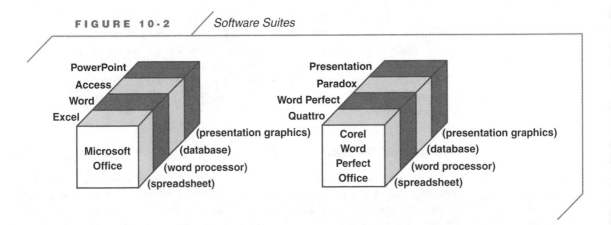

PROCESS DESIGN FUNDAMENTALS

Historically, enterprise computing, as opposed to scientific or statistical research computing, is characterized as "doing a little computing to a lot of data." That is to say, you spend most of your resources moving data from place to place, with relatively simple data manipulation and computation performed along the way. Payroll and inventory processing are good examples of this type of information processing. As discussed in Chapter 8, both the input and output from such operations are retained in large electronic data stores called files. Because file processing is such a large part of enterprise computing, efficient file processing is essential to the overall effectiveness of any information system.

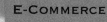

TECHNOTE 10-1

E-COMMERCE

In the broadest sense, electronic commerce describes three types of transactions: business-to-customer (B2C), business-to-business (B2B), and business-to-employee (B2E). Business-to-customer e-commerce is the most common kind of transaction, but the other types of transactions have long histories of success as well. Business-to-business e-commerce has its roots in the electronic data interchange (EDI) and electronic funds transfer (EFT) technologies. Business-to-employee transactions have followed the evolution of telecommunications from telephone to fax to e-mail and now the Web. The common advantage to all e-commerce transactions is that it offers more efficient service to an expanded client base.

In addition to concerns about hardware and software, there are several process design elements to address when considering using e-commerce in your system design.

Access: Who can access the service? Internet access is universal unless special password or account verification is employed. What safeguards exist to ensure uninterrupted access? How can the enterprise ensure that access speed will be maintained in the face of increasing Internet traffic? What will visitors be allowed to access?

Security: How can other enterprise information systems be secured?

Data Capture: How will transmissions to the enterprise be collected? How will data be routed to the proper internal information system or subsystem? How will data be archived for auditing purposes?

Payment Method: If payment is required, how will it be transmitted?

Product Delivery: If products or services cannot be delivered electronically, how will product or service handling systems be informed of the transaction? How will deliveries be tracked and verified?

Product Returns: How will product or service return transactions be transmitted and accounted for?

These are but a few of the concerns surrounding e-commerce. Fortunately, there is a growing body of software to help. Front-end e-commerce software is designed to manage Web-based order processing, whereas back-end e-commerce software handles the after-purchase accounting chores.

Advances in computing hardware and software offer the analyst and the user many specialized options to file processing. Word processors, spreadsheets, and databases are a series of productivity packages that make independent file processing easier and more efficient. However, this text includes efficient file sharing as another essential component of process-design work. *Efficient file sharing* describes software products that are capable of sharing files without a lot of user intervention, data transformations, and other inconveniences.

Of course, the user's previous experience and preferences also influence the choice of software, as well as the overall process design. Although the joint application design process eventually gives way to the prototyping design process, the user is still actively involved in design problem solving and evaluation. Two system models, the menu tree and the system flowchart, are particularly helpful as process design tools and as user-analyst communication devices.

The Internet presents several challenges to process design work. The analyst must consider the extent to which enterprise personnel and customers have access to processing logic and data store content via the Internet. As Internet transmission speeds and security protocols improve, more businesses have begun to rely on *electronic commerce*, or *e-commerce*, the buying and selling of goods online. Consumers use the Internet to purchase everything from airline tickets to corporate stock options.

Building on the DFD: The Menu Tree and System Flowchart

Figure 10-3 illustrates a partially completed menu tree for Sunrise Systems. Notice that the five processes on the data flow diagram introduced in Chapter 7 (Figure 7-9) compose the first tier of menu options, with a portion of the further detailed options shown on the second tier. In this case, three of the four second-tier options correspond with data flows illustrated in Chapter 9 on report and query design (Figures 9-1 and 9-2). The fourth option (Portfolio Notes) is closely associated with a data entry function designed to capture investor instructions and market news.

FIGURE 10-3 / *Sunrise Systems Menu Tree*

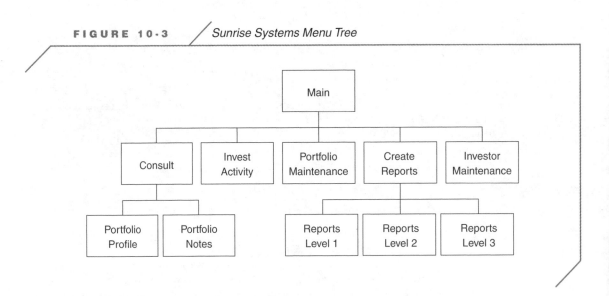

The menu tree, first described in Chapter 6, reflects the system design by incorporating the processes and data flows (inputs and outputs) into a hierarchical view of the system. The system flowchart, also introduced in Chapter 6, reflects the design in a different fashion by showing the relationship between and among processes and data stores or files. Figure 10-4 presents the *composite system flowchart* for Sunrise Systems. A composite system flowchart reduces several DFD subsystems or menu options into one processing rectangle.

This composite system flowchart also identifies several peripheral hardware devices. Thus, you begin to develop what is called the physical system design.

FIGURE 10-4 / *Sunrise Systems Composite System Flowchart*

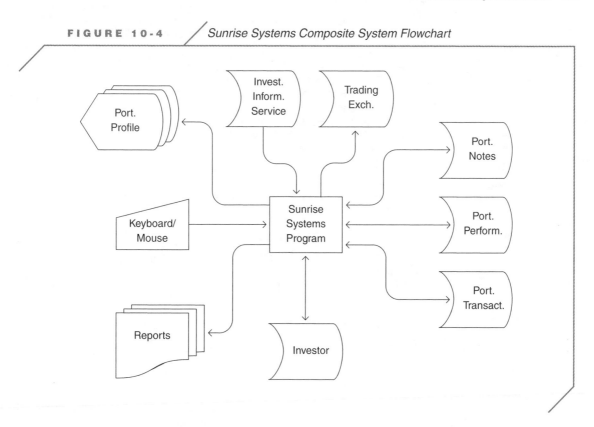

Remember, the logical design, as presented in the DFD, does not show the real-world implementation of the design. Now, as system development activities draw near, you must decide how the system will perform its functions.

To be really effective in the design process, the system flowchart should correlate closely with the different processes involved in the system. As mentioned in Chapter 7, one guiding principal in modern systems work is that big problems should be broken down into smaller, more manageable problems. Figure 10-5 presents the Sunrise Systems information system design in a way that emphasizes that problem-solving strategy. Here, you again see the five processes, data flows and data stores, but this time they are associated with 4GL application products and peripheral devices. Perhaps for the first time, the analyst and user can see exactly how the system will be constructed.

A *detailed system flowchart* shows the subsystem annotated with the 4GL application and associated file types that the analyst intends to use to implement the design. Notice that this system employs four of the major horizontal software products (word processor, spreadsheet, database, and graphics) as it processes data from one file to another. From this flowchart, can you speculate on the file usage types (master, transaction, history)?

FIGURE 10-5 / *Sunrise Systems Detailed System Flowchart*

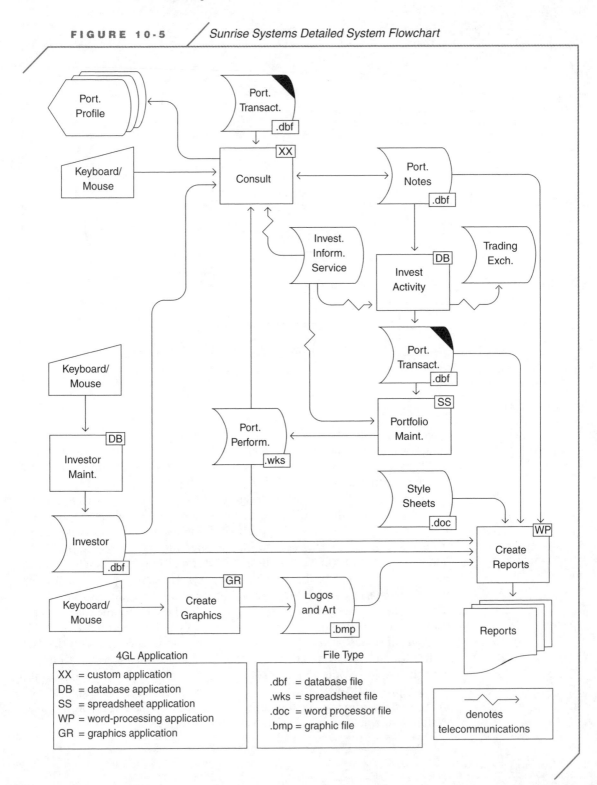

Notice the custom application specified for the Consult subsystem. This subsystem interacts with several file formats and peripheral devices and requires the use of powerful file-sharing techniques to assemble a custom document. To a lesser extent, the Portfolio Maintenance subsystem uses the import-export file-sharing method to convert database information into a spreadsheet format. The document processing evident in Create Reports is also noteworthy in the way that it uses a word-processing application to create a complex document. Figure 10-6 illustrates one such report and identifies the different source applications that contribute to the final product. Compare this report with the multisourced softcopy report illustrated in Chapter 9 (Figure 9-3). Both outputs demonstrate a subtle, but very noticeable, movement away from *data* processing and toward ***document processing***, in which the emphasis is on product assembly rather than on data manipulation.

The file-sharing interfaces between these products can be implemented in many ways. The specific techniques required to do this will vary, depending on the products and methods you select. At this point in the SDLC, you do not need to develop an exact sequence of commands. What is most important is that you understand the concept of file sharing sufficiently to incorporate it into your process design.

Low-Tech File Sharing

Before the computer became a common business tool, scissors, glue, and copy machines were literally the only file-sharing tools available. Incompatible file formats and the absence of a convenient transportation mechanism made it impossible to easily integrate separately defined and developed information segments. Today, *low-tech file sharing* describes clipboard-based and import-export file-sharing methods.

For example, cut-and-paste functions certainly improved file sharing. A temporary storage space, known as the ***clipboard***, serves as a transporting medium upon which data is copied to and from. Thus, it is possible to introduce one file format onto the clipboard and then paste it into another file format. This is possible because of the automatic file conversion process that is embedded into the operating environment. However, even this simple method of combining text and graphic files requires special user intervention to assemble the finished product, which makes it impractical when the analyst's intent is to design an information system that requires minimal user expertise.

File import-export represents a more sophisticated and product-specific technique for sharing files. This method allows a 4GL product such as WordPerfect to import text developed in Microsoft Word, or even to import spreadsheet information from Lotus 1-2-3. To accomplish this, the importing product must exercise special file-translating instructions, after which the imported file can be manipulated as if it were a native file. Import-export operations are usually activated through a series of menu selections or keystroke sequences. By collecting these commands into batch files or macro instructions, the analyst can provide the user with file-sharing options that require little user intervention.

FIGURE 10-6 / *Sunrise Systems Investor Report*

Source: ──▶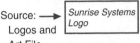
Logos and
Art File

July 14, 2004

Mr. Ronald Allen

Source: ──▶ *2727 Post Street*
Investor   *Albuquerque, NM 87100*
Database
File   *Dear Mr. Allen:*

*This report summarizes the performance of your investment
portfolio during the second quarter of 2004. The information
is taken from several sources: your consultations with your
investment advisor, the market transactions specific to your
portfolio, current market prices, current market news, and the
professional judgement of the staff at Sunrise Systems.*

Source:
Style Sheet
File

Market Value

*Taken as a whole, your portfolio performed reasonably well
during this period of time. As you can see from the performance
chart below, the current market value of the portfolio is much
higher than at the beginning of the period. Of course, this does
not take into account the numerous transactions that occured
during the last three months. This information is detailed in
the next section of this report*

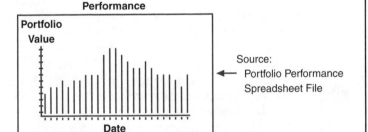

Source:
◀── Portfolio Performance
Spreadsheet File

Page 1 of 4

In spite of their clumsiness, all these methods are still used to some degree, depending upon the available computer resources and skills of the analyst and user.

High-Tech File Sharing

Object embedded and linked file-sharing methods describe *high-tech file sharing*. A method in which a copy of the source data is placed into a destination file is called *object linking and embedding (OLE)*. Figure 10-7 presents these two approaches as they might apply to the multisourced output that was introduced in Chapter 9 (Figure 9-3).

Using a pointer, or path, object linking connects the source file to the destination file. The actual source data is never really copied into the new composite document. However, the *linked object* includes a reference to the source program, which allows the user to edit the source file from the destination document. Further, any changes made to the source document are automatically available to the destination document.

Object embedding actually stores a copy of the source data in the destination file. As with object linking, the *embedded object* can be edited from the destination document. However, one important difference between the two methods is that changes to the original (source) document are not available to the destination (embedded) document.

In practice, both methods would produce similar looking destination documents. With object linking, however, the user can start the source program from the destination document by double-clicking on the embedded object. Thus, if the process design calls for user access to the original source files, linking offers a very user-friendly solution. For example, while discussing the portfolio profile with the investor, the analyst could enter investor instructions into the Notes file directly from the profile screen.

The analyst should exercise care in applying this technology. Some files, such as payroll or permanent history files, should be protected from update through nonnative operations. It would be improper to allow an investment analyst to initiate buy and sell orders from a screen that did not include the appropriate password protocol.

Although this difference may seem subtle, its significance will become clearer as you gain more experience with these high-tech methods. What should be emphasized, however, is the necessity of the automatic update feature for the investment example. Assume that the three source files are maintained via separate subsystems. Once the Investment Profile output (hardcopy or softcopy) is designed and implemented with OLE, the user can always see the latest composite information about a portfolio without performing any cumbersome file-sharing command sequences.

FIGURE 10-7 / *Object Linking and Embedding*

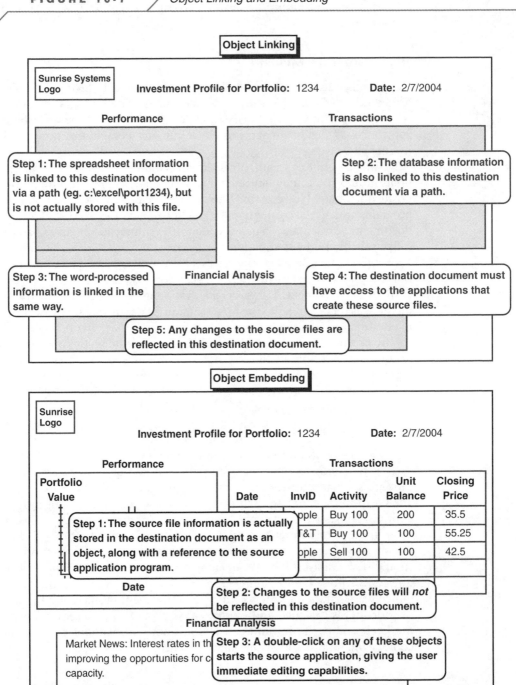

Internet File Sharing

Sharing files between computers connected to the Internet has long been possible with e-mail and *File Transfer Protocol (FTP)* software. Both of these methods require detailed information about the source or destination computer. This makes the file-sharing process very specific.

More recently, another class of Internet file-sharing software has been inspired by the success of software that enables audiophiles to swap digital music files without prior knowledge of the source or destination computer. Any computer with this type of software installed and open is automatically hooked into a special type of network where computers can communicate directly with one another without going through a centralized computer.

This is a very different kind of peer-to-peer network from the one briefly mentioned in Chapter 1. Recall that such a network is traditionally defined as one or more computers connected via a direct cable. In this case, the computers are connected via the Internet. The potential for easy-to-use, worldwide file-sharing applications is staggering.

For example, a far-flung enterprise might use such software to allow its intranet clients to locate and share any type of file (audio, video, image, print document, etc.) without knowing where the files exist. Thus, as illustrated in Figure 10-8, a doctor, connected to the Center for Disease Control's intranet, might simply request an image file named tb_23.jpg and expect to download that file from some other (unknown) computer in the intranet.

FIGURE 10-8 / *Internet File Sharing*

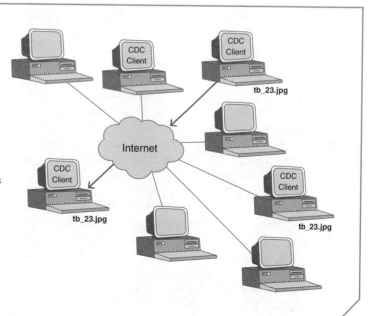

> Doctor(1) logs into the CDC intranet, (2) launches file-sharing software, and (3) requests a copy of **tb_23.jpg** from anyone in the CDC intranet.

>The file-sharing software locates the file, perhaps on more than one CDC client station.

>The file-sharing software downloads a copy of the file in the most efficient manner possible.

File Sharing with Middleware

File sharing is made more complicated when newly designed information system components must interface with one or more existing legacy systems. A *legacy system* is an ongoing, existing system that uses older technologies that are incompatible with newer 4GL software. For example, consider a project to develop a new data collection and retrieval system for student records. Cost constraints might cause the client and analyst to decide to retain, rather than replace, an existing nonrelational database of archived student transcript data. To ensure the integrity of the overall student records information system, the new system must be able to access data efficiently from the legacy system. Such a task is well beyond the file-sharing technologies previously discussed.

Middleware describes a diverse set of software that functions as a bridge between incompatible technologies. Microsoft open database connectivity (ODBC) software is a good example of middleware. It allows otherwise incompatible database management systems to work together. For example, with the proper ODBC driver, Access can work with data maintained in an Oracle database.

Process Design: Step-by-Step

Figure 10-9 suggests a sequence for the analyst to follow when working through the process design for a new system.

FIGURE 10-9 / *Process Design Steps*

1. List the computer processes in the DFD, USD, and menu tree.
2. If absent, add to this list a maintenance process for each master file identified in the ERD.
3. If absent, add to this list an updating process for each transaction file identified in the ERD.
4. Review the system output list to be sure that every output can be identified with a process on your process list. Add processes where necessary.
5. Evaluate each process to determine the best implementing software to use.
6. Write a summary for each process, briefly explaining how you expect the software to manipulate the data inputs into information outputs.
7. Develop a system flowchart, associating all the processes with specific software and all the I/O with physical devices and media.
8. Develop a program structure chart for those processes that require some custom programming. For example, this would not apply to a mail-merge process, but it would be necessary in a complex revenue projection or simulation process.
    Note: Program structure charts and flowcharts are discussed in the next section.
9. Update the system resource requirements to include the software and hardware needed to support the process design.
10. Update the data dictionary to include the processes that you have designed as well as the system flowchart.

PROGRAMMING PERSPECTIVES

The final design may involve elements from all the software options, with a considerable amount of file processing and file sharing. The analyst must carefully combine this collection of software to create a seamless user environment. More often than not, this involves programming.

Although the use of 4GL products does not require the skills of a systems programmer, the analyst must understand enough about programming to take full advantage of the products. Further, the analyst must bring diverse software solutions together as a user-friendly, unified package. This inevitably requires the use of macros, batch files, command files, and even some small programs. Today, such programming requires a working knowledge of the fundamentals of programming.

Chapter 12 is devoted to 4GL programming. To complete the process design, however, you need to use many of the programming tools presented in the following material.

Programming for the Nonprogrammer

The word "programming" has many negative connotations associated with it. Most people believe that computer programming is something totally out of their realm of experience, requiring extraordinary math skills. To the contrary, the mathematical manipulations required in programming are usually quite elementary—you have all programmed something in your everyday lives. What is most important in programming is the ability to solve problems in a logical manner. If you can figure out how to use the automatic timer recording feature on your VCR, you can learn enough about programming to put 4GL products to work in a small enterprise. To illustrate, Figure 10-10 shows part of the recording instructions on a typical VCR.

Notice that a picture of the VCR control panel supplements these instructions. This gives the user two methods of learning how to use the product. Visual representations are sometimes much easier to follow than written instructions, but are especially useful in combination with good written instructions. This text uses the same principle to describe the logical processing flow of a computer program. Not surprisingly, such pictures are called program flowcharts.

FIGURE 10-10 / *VCR Automatic Timer Recording Instructions*

1. Check that the power cord is plugged in.
2. Insert cassette.
3. Slide the Clock/Clock Adjust/Program switch to the right (PROG) position.
4. Press SET + and select program number. Press SELECT to finalize selection.
5. Press SET + to select START DAY. Press SELECT to finalize selection.
6. Press SET + to select START TIME (HOURS AM/PM). Press SELECT to finalize selection.
7. Press SET + to select START TIME (MINUTES). Press SELECT to finalize selection.
8. Press SET + to select STOP TIME (HOURS AM/PM). Press SELECT to finalize selection.
9. Press SET + to select STOP TIME (MINUTES). Press SELECT to finalize selection.
10. Press SET + to select CHANNEL NUMBER. Press SELECT to finalize selection.
11. To program several events, press SELECT and repeat steps 4–10.
12. To review selections, press SELECT.
13. Select recording speed SP/EP.
14. Slide Clock/Clock Adjust/Program switch to the left (CLOCK) position.
15. Activate Timer Record mode by pressing TIMER button.

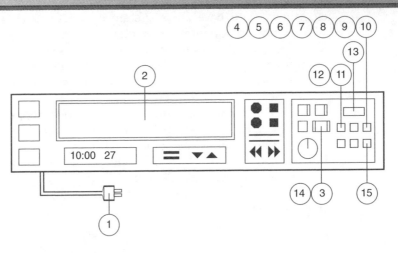

Program Flowcharts

Figure 10-11 presents a different visual description of the VCR instructions. A *program flowchart* uses standard shapes, flowlines, and text to give the user a two-dimensional view of the process. Furthermore, you can use the flowchart as an introduction to computer programming because it illustrates three fundamental programming concepts: instructions are always sequenced, some instructions are executed repeatedly, and some instructions are executed selectively.

FIGURE 10-11 / *VCR Automatic Timer Recording Flowchart*

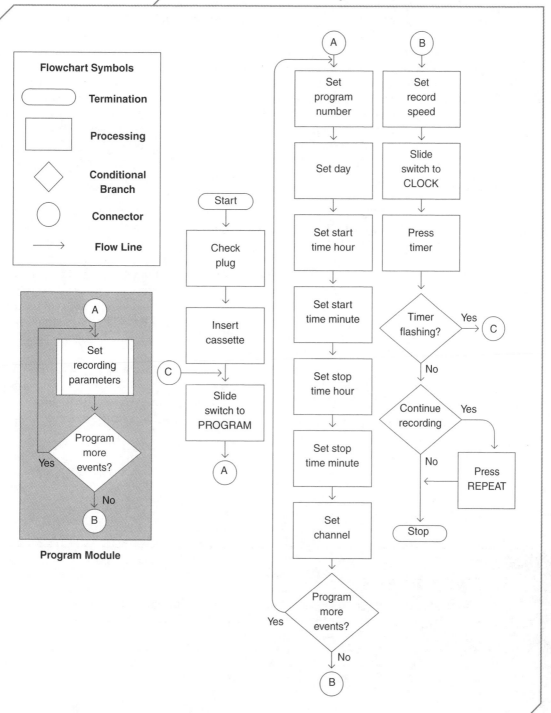

The sequential nature of programming is rather obvious. Everyone has learned that if you follow instructions, one after another and in some order, you are more likely to reach your goal. Of course, if the instructions are flawed, you will not be as successful. Computers follow instructions meticulously. For the past 50 years, computers have executed one instruction after another. Although mainframe and supercomputers can execute more than one instruction at a time, PCs are generally limited to executing one instruction at a time.

Many solutions involve the repetition of certain actions. In the VCR example, you are instructed to repeat steps 4 through 10 for each recording segment you desire. Good problem solvers look for patterns of instructions that can be reused, or repeated. In this case, the user supplies the specific data (day, time, and channel) for each repetition of the instructions, but the basic steps are the same each time. The ability to repeat instructions saves the programmer time and makes the solution much more versatile.

To make programs even more versatile, computer languages have instructions that can alter the normal sequence of program execution, based on the results of a test condition. Step 14 in the figure illustrates this concept. If the timer light is flashing, you must return to step 3 rather than continuing to step 15. Programmers call this a "branch" and use the diamond-shaped flowchart symbol to convey this meaning. One important restriction to such instructions is that the questions, or test conditions, can be answered only with yes or no. The ability to include instructions that are executed only in certain situations is a great advantage to the user. Users can effectively change program behavior by changing the data they provide the program.

Programming flowcharts vary in detail to suit the needs of the user. In the process-design phase, the analyst must create the flowcharts that describe the programs that tie the information system together. Usually, these programs are best described by uncluttered, high-level flowcharts called structure charts.

ThinkingCritically

ProgrammingExpertise

How would you respond to the statement, "Regardless of how much the analyst relies on 4GL technology, there will be some programming to be done when developing a computer information system"?

Structure Charts

A *structure chart* shows the relationships between and among the different functional components of a complex software product. In this sense, it is much like a hierarchical menu tree, with each menu option shown as a separate functional component. A structure chart may be composed of several levels of functional components and is read from top to bottom, left to right. The conditional execution symbol shows that the user can choose which functional component to activate.

As a computer information system is a collection of programs, so a program is a collection of modules. A *module* is a collection of instructions that perform a specific function. A structure chart shows how the modules work together in a program. For example, you could combine steps 4 through 10 in Figure 10-10 to form a single module called Set Recording Parameters. This allows the programmer to replace the middle section of the flowchart with the shaded portion to the left in Figure 10-11. The double vertical line indicates a module. Taken to the extreme, a program can be constructed from nothing but a series of modules. The flowchart for such a program is called a *structure chart*. Figure 10-12 is the structure chart for our VCR example.

Structure charts are extremely important to the modern programmer because they reflect a modularized approach to programming. Figure 10-12 illustrates the level of processing design detail required at this stage of the SDLC—remember, you do not need to get bogged down in detailed program flowcharting.

FIGURE 10-12 / *VCR Structure Chart*

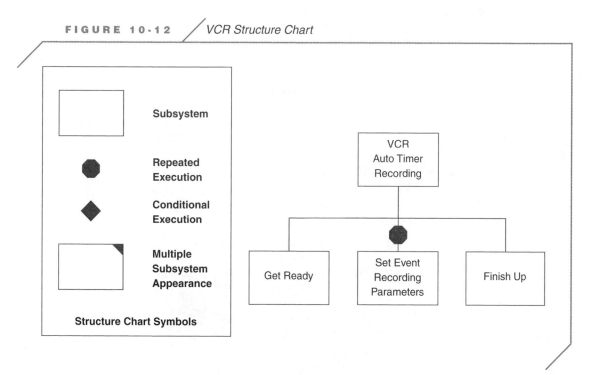

Menuing Software

The menu tree is described in Chapter 6 as the popular way to present information system options to the user. Menus are used in most of the 4GL products to provide the user with an electronic quick-reference guide. Users can step their way through various command sequences, even if they do not recollect the exact wording of a particular command.

Figure 10-13 presents a menu tree for the operation of a VCR. Notice that Recording is merely one of the many first-level options. Furthermore, when the user selects this option, the system immediately presents another menu. This process can continue until the user has, in effect, built a very specific, detailed command. Notice the familiar Automatic Timer Recording option nestled deep within the tree.

The beginning of this section alerts you to the importance of creating a seamless user environment. The program that is most effective in doing this is *menuing software*, which helps the analyst construct menu sequences and attach application software to the menu options. Your process design should include a menu program diagram that describes exactly how the user will move within and between the subsystems of your information system. Do not be overly concerned about the actual programming at this time. Several user-interface software products are available to choose from if you do not want to write the code yourself. The next two chapters offer considerable instruction on both these development stage options.

FIGURE 10-13 / *VCR Menu Tree*

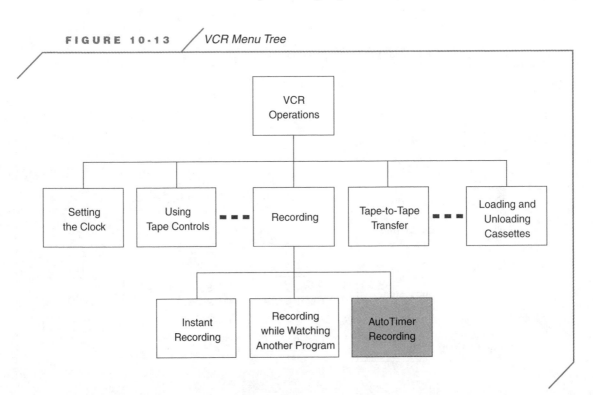

THE CORNUCOPIA CASE

The owner is delighted at the prospect of a point-of-sale/inventory/reorder type of system, but nervous about the way it will be implemented. She wants to make the transition to computerized sales as easy as possible on herself and her employees. The system will use a bar code scanner to capture the universal product code (UPC) of the CDs that are sold. Data capture procedures for other products will be added later. The actual computer system will be installed with a very low profile at the checkout counter, with only the monitor, keyboard, mouse, and scanner on the desktop. After the initial boot and password processes are complete, the system displays the Cornucopia switchboard screen until the user selects a subsystem option. Upon exit from a subsystem, the switchboard reappears.

The four primary areas of concern (customer record keeping, reordering system, customer communications, and sales trending) are all addressed in the design presented in the revised data model and menu tree. All eight processes in this system are listed in Figure 10-14.

FIGURE 10-14 / *Cornucopia Processing Requirements*

Process	Implementing Software
Customer update	Database
CD update	Database
Supplier update	Database
Sales transactions	Database
Reordering	Database
Sales trends	Database, spreadsheet
Correspondence	Word processing, database
Internet	Web browser, Web page construction tools

Revised New System Data Model

Figure 10-15 presents the table relationship diagram produced by Microsoft Access. As advised by the earlier discussions on file, form, query, and report design, the analysts created the files in order to create the necessary servicing screen forms; therefore, the relationship diagram is available. It displays the data attributes of each table object, as well as the cardinal relationships between the tables. Three new tables are added because of the process design activities: AutoOrder Temp, SalesSum, and SalesTrans Temp.

FIGURE 10·15 / *Cornucopia Table Object Relationships*

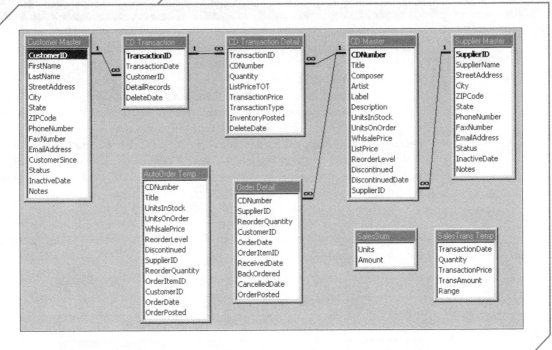

The AutoOrder Temp table is created each time the user selects the AutoOrder option from the Order Subsystem dialog box. It contains a potential order record for CDs that have an inventory amount (UnitsOnHand + Units OnOrder) less than the ReorderLevel. The user can access these records through the AutoOrder Form, choosing which orders to process and which orders to ignore.

The SalesTrans Temp table contains records used in the Sales Trends subsystem. The SalesSum table, which is derived from the SalesTrans Temp table, contains summary records that are eventually exported to a spreadsheet template. Both of these tables were added to the project when it became obvious that the original sales transaction tables would be too cumbersome to work with using the Sales Trends subsystem.

This exercise illustrates that system design does not always proceed in a sequential fashion. If the analysts have carefully documented the design at each stage of its development, the redesign is much easier. For example, the original new system ERD introduced in Chapter 8 (Figure 8-13) can be revised to include the new files. In this case, the analysts choose to use the table relationship diagram rather than revise the original ERD.

New System Switchboard

Figure 10-16 presents Cornucopia's switchboard and subsystem option dialog boxes. The switchboard buttons closely match the processes defined in the data flow diagram. Five of the subsystems are further defined by dialog boxes, which request users to make another selection. Together, the switchboard and the dialog boxes present users with all of the options detailed on the menu tree (see Figure 10-14).

FIGURE 10-16   *Cornucopia Switchboard*

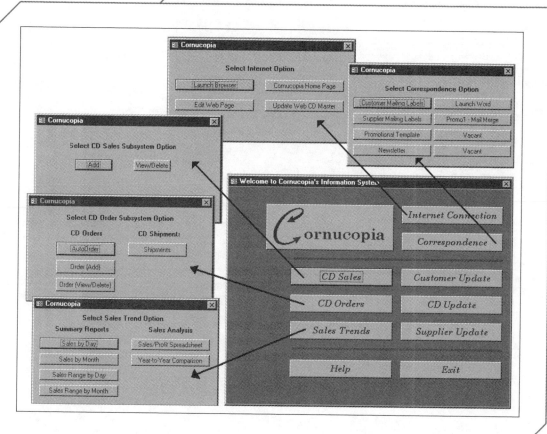

New System Interface Objects

Figures 10-17, 10-18, and 10-19 present subsystem structure charts for the master file maintenance, sales transactions, and CD-ordering process logic. These illustrations associate the screen forms with the tables they service. The diagram is similar to the data flow diagram, in that its focus is on system processes or functions. But, there is an important addition to consider: the annotated flow lines reveal the object-oriented nature of 4GL software.

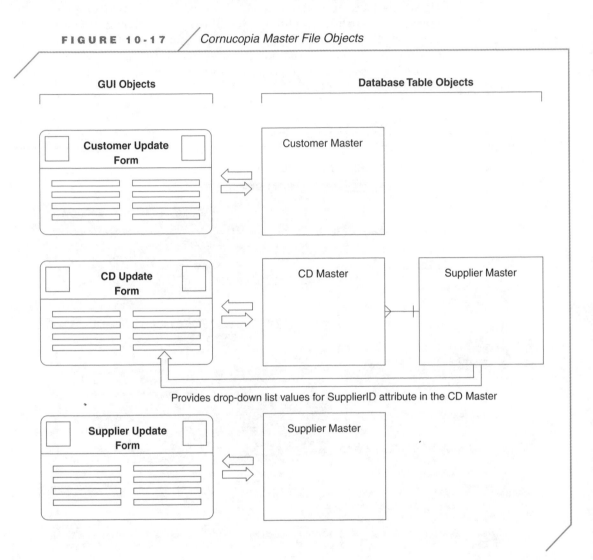

FIGURE 10-17 / *Cornucopia Master File Objects*

The master file maintenance diagram indicates that each of the master tables is serviced by one interface object (i.e., screen form). The CD Update Form requests information from the Supplier Master table through a drop-down list box object. Each form provides the basic service methods (create, retrieve, update, and delete) identified on the object model introduced in Chapter 8 (Figure 8-14). In other words, the analysts can begin to implement the object model by developing tables (object instances of the table class) and forms. Notice that queries and macros permit further implementation, with little or no programming.

FIGURE 10-18 / *Cornucopia Sales Transaction File Objects*

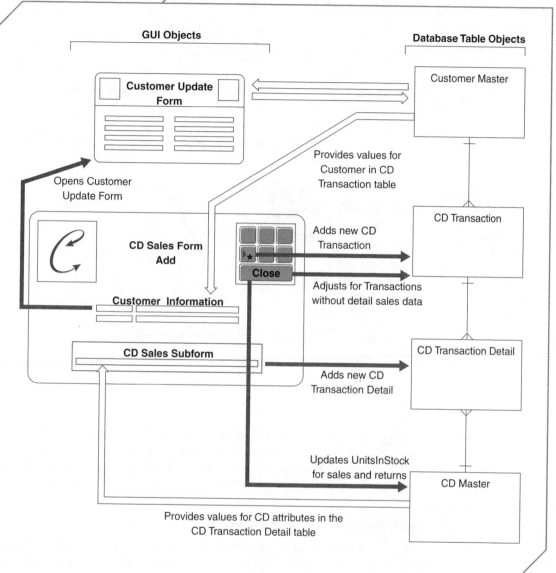

The CD Sales Form is much more complicated. This form is tied to the CD Transaction table, so that when the New button is clicked, a new CD Transaction is written to the table. When CD sales or returns are entered, new records are written to the CD Transaction Detail. When the Close button is clicked, the CD Master table UnitsInStock value is changed to reflect the sale or return. If the user has created a new CD transaction without any sales or return transactions, the close event sets the DetailRecords value to false. The CD and Customer master files provide drop-down list box values to make data entry easier. Finally, the user can add a new customer by clicking the Add New Customer button.

FIGURE 10-19 / *Cornucopia CD Order Objects*

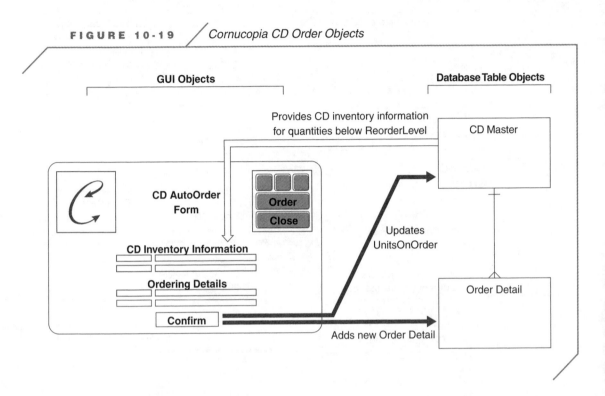

Figure 10-19 shows that the CD AutoOrder Form is tied to the CD Master and Order Detail tables. Actually, this form uses the AutoOrder Temp table, which is created when the user selects the AutoOrder option. When the user clicks the Confirm button, several actions occur, as explained next.

New System Structure Charts

Figure 10-20 presents three views of the processing logic for the Confirm button on the AutoOrder Form. The event procedure structure chart lists seven actions, two of which execute queries with the DoCmd statement. The Update UnitsOnOrder from AutoOrder query is depicted in pseudocode and a flowchart.

These logic diagrams may seem to present a piecemeal view of the system. This is the nature of 4GL software, where information products are assembled from predefined objects rather than constructed from scratch. In this case, table and form objects come equipped with a set of useful behaviors to which the analysts add specific features. The analysts only need to diagram custom features, and sometimes even those require no explanation other than a descriptive name.

FIGURE 10-20 / *Cornucopia AutoOrder Form-Processing Logic Charts*

Event Procedure Structure Chart

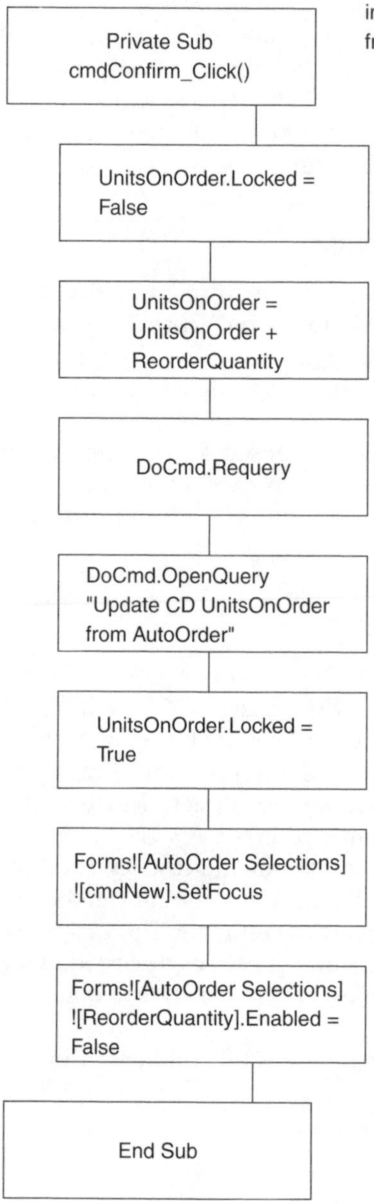

Pseudocode for Update CD UnitsOnOrder Query

Find the record in the CD Master table that matches the CD information on the AutoOrder form. Then, add the ReorderQuantity from the form to the UnitsOnOrder attribute in the master record.

Flowchart for Update CD UnitsOnOrder Query

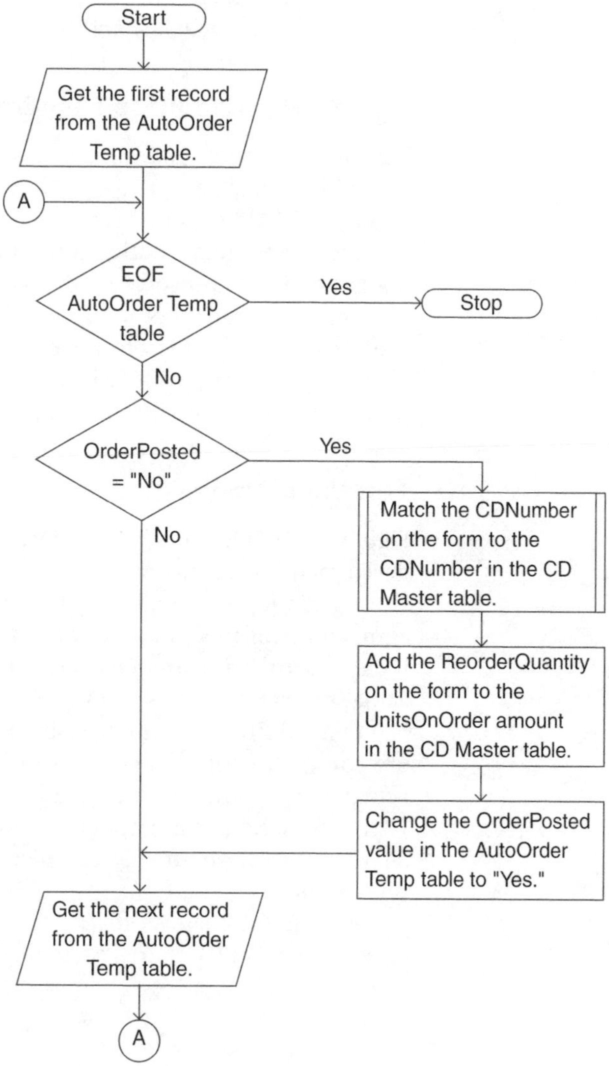

General Process Design

The CD Sales and CD Orders subsystem define Cornucopia's online transaction-processing system. The Sales Trends, Internet, and Correspondence subsystems use data supplied by the online transaction processing (OLTP) system to provide valuable information about the operation of the enterprise. The analysts envision using all of the applications in Microsoft Office to implement this design, with the switchboard as a convenient interface to everything. Most of the work will be done with the relational database package, which provides the analysts with ample object-oriented features, automated query building, a rich set of predefined macros, and Internet-compatible file type options.

Processing Resource Requirements

The processing design decisions confirm that the hardware specified in the I/O System Resource Requirements Notes introduced in Chapter 9 (Figure 9-22) is appropriate. No new processing resource requirements were identified during process design because file, form, query, and report design decisions dictated many hardware and software decisions.

One question persists, however. The proposed keyboard-based update process for the CD Master table promises to be very tedious and error prone. The analysts' earlier research into vertical solutions indicates that there is a possibility of downloading CD updates from suppliers.

Time and Money

Figure 10-21 illustrates the Project Status Report as of Week 8. The process-design activities consumed seven hours of the analysts' time during Week 7. During Week 8 the analysts spent six hours on design, bringing the estimated completion of this phase to 95%. Notice that even though the analysts' estimate was high for analysis activities and low for design activities, the total actual hours is very close to the total estimated hours.

One additional hour was spent on development activities, resulting in a 5% completion on that phase. This latter item is of some concern because the actual completion percentage falls far below the estimate. The analysts show a 70% completion of the design review session because, as discussed in the next chapter, they have already completed most of the materials required for the presentation.

Figure 10-22 shows the current progress toward completion in terms of Cornucopia's PERT chart.

FIGURE 10-21 / *Cornucopia Status Report – Week 8*

	A	B	C	D	E	F	G	H	I	J	K	L	M	N	O	P	Q	R	S	T
1	Date:	Cornucopia Project Status							As of: Week 8											
2																				
3	Activity	% Comp.	Status	1	2	3	4	5	6	7	8	9	10	11	12	13	14	15	16	Total
4	Analysis - Estimate	100%		4	5	5	4													18
5	Actual	100%	ok	2	4	2	3													11
6	Design - Estimate	100%					4	4	5	5	4									22
7	Actual	95%	ok			1	3	8	5	7	6									30
8	Develop - Estimate	20%									4	4	5	5	4					22
9	Actual	5%	--								1									1
10	Impl. - Estimate	0%													4	5	5	5	5	24
11	Actual	0%	ok																	0
12	Total - Estimate			4	5	5	8	4	5	5	8	4	5	5	8	5	5	5	5	86
13	Actual			2	4	3	6	8	5	7	7	0	0	0	0	0	0	0	0	42
14																				
15	Contract	100%	ok	C																
16	Prelim. Present.	100%	ok			C														
17	Design Review	70%	ok							S										
18	Prototype Review	0%	ok										S							
19	Training Session	0%	ok															S		
20	Final Report	0%	ok																S	
21																				

Status8.xls — Status / Sheet2 / Sheet3

FIGURE 10-22 / *Cornucopia PERT Chart – Revisited*

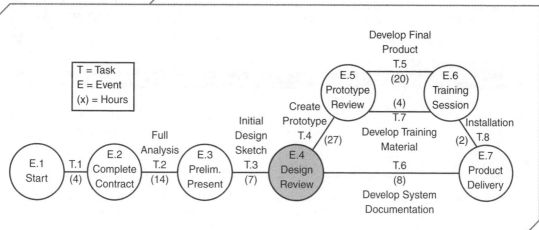

Cornucopia with Visible Analyst

As demonstrated in this project, design revisions are commonplace. A CASE tool can facilitate these revisions because it allows the analysts to easily manipulate the numerous models developed throughout the design phase. Further, the data dictionary features of a CASE tool help to incorporate the same change in each of the models.

As the development phase approaches, you may be concerned about the actual programming skills required of you. 4GL products assume many programming responsibilities. Figure 10-23 illustrates how Visible Analyst can help the analysts construct the subsystem structure chart. The Repository menu provides an option to enter data dictionary information, such as a description of the module's purpose. The cross-reference between the repository and structure chart gives the analysts complete access to the design documentation.

FIGURE 10-23 / *Cornucopia Structure Chart with Visible Analyst*

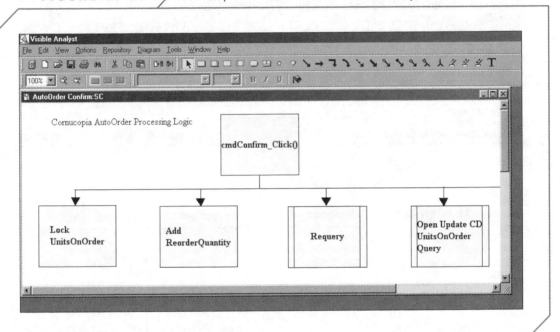

SUMMARY

This chapter presents detailed information on how to develop the process designs of a small-enterprise information system.

Several 4GL options are available (vertical, horizontal, or integrated software) as well as the traditional customized programming option. The analyst may find that the best design is one that incorporates some of each of these options. If this is the case, the file-sharing requirements within the system can be considerable.

Regardless of how much the analyst relies on 4GL technology, some programming must be done. Program flowcharts, structure diagrams, and menu trees are a good high-level introduction to programming. Later on, these tools help you build the programs that control the user interface to the information system.

TEST YOURSELF

Chapter Learning Objectives

1. Which prewritten software products are appropriate to incorporate into the process design?

2. What elements of data flow diagrams correspond to elements of system menu trees and system flowcharts?

3. Under what circumstances is it appropriate to incorporate the various file-sharing techniques into the process design?

4. How are a subsystem structure chart and program flowchart similar and dissimilar?

5. How much programming expertise is required to complete a 4GL-based, small-enterprise information system?

Chapter Content

1. Vertical software is designed to solve a narrow range of problems within a specific industry or information-processing setting. True or False?

2. Horizontal software is designed to solve a narrow range of problems within a specific industry or information-processing setting. True or False?

3. Integrated software combines several horizontal software elements into a common user interface. True or False?

4. The data flow diagram identifies the major process elements in the system. True or False?

5. The menu tree is a good place to start designing the process elements of the system. True or False?

6. A detailed system flowchart is annotated with 4GL applications and file types. True or False?

7. File sharing can only be accomplished with object-oriented programming languages. True or False?

8. Middleware is defined as software that is a combination of vertical and horizontal software. True or False?

9. A programming flowchart presents the instructional logic in a two-dimensional format. True or False?

10. A structure chart illustrates how the functional components, or modules, of a computer program work together to solve the information system problem. True or False?

11. A good process design invariably involves efficient file-processing and file-sharing methods. True or False?

ACTIVITIES

Activity 1

Compare Sunrise Systems' composite system flowchart and detailed subsystem flowchart (Figures 10-4 and 10-5). Are all the files, screens, and reports represented on both charts? If not, make the necessary corrections.

Activity 2

Use Sunrise Systems' detailed system flowchart (Figure 10-5) to identify the source files and source applications for each of the of the OLE outputs illustrated in the Investor Report (Figure 10-6). Modify Sunrise Systems' detailed system flowchart to accommodate an investment analyst's on-demand query for output regarding portfolio performance.

Activity 3

Prepare a detailed flowchart describing how to program a DVD player or some other electronic device.

DISCUSSION QUESTIONS

1. Assuming that your design incorporates horizontal software, how does the frequent updating of this software influence functional and operational obsolescence?

2. How do the various file-sharing options described in this chapter complicate and simplify processing design?

3. What are the processing design advantages of using a relational database as the source table for a mail merge operation?

4. Consider the file sharing involved in the Sunrise Systems example. What file-sharing method would you recommend in this situation? Defend your choice.

5. How would you define the term "4GL software"?

6. Under what circumstances does Internet file sharing become an infringement on copyright owners' rights?

7. How would you go about verifying the performance claims of a vertical software product?

8. In what ways do small-enterprise information systems resemble vertical software?

9. When you program your car radio, are you a programmer or a user? Defend your answer.

10. How does an ATM welcoming screen function as a switchboard to the ATM's functional modules?

PortfolioProject

Team Assignment 10: Process Design

As processing design nears completion, you are poised to begin building the system you have designed. Cornucopia's PERT chart (Figure 10-22) pinpoints your position in broad terms, and the detailed task list (Figures 6-14, 6-15, and 6-16) identifies many tasks that remain. Remember, while the PERT chart and detailed task list are linear in appearance, the blurred-line version of the enhanced SDLC (Figure 1-8) provides for the possibility that you may return to design activities should the need arise.

Several Cornucopia process-design illustrations focus on detailed relationships between database forms and tables (Figures 10-17, 10-18, and 10-19). The next three chapters show you how these designs are implemented. Don't be alarmed if your design images are much more general. As your experience with relational database software grows, your designs will become more detailed. At this point, it is sufficient that you identify the 4GL product you will use to implement each processing component of your system. Over the next few weeks, you may find yourself moving back and forth between design and development in order to achieve the necessary level of detail your project requires.

As always, it is important for team members to communicate about their project work. The PERT chart provides some evidence that communication is especially important from here to project completion. Notice that work on several tasks can proceed independently. One team member can work on project documentation, while another team member works on a prototype. Good team communication makes it more likely that all the independent pieces eventually fit together to create a uniform system.

In order to develop a workable process design and prepare materials for the upcoming design review session, you must complete the following tasks:

1. Develop a list of processing requirements, along with the hardware and software resources required to implement your processing design. Submit a copy of this list to your instructor.
2. Develop composite and detailed new system flowcharts. Submit a copy of these flowcharts to your instructor.
3. Write a summary for each process, briefly explaining which 4GL software features you intend to adapt to manipulate data inputs into information outputs. Submit a copy of these summaries to your instructor.
4. Develop any revisions to your new system DFD, ERD, and menu tree. Submit copies of these revisions to your instructor.
5. File a copy of the new system-processing requirements, flowcharts, process summaries, and model revisions in the Portfolio Binder behind the tab labeled "Assignment 10."

Development

prototyping

When you complete this chapter, you will be able to:

OVERVIEW

This chapter begins by addressing the final two pieces of the design puzzle: the resource specifications and the cost/benefit analysis. The design review session begins the transition from design to development. A discussion on the fundamentals of prototyping moves the project clearly into development. Finally, the chapter offers a plan on how to build upon the prototype in order to move into full-scale product development. You will learn to define the term analyst-programmer, which reflects the expanded role of a small-enterprise systems analyst.

- Develop the resource requirements for all six components of a computer information system.

- Develop a cost/benefit analysis chart.

- Prepare a design review document.

- Identify the information system design elements that are good candidates for prototyping.

- Develop a prototype for your project.

SYSTEM RESOURCE REQUIREMENTS

A list of *system resource requirements* describes all the hardware, software, data-handling, procedure-handling, and personnel resource needs associated with the SDLC. Obviously, the costs of most systems include the purchase and assembly of hardware and software. These are usually what we think of first when explaining to the user that certain acquisition costs occur well before we implement the system. Our study, however, shows that computer information systems are composed of six components, each of which the analyst should evaluate in terms of its resource requirements. The small enterprise is particularly sensitive to expenses for unusual resource requirements. You must conduct a thorough analysis of all components of the system to ensure that the necessary resources are available when you need them to develop, implement, and maintain the system you have designed.

Data, People, and Procedure Resource Needs

Of these three components, only data surfaced in the previous discussions about resource needs. The data-handling and file-processing activities of the system may require special resources over and above the typical hardware and software requirements. For example, the market news files used in the Sunrise Systems example not only must be purchased, but they also have associated data transmission costs.

The people who use the system may require special skills that would not normally be included in the standard training package. Special hardware, such as audio or remote-controlled devices, might be required. Procedures may involve online help or other visual aids, as well as someone on staff to keep the procedures up to date.

Figure 11-1 suggests some areas to consider as you begin to itemize the list of system resource requirements that will eventually be included in a comprehensive and detailed cost summary.

FIGURE 11-1 / *Potential Data, People, and Procedure Resource Needs*

Data	People	Procedure
Data collection hardware	Specialized I/O for people with special needs	Online help
Purchase and/or subscription of data files	Custom training for high-tech computer systems	Ongoing analysis for procedure maintenance
Data storage facility	Special work area needs	Records storage requirements

Hardware, Software, and Networking Resource Needs

The design work outlined in the preceding three chapters includes detailed specifications of hardware and software for file, form, report, query, and process design. Although the extended discussion of networking appears in Chapter 13, the following illustrations reflect a common situation in which a local networking specialist provides detailed networking resource needs. The analyst can now assemble these requirements into a coherent list, ready for additional product specification, cost, and vendor information. Figure 11-2 uses a simple format to illustrate some generic system resource requirements.

FIGURE 11-2 / *System Resource Worksheet*

Item Description	Vendor	Cost	Comment
Hardware:			
server	PDQ Ltd.	2000	single source
platforms (ea.)	PDQ Ltd.	1200	single source
monitors (ea.)	Vector Inc.	0	bundled
printer	HiRes Corp.	1275	bid
Software:			
operating system	Microsoft	1500	bundled
MS Office	CompSoft	450	mail order
Networking:			
network installation	Network Services	2500	bid
network maintenance	Network Services	125	monthly
Data:			
Internet fees	FastNet	100	monthly
People:			
graphic artist	LineArt Agency	250	contract service
Procedure:			
portfolios	ABC Supply	250	art storage

Product Research: Request for Bids and Proposals

Of the many ways to find out what products are available and how much they cost, no single method stands out as the best. What is important to remember is that some research is necessary to properly evaluate the product literature and marketing hyperbole. Monthly trade magazines, professional journals, the Internet, industry conventions, and software retailers can provide some perspective on product availability, cost, performance, and compatibility. However, sometimes using a formal process to collect this information is required.

When required by law or company policy, the analyst must prepare a formal document requesting potential vendors to submit proposals to fulfill some of the system resource requirements. Although no standardized format exists for these documents, they generally include the items illustrated in Figure 11-3. Often these documents are so complex that you should have them reviewed by a lawyer. Above all else, however, you must be sure to state clearly what you expect the resource to do. This gives potential vendors the opportunity to apply their problem-solving expertise as well, rather than simply supplying catalog numbers and price quotes.

When you have already decided on a particular product that more than one vendor can supply, you should prepare a *request for bid (RFB)* document. When you have only the general product specifications, you should prepare a *request for proposal (RFP)* document. On the surface, the main difference between the two documents concerns the wording of the resource requirement specification. In practice, potential customers evaluate bids and proposals quite differently. With an RFB, price, delivery date, and warranty are normally what separate the bids. The RFP, on the other hand, may reveal significant differences in product options, manufacturing standards, and performance potential.

Making a decision is easy when all things are equal. Of course, they never are, especially when it comes to computer resources. For many years, purchasing decisions were easy to make because a few companies dominated the computer hardware business. This has changed, as hundreds of competing companies flood the marketplace, some with manufacturing as well as assembly operations. The desktop computer market is still split, albeit unevenly, between the PC and Macintosh. Small-enterprise computing analysts do have some recourse, however. Other than firsthand experience, the best way to begin your evaluation is with the product reviews that appear in several popular magazines. Some Internet sites provide user reviews as well.

BillableSoftwareExpense

Your work as an independent computer consultant has steadily grown over the past two years. What began as weekend work for friends and associates now consumes almost 20 hours a week, with referral business branching into small enterprises from neighboring communities. Recently, you negotiated a $15,000 contract to design, develop, and implement an information system for a local cellular phone company. You realize that this contract requires a professional approach to many of the activities and deliverables associated with the project. Further, you calculate that the project's time constraints demand that you dramatically improve your productivity if you are to complete the work on time. This motivates you to purchase a $2,500 CASE tool.

Under what circumstances is any or the entire CASE tool purchase price billable to your client? Given your answer to the preceding question, who owns the user's license to the CASE tool?

Thinking Critically

In addition to cost and performance, delivery is also a critical issue. The analyst is responsible for coordinating resource deliveries with site preparation, installation, development, testing, and training.

FIGURE 11-3 / *Request for Bids and Proposals*

To: Vendors
From: Sunrise Systems
Date: February 1, 2004

General System Description:
Sunrise Systems provides individual financial planning and investment analysis for 250 to 300 clients. These services include...

Resource Requirement Specification:
Sunrise Systems seeks to purchase the following computer equipment:

RFB language:	RFP language:
1. XYZ Business System (detailed specs attached)	1. High-performance desktop computer
2. QRS Laser Printer—Model 1234 with cables	2. 20 ppm color laser printer with cables
3. FGH high-speed modem	3. High-speed modem

Evaluation Criteria:
Sunrise will evaluate the bids based on delivered price, delivery date, and vendor product warranty.

Submission Instructions:
Vendors should submit bids to Sunrise Systems, Attn: G. Donovan, no later than March 12, 2004. Vendors may bid on all or part of the specified computer equipment.

Schedule:
Sunrise Systems will open bids on March 15, 2004, evaluate the bids the week of March 15, 2004, and respond to vendors by March 24, 2004.

COST/BENEFIT ANALYSIS

Everyone is interested in the bottom line, which, with information systems, translates to, "How long will it take for this system to pay for itself?" For the most part, calculating system costs is a straightforward task. Benefit calculations, on the other hand, pose a more difficult problem. Because most information systems are designed to meet internal information needs, we cannot rely on traditional free-market mechanisms to determine the system's dollar value or benefit. When the system actually generates an information product that is for sale, you can obviously consider potential sales revenues as offsets to system costs. However, when this is not the case, many of the cost and benefit items are difficult to value in terms of dollars and cents. Nevertheless, that is *cost/benefit analysis:* an activity that plots project costs and benefits over time to determine if, or when, the system becomes economically justifiable.

Cost Elements

Previous chapters address costs in terms of hardware, software, and labor, as reflected in the project budget. The resource specifications developed throughout the design process provide the basis for much more accurate cost estimates. In addition, when used, the RFB or RFP can supply cost information as well. However, the analyst must also consider other *intangible costs*.

Some of the potential intangible costs associated with a system are risk, opportunity cost, employee resistance, and future commitment to the technology. For instance, were you to apply these terms to your decision to enroll in this course, you might find it difficult to place a dollar value on the intangible costs. For example, you could fail this class. If doing so makes you ineligible for the track team, causing you to miss the Olympic trials and forcing you to get a job, you might be able to calculate that the intangible risk cost is sizable. But, you certainly can't report this as an expense on your income tax form. The point is that you need to know that potential costs exist that can become quite real, even though they never appear on a balance sheet.

Benefit Elements

Presumably, considerable benefits are to be gained from the computer information system. Why else would anyone undertake the process? Yet, benefit identification and valuation is a troublesome task. The project contract and the project preliminary presentation report should identify the measurable goals of the system. However, how do you put a dollar figure on goals such as "reduce input errors by 10 percent" or "increase client awareness of our non-profit services"? The continued participation of the user throughout the analysis and design process may provide some answers to this question, but generally, this job belongs to the analyst. This is why in Chapter 2 you were encouraged to establish baselines against which the new system performance could be measured.

As with costs, benefits can be tangible and intangible. Some potential *intangible benefits* are increased flexibility, improved product quality, and improved employee morale. As an example, the multisourced Investment Profile output designed for Sunrise Systems greatly improves information access, thus allowing users to piece together trends or otherwise infer new information. This competitive advantage is likely to lead to increased investment activity and, in turn, increased commissions and profits, all of which can be estimated and tallied into the tangible benefit column. Nevertheless, intangible benefits may occur as well. Users may see this high-tech file sharing as an example that Sunrise Systems is on the cutting edge, not only with computing technology but also with investment analysis. Because it is difficult to say exactly how and when this attitude will translate into dollars, we include this as an intangible benefit.

Figure 11-4 presents a cost/benefit worksheet that helps sort out the various costs and benefits of a new information system. One sample entry on this worksheet requires some explanation. It is tempting—but dangerous—to look for benefits in the form of information system-induced personnel cuts. Although personnel cost savings accrue very quickly on paper, you will fight an uphill battle to realize those savings, especially in the small enterprise. Workers, feeling their job security threatened, have a way of making themselves useful, if not indispensable. Often, the introduction of a new information system requires more personnel, namely someone to run the computer.

FIGURE 11-4　／ *System Cost/Benefit Worksheet*

System Cost Elements

Contract Costs

Contract Costs	Amount	Comment
Hardware	$ 3,000	
Software	$ 2,000	
Labor	$ 5,000	

Noncontract Costs

Noncontract Costs	Amount	Comment
User Hours - Participation	$ 1,000	20 hours
User Hours - Training	$ 500	10 hours
Office Remodel	$ 500	
Data Conversion	$ 500	10 hours

Ongoing Costs

Ongoing Costs	Amount	Comment
System Maintenance	$ 100	monthly
Network Connection	$ 100	monthly

Intangible Costs

Intangible Costs		
Risk of Failure		
Reduced Cash Balance		

System Benefit Elements

Contract Objective

Contract Objective	Amount	Comment
Reduce Workforce	$ 1,000	monthly
Increase Sales Revenues	$ 500	monthly
Reduce Inventory Expense	$ 500	quarterly

Intangible Benefits

Intangible Benefits		
Improved Customer Service		
Improved Employee Morale		

PROJECT DELIVERABLE: THE DESIGN REVIEW SESSION

Joint application design activities and user-interface prototyping create a consistent working relationship between the analyst and the user. As the design phase draws to a close, the analyst should schedule a formal *design review session* to present the results of these activities in a well-coordinated design document and oral presentation. This session should include an overview of the new system design and resource requirements, as well as a description of how the system meets the objectives set forth in the project contract. Furthermore, the user needs to understand how the incremental design decisions made during the collaborative analyst–user activities affect the overall design. Finally, it is critical that the user appreciate the consequences of any major changes to the design from this point on. Figure 11-5 provides general guidelines for the report and oral presentation.

FIGURE 11-5   *Contents of the Preliminary Presentation Report*

Written Material:

1. Overview of the new system design

2. Models: USD and menu tree

3. User interface designs: switchboard, forms, reports

4. Web site homepage design

5. Resource requirement specifications

6. Cost/benefit analysis

7. Project status report

8. Project budget

9. Appendix (DFD, ERD)

Oral Presentation:

1. Visuals for items 1-8 above

2. Question-and-answer period

3. Preview of the next step

In some situations, in conjunction with this session, the original contract may provide for another *"go/no go"* on whether to continue, suspend, or abandon the project. Even though considerable resources have been expended already, the user may decide to cancel or postpone the project for a variety of reasons. The financial and competitive climate in which the enterprise operates can change rapidly, as can the proposed system's implementing computer technology. The cost/benefit analysis may be disappointing, or the user may have lost confidence in the analyst.

Previewing the Prototype Report and Review Session

The design review session is another transition event in the project. The design documents, along with any user feedback during the session, provide the blueprint for product development. As the first step in product development, the analyst creates a prototype of the major system functions. The remainder of this chapter is devoted to descriptions of prototyping and the need for continuing user participation during development activities. When the prototypes are complete, the analyst conducts a *prototype review session*, which provides another opportunity for formal review and discussion of the information system while it is still possible to introduce significant design changes.

PROTOTYPING FUNDAMENTALS

A *prototype* is a model or pattern of a product. Computer information system prototyping is a relatively new phenomenon, brought about by changes in user expectations and advances in computer hardware and software technology. Computer information system prototypes function on different levels and serve different purposes. For example, the screen designs created in earlier chapters are user-interface prototypes that enhance understanding of the data flows identified on the data flow diagram. An analyst can use processing prototypes, on the other hand, to test complex subsystem designs. Although you have already experienced prototyping to a limited degree, you need a good grasp of some of the fundamentals.

Prototyping and the SDLC

The enhanced SDLC (Figure 1-11) identifies prototyping as a design-development methodology that facilitates continued user involvement in the project. Some prototyping occurs naturally, as the design elements for input/output and processing are prepared through joint application design activities. Such informal prototypes might involve pen-and-paper sketches, practice walk-throughs, or other products generated by the preliminary review and design review session.

The combination of powerful 4GL software and sustained user participation in the process, however, creates the potential for a much more formal prototyping methodology. An example of this trend is *rapid application development (RAD)*, which shortens the process by combining joint application design, 4GL software, CASE tools, and prototyping to deliver the most important portions of the system quickly. This text incorporates prototyping into the traditional SDLC model, offering a way to mix the old and new methodologies gently.

Prototyping Levels and Types

There are two distinct prototyping types. An analyst develops a *reusable prototype* with the intention that it will be transformed into the final product. A *throwaway prototype*, as the name suggests, is discarded after it has served its purpose. One of three prototyping levels can further distinguish each type: input/output, processing, and system. The *input/output prototype* is limited to the user interfaces. The *processing prototype* includes basic file maintenance and transaction processing in the prototype. The *system prototype* offers a complete working model of the system.

TECHNOTE 11-1

RAPID APPLICATION DEVELOPMENT

Rapid application development (RAD) is a close relative of prototyping. Its promise is that the combination of modern analyst productivity tools, with the collaborative user-analyst relationship, allows dramatic savings in product development time for projects of any size.

Formally, RAD is a process that combines the use of four elements already presented as part of our enhanced SDLC:

1. Joint application design
2. Analysts with 4GL product expertise
3. CASE tools
4. Design-development prototyping

Joint application design is a formal process that guarantees the active participation of users in the design activity. Although less formal, prototyping has the same effect when used properly. In medium-size to large projects, such techniques are required in order to build the necessary collaborative structure into the systems project.

The analyst productivity tools consist of 4GL products that help the analyst with each step of the project. The acronym SWAT was coined to describe "specialists with advanced tools." Thus, the analyst would call on the **SWAT team** to work with users to develop the project quickly.

In some ways, this approach is ideal for small-enterprise projects. This text recommends the use of all four components of RAD, along with the more traditional methodologies presented in previous chapters.

In the small-enterprise setting, you probably do not need to create a throwaway prototype of any kind because the 4GL products that are used to create the model are the same products that will be used to create the final system. Likewise, it would be very unusual for the small-enterprise analyst to create a system-level prototype of any kind. The model and the real thing would be virtually the same. The reusable, input/output, and processing prototypes are most appropriate for projects of the size and scope discussed in this text. Figure 11-6 shows the likelihood of using a particular prototyping approach in the small-enterprise setting.

FIGURE 11-6   *Prototyping Matrix for Small-Enterprise Projects*

Level	Type	
	Reusable	Throwaway
Input/Output	Frequently	Sometimes
Processing	Sometimes	Rarely
System	Never	Never

This translates into a straightforward list of system elements that are candidates for prototyping. Many of the input/output forms and reports and much of the basic master file maintenance processing can be effectively prototyped. Even some of the basic transaction processing can be included in this list.

Figure 11-7 illustrates the reusable prototyping process as it might work for small-enterprise projects. Those design elements shown with shaded, bidirectional arrows are good candidates for prototyping. The user-analyst collaboration and prototype testing activity serves to perfect the design and make the development activities much easier. The other design elements are much more difficult to prototype, and therefore are not normally included in this process.

FIGURE 11-7   *Design Element Prototyping in Small-Enterprise Projects*

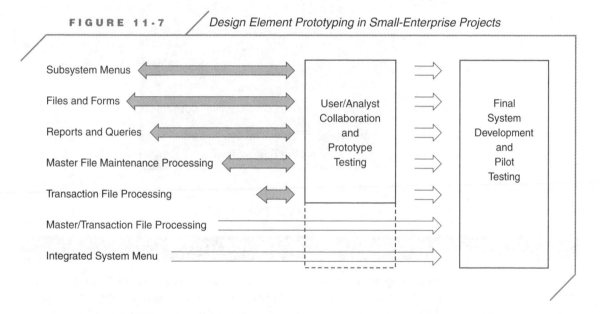

Prototyping Tools

Many of the tools used to create the earlier system models can also be used to create prototypes. Word processor, spreadsheet, and graphics software provide a foundation for any computer professional. The *relational database management system (RDBMS)*, however, is the mainstay for small-enterprise prototyping, in that it enables the user to relate multiple files within the database. Modern relational database products possess user-friendly interfaces, full-featured command languages, powerful GUI builders, and code generators.

Figure 11-8 provides a list of design-development tools and suggested prototyping applications.

FIGURE 11-8 / *Prototyping Tool Kit*

Tool	Uses
Word processor	Sample correspondence
	System documentation
Spreadsheet	Tabular output
	Charting output
Graphics package	System diagrams
	Presentation overviews
RDBMS	File creation
	Form design
	Report design
	File maintenance
GUI builder	Menu design
	Form design
4G languages	Processing code generation
Desktop manager	System menus
CASE	System diagrams

ADVANTAGES AND DISADVANTAGES OF PROTOTYPING

Prototyping activities consume valuable time and resources. Thus, the analyst should approach prototyping only after serious consideration of the needs the prototypes will serve. One way to do this is to evaluate the advantages and disadvantages of prototyping, as summarized in Figure 11-9.

Several of these advantages revolve around a very simple premise: at some point during the design-development activities, the analyst must close off further changes to the information system. Prototyping permits the user to evaluate the input/output interfaces very early in the process. Therefore, fewer design revisions are required during product development, when such changes are more difficult and expensive to incorporate. Furthermore, to the extent that prototyping shortens the time from design to implementation, there is less chance that the information needs of the enterprise or the preferences of the user will significantly change.

Likewise, the disadvantages of prototyping relate to the same basic concept. Because it is so easy to create an illusion of the information system, analysts sometimes forget that they must eventually convert the prototype into a real-world product. The term *industrial-strength information system* refers to this potential dilemma. An industrial-strength information system must be able to withstand inconsistent or exceptional data entries, improper or illegal user input requests, and unusual or erratic processing demands. Prototypes do not normally satisfy these criteria.

FIGURE 11-9   / *Prototyping Advantages and Disadvantages*

Advantages:
  1. Improves analyst/user communication
      Helps the user to define needs
      Helps to train the user early
  2. Concentrates attention on information products rather than
      on processes
  3. Encourages design change decisions to be made early in the
      project rather than late in the project
  4. Helps to test design performance
  5. Saves money
Disadvantages:
  1. Discourages careful documentation
  2. Process design problems may be underestimated
  3. Encourages oversell of system performance

FOR EXAMPLE: THE POLITICAL RESEARCH CORPORATION

The Political Research Corporation (PRC) is introduced here and used throughout this section of the text to illustrate the project development activities. PRC is an independent enterprise that designs, distributes, collects, organizes, and summarizes voter opinion research questionnaires. In other words, this company is in the information business to serve politicians, governmental agencies, lobbyists—even other information providers. The fact that PRC produces information for profit does not, and cannot, change the way the analyst approaches the information system designed to generate this product.

To familiarize you with this particular portion of PRC's information system, Figures 11-10 through 11-13 present the context diagram, data flow diagram, system flowchart, and some output prototypes. The context diagram identifies three external entities to this system: clients, voters, and the registrar of voters. The voter survey process is initiated when a client presents a series of questions to PRC. In some cases, the client might simply outline the general topic of interest, leaving the formulation of the specific questions to the PRC staff.

FIGURE 11-10 / *PRC Context Diagram*

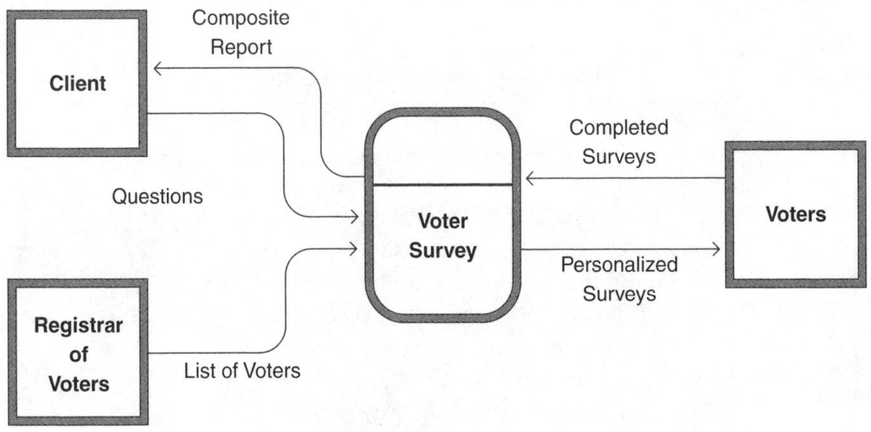

The data flow diagram (Figure 11-11) provides much more detail about the seven subprocesses of the system. The three external entities reappear as either data sources or information sinks to individual subprocesses. The data stores and data flows add detail, to show more specifically what the system does. Notice that this is a purely logical rendering of the process model: there is no reference to the physical devices, application programs, or file storage media involved in the system implementation.

The system flowchart (Figure 11-12) presents much of the physical design of the system, along with shaded areas to denote the prototyped outputs. Also, the type of processing software and the type of file storage are identified. This diagram is read from top to bottom, left to right, with numbered, circular connecting symbols to show how the outputs from one process are used as inputs to another process. Once again, notice that the system employs four major horizontal application software types (word processor, spreadsheet, database, graphics) integrated into a single information system.

Finally, the output prototypes (Figure 11-13) illustrate the personalized survey, the analysis softcopy view, and the composite report. Another obvious prototyping candidate is the system menu. Can you visualize what such a prototype would look like? What software product might you use to develop such a prototype? Would you recommend any other input/output interfaces or processes for prototyping?

FIGURE 11-11 /*PRC Data Flow Diagram*

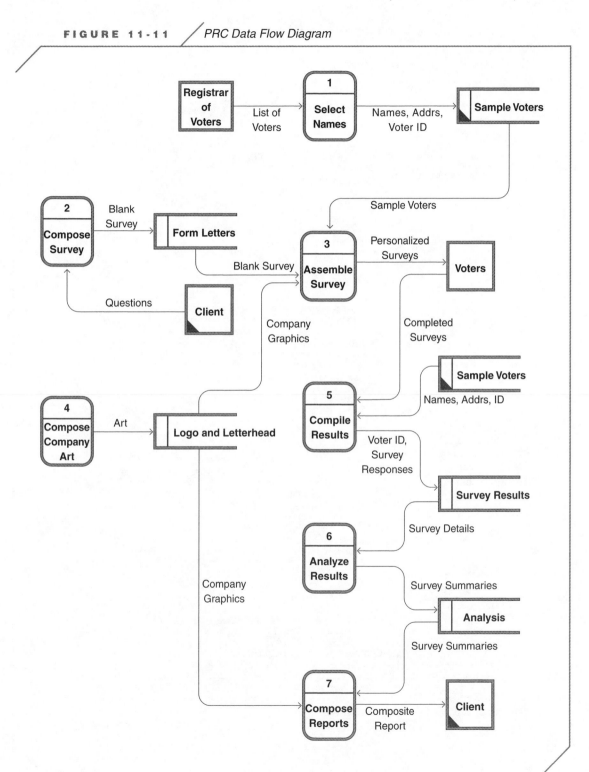

FIGURE 11-12 / *PRC System Flowchart*

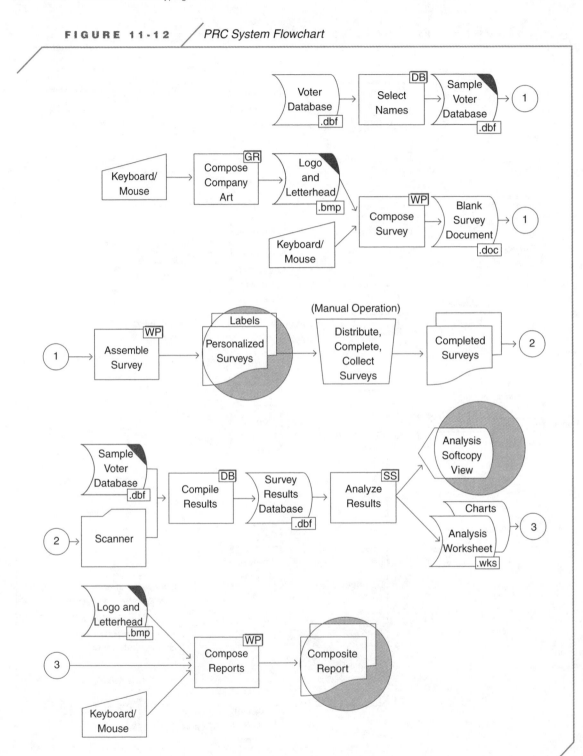

FIGURE 11-13 / *PRC Output Prototypes*

  Political Research Corporation

January 12, 2004

Ms. Beverly Duane
123 Post St.
San Francisco, CA 99901

Dear Ms. Duane:

You have been randomly selected to
participate in the legislative survey for
2004. Our records show that you last
voted in Los Angeles in 1998, but that
you are now registered to vote in San
Francisco. The following questions are
particularly relevant to metropolitan
voters.

Your responses will be summarized
with those of other voters and reported
to the Legislative Analyst. All individual
reponses will be kept confidential. Please

Personalized Survey

PRC **Political Research Corporation**

April 1, 2004

Legislative Analyst
34 State Avenue
Sacramento, CA 92156

Re: 2004 Legislative Survey
Attn: R. Smith

The following summary data is submitted
per your instructions. Please continue to
Page 17 of this document, where our
written analysis of the reponses to these
issues begins.

Response Percentages

Issue	A	B	C	D	E
1	12	32	46	5	5
2	67	12	10	3	8
3	4	7	12	56	21

Composite Report

PRC **Survey Response Data**
2004 Legislative Survey

IdNo	Precinct	Issue1	Issue 2	Issue 3
123	P-472	A	C	A
124	R-178	B	B	C
125	K-945	B	D	A
126	F-637	C	A	B
127	P-261	A	B	C
128	T-454	D	B	C

Analysis Softcopy View

PROTOTYPING: STEP-BY-STEP

Prototyping should begin with two strategies in mind: do the easy things first
and engage the user early in the process. Master file maintenance prototypes
are relatively easy to build and usually spark the user's curiosity about forms
and reports. The system menu is another obvious choice for early prototyping.
Because many of these elements stand alone, with no supporting program,
they represent the first level of your reusable prototype (input/output). Later,
as you move into the second level (processing), you incorporate these elements into subsystems. As a convenience and planning document, the analyst
should note or highlight those elements of the system flowchart that are prototyping candidates.

To help the user understand exactly which elements of the system are prototyped, the analyst should prepare a prototype USD. A *prototype USD* is a
copy of the original new system USD, with highlighting superimposed on
those portions of the information system that are prototyped. Figure 11-14
offers a step-by-step process for prototyping.

FIGURE 11-14   / *Prototyping Steps*

1. Identify prototyping candidates.
    User interfaces (menus, dialogs, inputs, outputs)
    Master and transaction file maintenance
    Simple processing functions

2. Build the prototypes with "tools."
    4GL software (word processor, spreadsheet, database)
    Graphics/presentation software
    CASE software

3. Test the prototypes to ensure that they can be easily accessed for
    demonstration purposes.

4. Prepare a prototype USD to identify those parts of the system that are
    prototyped.

5. Collaborate with the user to evaluate the prototype and make changes
    where necessary.

6. Transform the prototype into a working system, as described in the next
    two chapters.
    Remove any unnecessary code created by the code generator.
    Build program "shells" around this code, using a modular program design.
    Add any necessary program controls and linkages.
    Conduct incremental program development and testing.

PROTOTYPING WITH CASE TOOLS

CASE products come in many varieties. Figure 11-15 differentiates the kinds
of CASE products and correlates them to the different activities of the enhanced
SDLC. In general, the CASE tool functions discussed up to this point in the text
are *front-end CASE* or *upper CASE* products, which assist the analyst with the
analysis and design activities of the project. Data dictionaries, along with process,

data, and system models—as well as some project documentation—can usually be created with such products. As we move into prototyping and development work, the *back-end CASE* or *lower CASE* products are required. These products include tools to help the analyst build menus, user dialogs, screen forms, reports, and processes. Often, these back-end CASE tools are presented to the analyst as a development workbench. This would include a graphic interface with convenient toolbars and menu options to assist the analyst with automatic code generation, program testing and debugging, and process documentation.

FIGURE 11-15 / *The Enhanced SDLC and CASE Tool Scope*

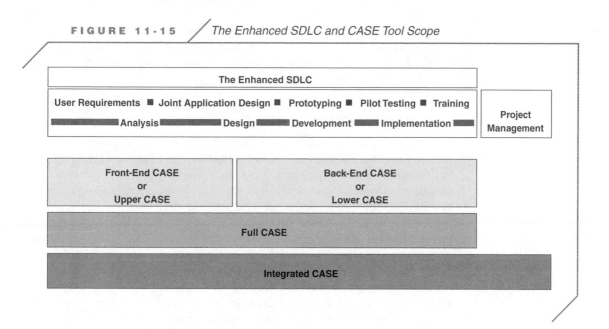

A *full CASE* product combines all the preceding features into one package. This usually includes a central repository of project materials, which is a well-coordinated, disk-based collection of project files, supporting documents, and the data dictionary. When such a repository is maintained on a local area network, the full CASE product can provide any of the relevant project materials to the individual analyst—or to any member of the analyst team—with the assurance that everyone is working with the same information. This is a tremendous advantage in medium-size to large information system projects in which the work may involve many analysts, a great deal of time, and different work locations.

Finally, the *integrated CASE* tool extends the full CASE approach one step further to include project management. Once again, this expansion of the CASE tool function becomes more important as the size and scope of the project increases.

THE CORNUCOPIA CASE

To perform a cost/benefit analysis the analysts must prepare two estimates. Cost projections are based on the resource requirement specifications and the future operating costs of the new system. Benefit projections are based on the goals and objectives set forth in the project contract. Although estimates of this nature are sometimes difficult to develop and justify, the analysts should prepare a rationale that explains the estimating process. This can lend some perspective to future evaluations of these estimates.

The analysts use the reusable prototype approach for the project mainly because of the small size of the project and the straightforward nature of the general system design. The prototyped input/output and processing elements can then serve as the basis for the full system development.

Resource Requirement Specifications

The processing design decisions confirm that the hardware specified in the I/O System Resource Requirement Notes introduced in Chapter 9 (Figure 9-22) is appropriate. No new processing resource requirement specifications were identified during process design because file, form, query, and report design decisions dictated many hardware and software decisions.

One question persists, however. The proposed keyboard-based update process for the CD Master table promises to be very tedious and error prone. The analysts' earlier research into vertical solutions indicates that there is a possibility of downloading CD updates from suppliers.

The hardware and software acquisition specifications are illustrated in Figure 11-16a and 11-16b. The data resource acquisition costs are not included because this is not part of the system design at this time. However, a modem is listed in the hardware and software specifications. The people and procedures costs are bundled into the labor costs associated with system documentation and training. At this time, there is no networking cost associated with the project.

FIGURE 11-16a / *Cornucopia Resource Requirement Specifications*

Hardware

Dell Precision Workstation 530:	Intel Xeon Processor, 2.20 GHz, 512 KB Cache
Memory:	512 MB PC800 ECC RDRAM (2 RIMMS)
Hard Drive:	120 GB 7200 RPM IDE Hard Drive - DataBurst Cache
Floppy Drive:	3.5" 1.44 MB Floppy Drive
Internal Zip Drive:	IOMEGA Zip 250 MB
CD-ROM, DVD, and R-W Drives:	48X/24X/48X IDE CD Read-Write
Modem:	V.90 PCI Data/Fax Controllerless Modem
Graphics Card:	ATI, Radeon VE, 32 MB, VGA (dual monitor capable)
Sound Card:	Creative Labs Sound Blaster Live! Value
Speakers:	harman/kardon 695 Speakers
Keyboard:	Entry Level Quietkey Keyboard, PS/2 (no hot keys)
Monitor:	17-inch Dell (16.0 inch vis) M782 Flat CRT Monitor M782
Mouse:	Dell, PS/2 (2-button, no scroll)
Printers:	Brother HL-1470N Laser Printer
Power Protection:	Belkin Components Surgemaster Gold 9 outlet
Hardware Support Services:	3Yr Parts + Onsite Labor (next business day)

Software

Operating System:	Microsoft Windows XP Professional
Productivity Software:	Microsoft Office XP Professional
Productivity Software:	ADOBE ACROBAT 5.0
Security Software:	Symantec Antivirus CorpEd 8.0

Specific product choices have required careful consideration of vendor Web sites, several industry magazines, direct calls to software manufacturers, and visits to software retailers and industry conventions. Based on this information and the highly competitive pricing policies in the market, the analysts and user have agreed that preparing a formal request for bids or proposals is not necessary.

FIGURE 11-16b / *Cornucopia Resource Requirement Specifications*

Bundled price for the above	$ 3,800
Vendor: Dell Corporation	
Other Sourced Hardware/Software	
APC Back-UPS BF 500	$ 100
Wasp Nest Professional Bar Code Reader	$ 350
Networking	
None	$ 0
Data	
None	$ 0
People	
Training (bundled with labor cost)	$ 0
Procedures	
Documentation (bundled with labor cost)	$ 0
Miscellaneous	
Paper products (paper, labels)	$ 100
Floppy disks	$ 25
LaserJet cartridge (spare)	$ 75
Zip cartridges (spare)	$ 50
Cleaning materials	$ 50
Reference materials	$ 75

Summary	
Bundled Hardware/Software	$ 3,800
Other Sourced Hardware/Software	$ 450
Data	$ 0
People	$ 0
Procedures	$ 0
Miscellaneous	$ 375
Subtotal	$ 4,625
Tax and Shipping	$ 275
Grand Total	**$ 4,900**

Figure 11-17 compares the actual costs to the budgeted costs. The budget surplus is just a little more than 10 percent.

FIGURE 11-17 / Cornucopia System Resource Budget Surplus

Budget Category	Actual Cost	Budgeted Cost
Hardware and Software (Bundled)	$3,800	$4,000
Hardware and Software (Other)	$450	$1,500
Data	0	0
People	0	0
Procedures	0	0
Networking	0	0
Miscellaneous	$375	$400
Tax and Shipping	$275	0
Total	$4,900	$5,500

Cost Projections

The cost projections for the delivery of the information system are identical to those on the original project budget. Even though the latest update shows that, at this time, actual expenditures are less than estimated, this is a result of the timing of hardware and software purchases and some minor inaccuracies in labor estimates. Therefore, the original budget remains unchanged.

The cost projections for the initial system implementation and continued maintenance are based on the original contract. The 12 hours of user training were projected at $25 per hour and spread over several weeks just prior to, and after, the system becomes operational. The maintenance costs are estimated at 1 percent per month of the $10,000 system price.

These projections are summarized in the cost/benefit projections (Figure 11-18). Notice that the intangible costs are not included. Certainly, the risk is very real that the system will not deliver the services it promises, or that the user will not be able to use the Sales Transaction subsystem without degrading customer service. Nevertheless, because quantifying these potential costs is virtually impossible, they are simply noted and ignored in the official cost projections.

Benefit Projections

The benefit projections are associated with the project contract introduced in Chapter 2 (Figure 2-10). The estimated benefits and rationale are as follows:

Decreased Reorder Costs: Contract Objective 3
    The new reordering system will allow the owner to shift this responsibility to a $6-per-hour clerk, thus freeing the owner's time, which she values

at $20 per hour. Assuming that reordering time will be reduced to 10 to 15 hours per month, a saving of $200 per month is conservative.

Increased Repeat Customers: Contract Objective 4

The customer correspondence system will increase business, but it will take six months for this to show results. A conservative estimate that this will result in 10 more sales per month, at an average profit margin of $20 per sale, yields a $200 monthly benefit.

Decreased Out of Stock: Contract Objective 5

The CD reordering system will provide a more accurate record of which CDs have been sold. The owner will be able to scan this report quickly and note the items she wants the clerk to reorder. A conservative estimate is that this will eliminate five lost sales per month. These benefits will begin to accrue three months after the new system is in place, which should be sufficient time for the most active inventory items to turn over. At the $20 profit margin per sale, this produces $100 per month in benefits.

Cost/Benefit Analysis

Figures 11-18 and 11-19 present the cost/benefit projections in a spreadsheet and graphic format. The graph shows the payback point to be somewhere in the 32nd month of the system. Given that the analysts estimate the system life to be three to five years, they can reasonably assume that considerable benefits will accumulate during that period. Such benefits might be directed at further system enhancements, thus further increasing the benefits and prolonging the useful life of the system.

FIGURE 11-18 *Cornucopia Cost/Benifit Projections*

CostBenefit.xls

	A	B	C	D	E	F	G	H	I	J	K	L	M
1	**Date:**		**Cornucopia Cost/Benefit Projections**										
2													
3		**1**	**2**	**3**	**4**	**5**	**6**	**7**	**8**	**9**	**10**	**11**	**12**
4	**Costs**												
5	**Initial System**	1100	6100	1600	1000								
6	**Training**				100	200							
7	**Maintenance**					100	100	100	100	100	100	100	100
8													
9	**Benefits**												
10	**Decrease reorder costs**					200	200	200	200	200	200	200	200
11	**Increase repeat customers**										200	200	200
12	**Decrease out of stock**							100	100	100	100	100	100
13													
14													
15	**Cumulative costs (Y1)**	1100	7200	8800	9900	10200	10300	10400	10500	10600	10700	10800	10900
16	**Cumulative benefits (Y1)**	0	0	0	0	200	400	700	1000	1300	1800	2300	2800
17													
18	**Cumulative costs (Y2)**	11000	11100	11200	11300	11400	11500	11600	11700	11800	11900	12000	12100
19	**Cumulative benefits (Y2)**	3300	3800	4300	4800	5300	5800	6300	6800	7300	7800	8300	8800
20													
21	**Cumulative costs (Y3)**	12200	12300	12400	12500	12600	12700	12800	12900	**13000**	13100	13200	13300
22	**Cumulative benefits (Y3)**	9300	9800	10300	10800	11300	11800	12300	12800	**13300**	13800	14300	14800
23													

Projections / Graph / Worksheet

FIGURE 11-19 / *Cornucopia Cost/Benefit Graph*

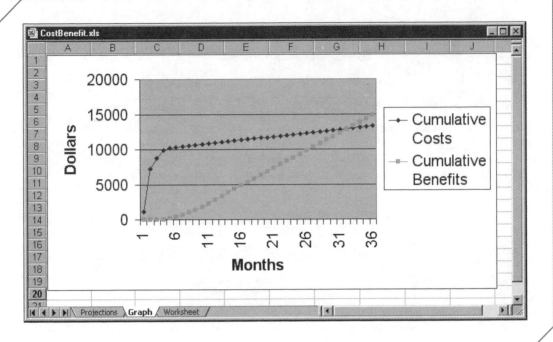

Design Review Session

During the design review session the analysts present several of the system design models, the most discussed of which are the menu tree and the user's system diagram (USD). The analysts demonstrate the user interfaces on a large-screen projection system. Figure 11-20 illustrates the slide used to present the switchboard interface.

At the conclusion of the session, the analysts present a rough schedule of activities for the development and implementation of the system. The user understands that the development phase continues the prototyping methodology used with some of the user interfaces. She also agrees to continue to participate in the project as time permits.

FIGURE 11-20 / *Cornucopia Design Review Session Slide Show*

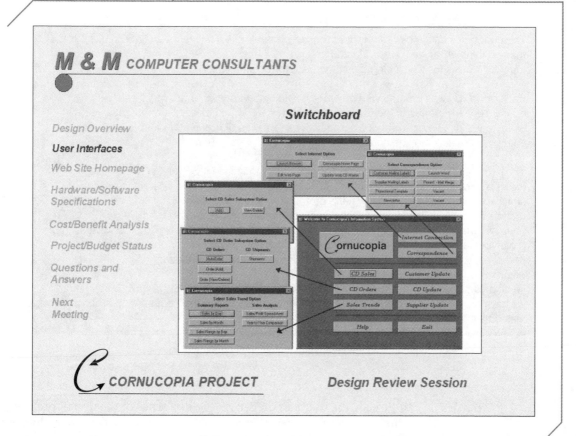

System Models: Highlighted for Prototyping

The analysts prepared a narrative to familiarize the user with prototyping and to encourage her active participation in the evaluation and revision of the prototypes (Figure 11-21).

FIGURE 11-21   Cornucopia User's Introduction to Prototyping

The prototypes developed for this information system project are models, or facsimiles, of the real thing. After we have evaluated and revised these models, the analyst team will use them as a basis for developing the complete information system.

Each prototype was created with the understanding that revisions are more easily made during the early stages of the design-development work. Therefore, we encourage you to imagine working with the various screens and interfaces of this system and to suggest any changes you think may be necessary.

These models are designed to provide you with a realistic view of many of the important inputs and outputs of the new system. In some cases, you may even be able to manipulate small amounts of sample data with the prototype product. However, you should realize that models do not always behave as their real-world counterparts will. Under extreme or unusual data and processing conditions, the prototype may break down. In other words, prototypes are not substitutes for fully developed products. Even after the prototypes are perfected, a great deal of work will remain before the product is ready for use.

To help you understand exactly which parts of the information system we have prototyped, we have modified the system diagram by shading the affected input, output, and processing elements of the system.

To provide the user and the analysts with a sense of the overall prototype plan, the system menu tree (Figure 11-22) and the user's system diagram (Figure 11-23) were adapted by shading the affected elements. However, why were only certain portions of the system chosen for prototyping? Remembering that prototyping costs money, the analysts chose those elements that would most rapidly advance the design and allow the user to react to a computer-based product.

FIGURE 11-22 / *Cornucopia Prototype Menu Tree*

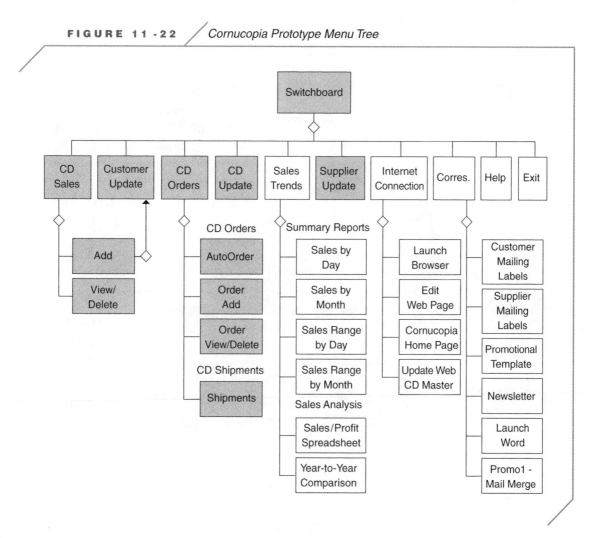

FIGURE 11-23 / *Cornucopia Prototype USD*

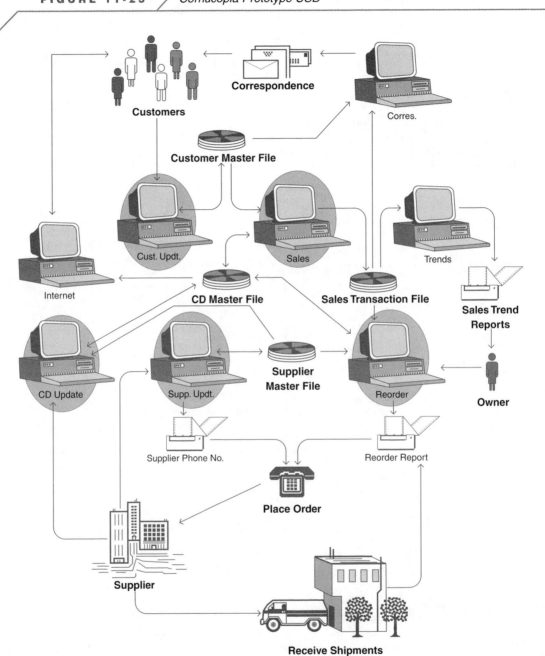

Menu Prototypes

The menu prototype can be constructed with a word processor, a paint program, or a relational database program's screen painter and code builder functions. The switchboard developed during the earlier design activities introduced in Chapter 10 (Figure 10-16) was constructed with a paint program and screen shots from several database forms. You will recall that one purpose of these early renditions of the menu tree is to help prepare the user for the JAD working sessions. During the prototyping activities, the menu tree serves a different purpose. Namely, it must help the user visualize how the system will work on a computer. Thus, the analysts must create a model that is easy to access and compatible with the way the rest of the prototypes work. Paint program output is too unlike the database implementation the analyst envisions for the system.

Figure 11-24 illustrates the design features of the CD Sales button on the switchboard prototype. This form was created with Microsoft Access. Beginning with a blank form, unassociated with any particular table, the analysts placed several objects on the form (one image, eight buttons, and two lines), changed several appearance properties (object name, button captions, fonts, picture reference, and color), and changed the button click event property to specify an action. The complete set of properties for the command button named "cmdCDSales" is displayed.

Of particular interest is the On Click event, which calls for a single statement macro named Open Sales Options to execute. This macro opens the Sales Form Options form, which is illustrated in Figure 11-25. The Sales Form Options form permits the user to open the CD Sales Form in either Add mode or View/Delete mode. The macro associated with the click event of the button captioned Add is included in the figure. This is a relatively complicated macro, especially when compared to the simple macro used to open the Sales Option form. This macro is discussed in the next chapter.

FIGURE 11-24 / *Cornucopia Switchboard Prototyping*

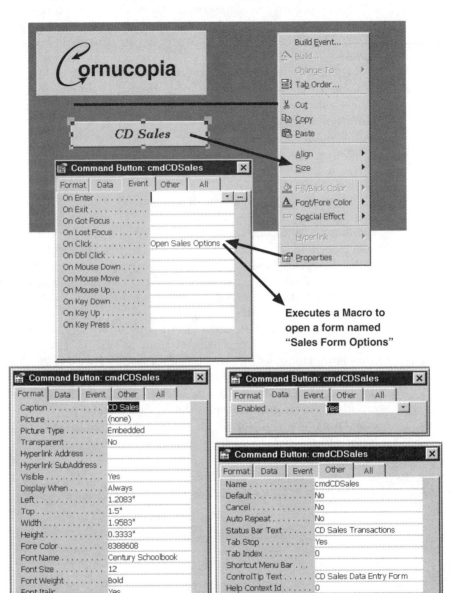

FIGURE 11-25   *Cornucopia Form Prototyping*

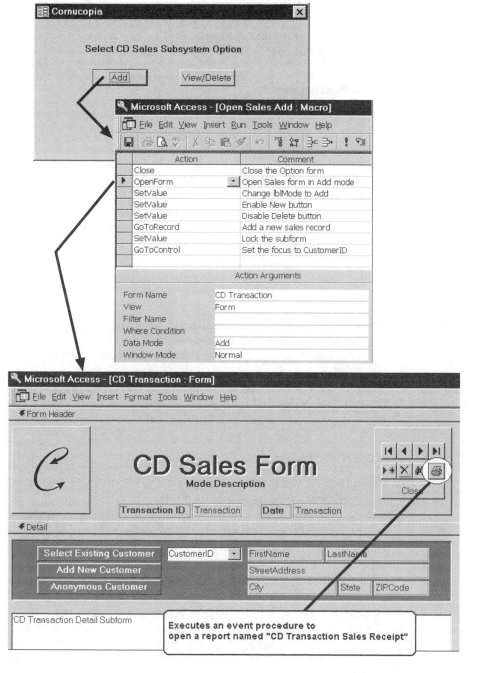

To review, the switchboard contains button objects. Each button has several associated events to which it is sensitive. Each event can signal an action. The action can be simple or complex. The action can be defined in one of three ways: directly in the event window, as code in an event procedure, or in a macro. Figures 11-24 and 11-25 illustrate how the click event is used to open up a series of forms using macros. A detailed discussion of event windows, event procedures, and macros appears in Chapter 12.

Form Prototypes

After developing the menu prototypes, the analysts chose the next easiest system elements to develop—the master file maintenance forms. The Customer Update and Supplier Update forms are very similar to the CD Update form introduced in Chapter 8 (Figure 8-16). Before these forms were created, the analysts developed a master template form to incorporate the form header design, which is used consistently in all of Cornucopia's data forms. The header contains the enterprise logo, form tiles, transaction record key information, and a custom toolbar. Form detail appears below the header.

Master file update forms were very easy to prototype. The analysts simply opened the template, attached a master table by changing the form's Record Source property, pasted the fields onto the form's detail section, and used the Save As option to save the new form.

The Sales Transaction subsystem posed a more formidable challenge to the analyst. However, because this subsystem is at the heart of the project design, including it in the prototype seemed appropriate. After creating a sample of the Sales transaction file, the analysts used the forms designer and a reiteration of the preceding process to create the screen form previously illustrated in Figure 8-17, diagrammed in Figure 10-18, and recapped in design view in Figure 11-25. A detailed account of the event procedures and macros associated with this form is presented in the next chapter.

Report Prototypes

Figure 11-26 shows the design view of the CD Transaction Sales Receipt report. This report was first created with the Report Wizard, and then modified to include subtotals, sales tax, and receipt totals. There are seven report sections. As with forms, the analysts established a consistent layout to use for all reports. The header and footer designs were copied into each new report layout.

Time and Money

The analysts spent eight hours transforming the design concepts into sample files and prototypes, which is reported under the development category on the project status report. This brings the estimate of completion up to 20 percent, which is still well below what the analysts had expected. It also appears that the labor hours saved during the analysis phase have been more than consumed during the design phase, with the prospect of more overruns during the development phase. When asked why this is so, the analysts reported that although no problems occurred while creating the master file prototypes, the sales transaction processing prototype was more difficult because of the complex file relationships involved.

FIGURE 11-26 / *Cornucopia Report Prototyping*

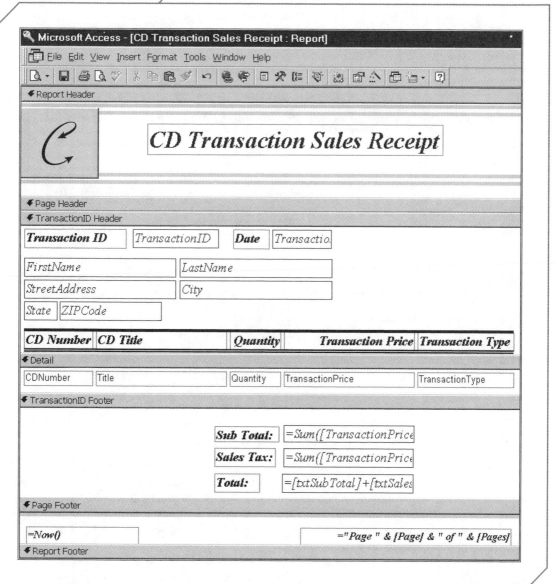

Figure 11-27 illustrates the project budget for Week 9. Notice that the expenditures for hardware and software seem to be complete and well within the estimates established for each category. On the other hand, the labor charges exceed the estimates.

The analysts are aware that the computer hardware and software performance-price relationship is volatile. Based on what is currently available, one of two things should happen to the project budgets and resource purchases. Either the budgets should be reduced or the products should be upgraded. In this particular situation, the labor cost overruns are offset by hardware and software cost savings.

FIGURE 11-27   / *Cornucopia Project Budget – Week 9*

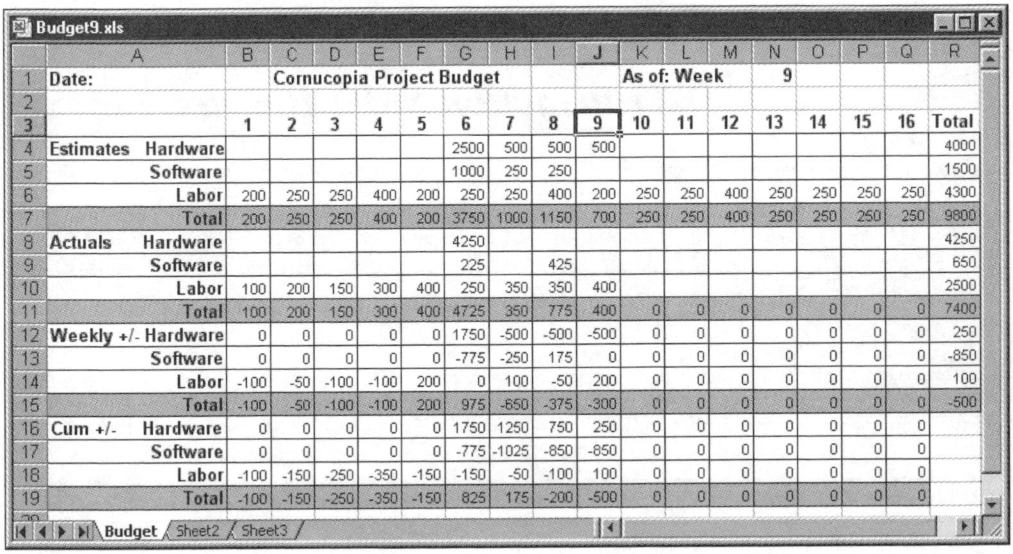

Budget9.xls

Date:		Cornucopia Project Budget								As of: Week	9							
	A	1	2	3	4	5	6	7	8	9	10	11	12	13	14	15	16	Total
Estimates	Hardware						2500	500	500	500								4000
	Software						1000	250	250									1500
	Labor	200	250	250	400	200	250	250	400	200	250	250	400	250	250	250	250	4300
	Total	200	250	250	400	200	3750	1000	1150	700	250	250	400	250	250	250	250	9800
Actuals	Hardware						4250											4250
	Software						225		425									650
	Labor	100	200	150	300	400	250	350	350	400								2500
	Total	100	200	150	300	400	4725	350	775	400	0	0	0	0	0	0	0	7400
Weekly +/-	Hardware	0	0	0	0	0	1750	-500	-500	-500	0	0	0	0	0	0	0	250
	Software	0	0	0	0	0	-775	-250	175	0	0	0	0	0	0	0	0	-850
	Labor	-100	-50	-100	-100	200	0	100	-50	200	0	0	0	0	0	0	0	100
	Total	-100	-50	-100	-100	200	975	-650	-375	-300	0	0	0	0	0	0	0	-500
Cum +/-	Hardware	0	0	0	0	0	1750	1250	750	250	0	0	0	0	0	0	0	
	Software	0	0	0	0	0	-775	-1025	-850	-850	0	0	0	0	0	0	0	
	Labor	-100	-150	-250	-350	-150	-150	-50	-100	100	0	0	0	0	0	0	0	
	Total	-100	-150	-250	-350	-150	825	175	-200	-500	0	0	0	0	0	0	0	

Budget / Sheet2 / Sheet3 /

Cornucopia with Visible Analyst

Although Microsoft Access is responsible for automatic code generation for the Cornucopia project, Visible Analyst can assist in creating SQL, C, and COBOL code, based on entries in the repository. Of course, the code is then submitted to traditional interpretation and compilation programs. Visible Analyst has an extensive online Help system. Numerous entries explain the code generation process in great detail.

The Corporate Edition of Visible Analyst is capable of generating SQL data definition language schema from one of the views or models contained in the repository. You can then export the code to a target relational database. Visible Analyst analyzes the repository and produces the schema, displaying the names of items it is examining on the status line. Errors are displayed on the screen, which you can save to a file, print, or ignore. The errors might be such things as entities without keys, entities with improperly specified keys, an invalid physical data type for a composing data element, or a number of other errors.

The Corporate Edition can also generate shell code in C or COBOL. The generated code includes global definitions, descriptive comments, function and perform statements, and passed parameters. The analyst can add code through the module description field of a module or macro.

SUMMARY

As the system design progresses, the analyst develops a more precise understanding of the actual system resource requirements, potential costs, and potential benefits. This information comes together in some form of cost/benefit analysis, or cost justification. Sometimes these figures do not add up, propelling the entire project into a reexamination. At this time, analysts and users still have an opportunity to significantly change the system objectives or design or even to abandon the project without incurring too much expense. The design review session is the place where such dialog may begin. This opportunity diminishes as the project moves farther into the SDLC, with the costly resource acquisition and prototype development activities.

Because prototyping spans both design and development, it is difficult to separate screen-building and code-generating 4GL tools from traditional computer programming. Although the distinction may not be apparent, the differences are significant: the former is associated with a mind-set that is free-form, experimental, and collaborative; the latter is methodical, pre-planned, and very much an individual activity. At this point in the project the nonlinear, back-and-forth process begins to flatten out. The project design becomes much more fixed, the user temporarily fades into the background, and the analyst-as-group-facilitator is transformed into the programmer-in-isolation, assuming a somewhat expanded identity as an analyst-programmer.

TEST YOURSELF

Chapter Learning Objectives

1. What differing challenges face the analyst in developing resource requirements for each of the six components of a computer information system?

2. In what ways does a well-prepared project contract make it easier to prepare cost/benefit projections?

3. Why are the USD and menu tree preferred over the DFD and ERD in a design review session?

4. What makes some information system design elements good candidates for prototyping?

5. How do 4GL products make prototyping easier?

Chapter Content

1. System resource requirements are developed during the design phase of the project. True or False?

2. System resource requirements only concern the hardware and software components of an information system. True or False?

3. The request for bids is unnecessary in a small-enterprise systems project. True or False?

4. In terms of brand name specifications, the request for proposals is much less specific than the request for bids. True or False?

5. A prototype is of no use beyond its ability to demonstrate how an information system product looks. True or False?

6. Rapid application development is only applicable to small projects. True or False?

7. Input/output products are good candidates for prototyping. True or False?

8. Transaction processing is difficult to prototype. True or False?

9. Small-enterprise project prototypes can be reusable or throwaway, and they can address input/output and processing elements. True or False?

10. It is best to begin the prototyping process with reusable, input/output prototypes because you cannot create the other elements without first creating reusable, input/output prototypes. True or False?

ACTIVITIES

Activity 1

Review the context diagram, data flow diagram, system flowchart, and output prototypes for the Political Research Corporation (see Figures 11-10 through 11-13).

    a. Correlate the images on the data flow diagram and system flowchart to identify the following matchups:

data flow diagram:	system flowchart:
processes	4GL software
data stores	file types
data flows	input/output documents

    b. Identify the master files and transaction files on the system flowchart.

    c. Describe the tools you would use to create the prototype products identified on the system flowchart and illustrated in the output prototypes.

Activity 2

Prepare a cost comparison, including three potential sources, for the purchase of a high-end desktop computer platform. Be specific about the platform configuration you are researching.

Activity 3

Prepare a cumulative cost/benefit chart that justifies your investment in your education.

DISCUSSION QUESTIONS

1. What advantage is there to preparing requests for bids or proposals for hardware and software, as opposed to simply using vendor Web sites to purchase system resources?

2. To what degree do the project contract's goals and objectives affect the effort to develop benefit projections?

3. Why is it useful to itemize intangible costs and benefits, even if they cannot be quantified in terms of dollar amounts?

4. How does the shrinking life span of computer information systems affect the cost/benefit analysis?

5. Under what circumstances might the user choose to abandon a project after participating in the design review session?

6. Describe a situation in which you have used a prototype to test the design of a product.

7. Speculate on why the throwaway prototype is more likely to be used in large system projects.

8. Explain how prototyping can save money by reducing the number of design changes that are introduced late in the project.

9. Explain how prototyping can encourage the analyst to oversell the performance of the information system.

10. Compare and contrast prototyping and computer programming.

PortfolioProject

Team Assignment 11: Design Review and Prototyping

This is a lengthy assignment, likely to challenge your team resources to the fullest. In previous team assignments, you developed many design review materials. The first part of this assignment guides you through the preparation of the remaining materials needed for your design review session. As with the preliminary presentation, you are encouraged to review Appendix C.

The second part of the assignment directs you to begin developing your prototype. You are required to modify the menu tree and USD to indicate which elements of your system are to be prototyped. Finally, you are required to prototype the GUID to maintain one of the system master files. For this part of the exercise you can rely on the switchboard and detailed form designs you have already developed to begin the prototyping process.

The design review session does not include these prototypes. When completed, they become the focus of the next project deliverable, the prototype review session, which follows in the next chapter.

In order to develop the materials for the design review session and begin the prototyping process you must complete the following tasks:

1. Develop resource requirement specifications for all six components of your system. Include as much detail as possible about products, vendors, prices, shipping costs, and taxes. Submit a copy of the specifications to your instructor.
2. Develop cost/benefit projections and a cost/benefit graph using the template file CostBenefit.xls on your data disk. Include a narrative supporting your benefit projections. Submit a copy of the projections, graph, and narrative to your instructor.
3. Develop a prototype menu tree and a prototype USD. Submit a copy of these two images to your instructor.
4. Develop a prototype of the form sequence required to maintain one of your system's master files. Prepare this GUID as a series of images in a slide show. Submit a copy of your slide show handout (three slides per page) to your instructor.
5. File a copy of your resource requirement specifications, cost/benefit materials, and prototypes in the Portfolio Binder behind the tab labeled "Assignment 11."

(continued)

Project Deliverable: Design Review Session

1. Submit a design review report containing the following items:
   a. Cover letter
   b. Narrative overview of the new system design
   c. Models (USD, menu tree)
   d. Screen designs (switchboard, forms, reports)
   e. Web site homepage design
   f. Resource requirement specifications
   g. Cost/benefit analysis (projections, graph, and supportive narrative)
   h. Project status report
   i. Project budget
   j. Oral presentation slide show handout (three slides per page)
   k. Appendix
      1. Revised DFD and ERD models
      2. Detailed system flowchart
2. Prepare a slide show and appropriate handouts to support a 20 to 30 minute oral presentation covering items 1b through 1i.
3. File a copy of your report in the Portfolio Binder behind the tab labeled "Design Review Session."

4GLprogramming

When you complete this chapter, you will be able to:

OVERVIEW

Eventually, almost every computer information system project requires some programming expertise. The nature of that expertise continues to evolve. In this chapter, you will learn how the various programming generations influence your work as a small-enterprise analyst. Although many of the necessary programming fundamentals are reviewed in Chapter 10, this chapter presents specific examples of programming code, with appropriate references to the control structures (loops, branches), modularization, and charting techniques presented earlier in the text. You will also learn the fundamentals of object-oriented programming, with an emphasis on the use of program libraries and reusable code fragments. You will certainly find yourself knee-deep in product help screens, computer magazine programming tips, and trade book sample program illustrations before you emerge with your working information system. To date, no product provides a clear detour around this meticulous and time-consuming task.

- Transform a prototype into an operational product.

- Use 4GL coding tools in program development.

- Develop graphical user interface dialogs (GUIDs).

- Blend 4GL-generated code with analyst-programmer code segments.

- Develop effective file sharing between applications.

THE EVOLUTION OF PROGRAMMING

Figure 12-1 illustrates the different programming language generations, beginning with machine languages in the 1950s and ending with the natural languages, a branch of computer science that, while actively and intensely researched, remains elusive. The VCR automatic timer recording operations discussed in Chapter 10 are very much like the *machine language* instructions of 50 years ago. The user enters instructions one at a time by pressing button sequences, and the program works only for a specific set of parameters and on a specific machine—no wonder the newer models offer a higher-level programming option that is keyed to preprogrammed television log codes.

FIGURE 12-1   / *Programming Language Time Line*

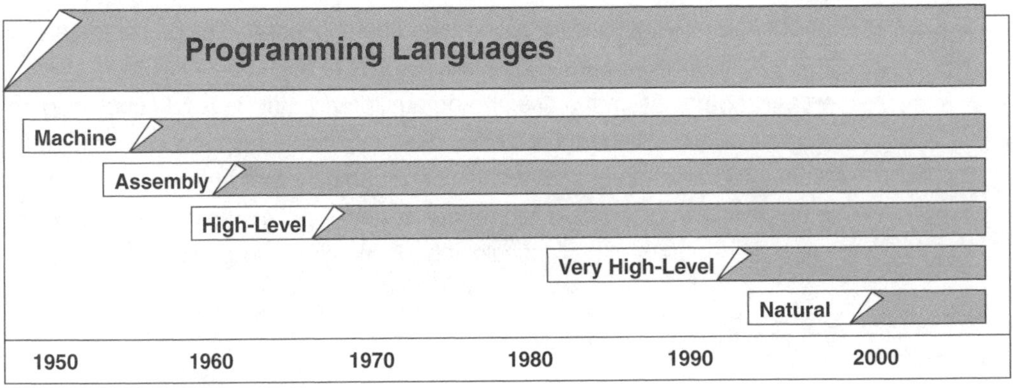

Desktop computer programming has evolved from the Intel 8080 microprocessor's *assembly language*, which is a low-level language tailored to fit a specific CPU architecture and characterized by its symbolic form. The evolution continued through early *high-level programming languages*, such as BASIC, Pascal, and C to the current fourth-generation language products that have become popular. The ever growing user population has naturally selected those tools that are easiest to use and offer the most productivity advantage. The billion-dollar software business, in response to such overwhelming market preference, continues to develop programming interfaces that resemble human language more and more. Visual Basic is an example of the trend.

Remember, however, that the older language generations still have a place in today's application development environment. They continue to persist for two reasons. First, a large base of existing applications are still viable and, for purely economic reasons, cannot be abandoned. Second, and perhaps more important, these older, more formal languages retain significant operational efficiencies because they require less system overhead. Of course, hardware advances tend to mitigate the overhead costs that go with the fourth-generation language development tools. Therefore, for our purposes, the very high-level languages offer the best tools for developing the design of our small-enterprise information system into a real product.

Procedural and Nonprocedural Programming

Early high-level programming languages were characterized by their procedural nature. *Procedural programming* requires the programmer to specify how to achieve a particular goal. The programmer must spell out each step in detail. In contrast, early very high-level languages were characterized by their nonprocedural nature. *Nonprocedural programming* requires the programmer simply to specify the goal, without concern for the specifics. For example, in the early 1990s a database programmer could supply one line of nonprocedural code ("List Sales for ProductId = 123"), whereas the COBOL programmer would need more than 25 lines of procedural code to achieve the same result.

The very high-level, *fourth-generation language (4GL) programming* discussed in this book embodies elements of both procedural and nonprocedural programming. Most of the general programming fundamentals covered in Chapter 10 apply to nonprocedural as well as procedural programming. The "List Sales for ProductId = 123" command works only within the context of a carefully planned operational environment. The database file must be active and properly indexed for efficient access. The output format must be tailored to the user's specifications. In addition, the command itself must execute at the user's direction. Thus, program structure charts and flowcharts are still useful tools for 4GL program design and development.

The newest very high-level languages, while continuing to include procedural and nonprocedural programming elements, are characterized by their objectlike nature. An objectlike language, as described more fully in the next section, incorporates most of the features of an object-oriented language. Many 4GL applications, such as Microsoft Access, include objectlike programming languages that make it possible to develop elaborate, customized solutions to fit a client's needs. To take advantage of 4GL applications, the analyst must develop a basic understanding of object-oriented programming and a well-developed appreciation for the objectlike nature of 4GL programming.

Object-Oriented Programming and 4GL Applications

The discussion on object modeling (see Chapter 5) introduced objects as an instance of a class. Recall that a class embodies data attributes and performs functions or methods and that classes communicate with one another via messages. *Object-oriented programming* languages are used to implement this model, and they require the programmer to apply a relatively new approach to data and the instructions that transform data into information. First, an object has two parts: *attributes* (e.g., field names, types, sizes) and *methods* (programming code segments). Second, objects respond to messages. Thus, the object

Sales might respond to the message "Display Amount for Product ID 123" with the same information generated by a traditional high-level program or a single 4GL command. From the user's perspective, the results are the same. From the programmer's perspective, however, the differences are significant.

Because objects are declared and classified as part of a huge library of objects, they can inherit attributes and methods from those classes to which they belong. In theory, the ever expanding library of objects means that the programmer can develop programs by selecting and customizing library objects. For example, if a program design calls for a window, the programmer can select a fully functional window object from the library to customize or use as is.

An *objectlike language* differs from an object-oriented language in an important way: the inheritance feature is somewhat diminished in objectlike products. For example, with an objectlike application, if an object's name is altered, the new object does not inherit the old object's code. While you can avoid this shortcoming through careful planning, it does provide a technical distinction between the two languages.

Perhaps more relevant to the small-enterprise project is the advent of **object-oriented databases (OODBMS)** and the addition of selected object-oriented features to popular relational database products. For example, Microsoft Access defines all the data files, queries, reports, and screen forms as objects, each with a long list of object attributes and access to many preexisting methods. Programmers develop applications by creating objects, modifying their attributes and behaviors, and then linking them together. Even though Access does not provide the full advantage of an OODBMS because important object-oriented features, such as inheritance, are missing, its objectlike characteristics greatly enhance the development process.

To illustrate, we return to the Political Research Corporation example. Figures 12-2 through 12-7 present the many facets of the switchboard form. The analyst used the Switchboard Manager tool in Access to create the PRC Switchboard (Figure 12-2) without writing a single line of programming code. The switchboard is a form object and the menu options displayed on the form are command-button objects. Each object comes from an object library with attributes and behaviors appropriate to its functionality. Object attributes and behaviors are then modified as needed. For example, the command button's Caption attribute is modified to describe a subsystem and its On Click behavior is modified to open a form or report. To summarize the object-oriented nature of the switchboard, form and button objects have two parts (attributes and behaviors), and each object responds to messages—the form reacts to the open message, buttons react to the click message. For future reference, in Access, an object attribute is called a property and a mouse click is called an event.

Figure 12-3 presents the switchboard form in design mode, along with the form property window and a command button property window. The Switchboard Manager automatically generated event procedures for the form's On Open and On Current events and each command button's On Click event.

Figure 12-4 shows part of that code. This window provides a relatively friendly interface to the code that customizes object behaviors and changes object property values. By selecting an object and event in the windows at the top of the work area, programmers can view and edit Visual Basic instructions as needed.

FIGURE 12-2 / PRC Switchboard

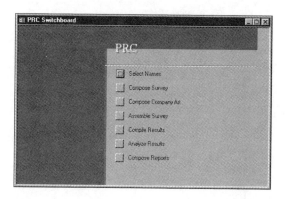

FIGURE 12-3 / PRC Switchboard Event Procedures

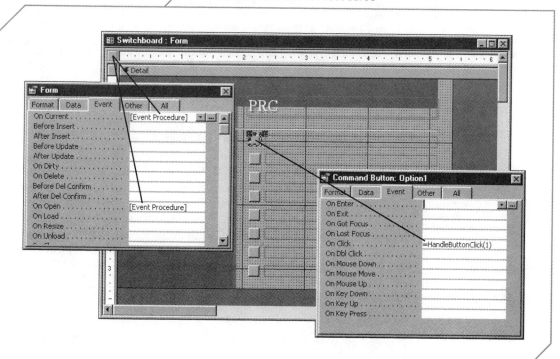

FIGURE 12·4 / *PRC Switchboard Visual Basic Code Window*

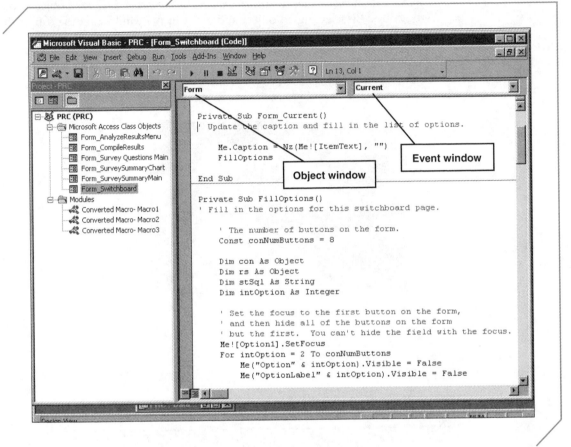

The switchboard form object behavior is supplemented by three subprocedures: Form_Open, Form_Current, and FillOptions. The first two subprocedures execute when the form opens. Form_Open minimizes the database window and initializes the form. Form_Current updates the form caption and then calls another subprocedure (FillOptions) to fill in the list of options. Figure 12-5 presents the form's subprocedures in a structure chart. As described in Chapter 10, structure charts show how subprocedures work together. This chart shows a fourth subprocedure (HandleButtonClick) that contains code to give the user access to subsystems described by the command-button captions.

FIGURE 12-5   /   *PRC Switchboard Structure Chart*

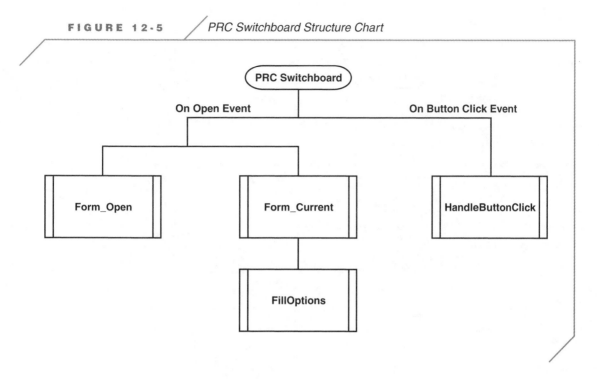

Figure 12-6 illustrates a two-dimensional diagram of the programming logic of FillOptions. This flowchart confirms the presence of procedural programming elements in the midst of a very high-level programming language subprocedure. The shapes labeled "Loop for each button" and "While there are more options" are examples of repetitive logic. The shape labeled "No options?" is an example of conditional logic. The Visual Basic code appears in Figure 12-7, where the numbered boxes associate specific instructions with the numbered symbols on the flowchart.

FIGURE 12-6   PRC Switchboard Subprocedure Flowchart

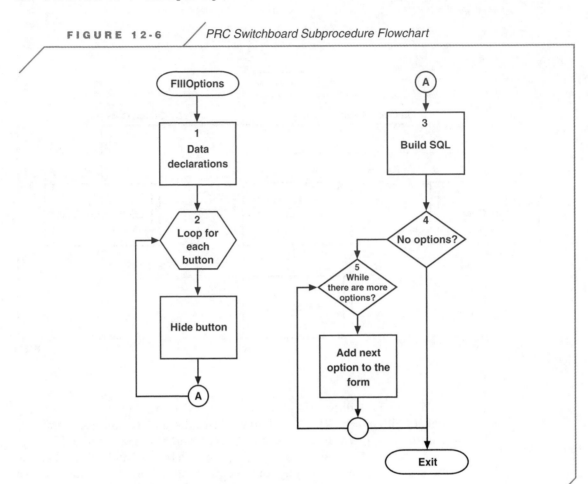

FIGURE 12-7 *PRC Switchboard Subprocedure Code*

```
Private Sub FillOptions()
' Fill in the options for this switchboard page.

    ' The number of buttons on the form.
    Const conNumButtons = 8                                              [1]
    Dim con As Object
    Dim rs As Object
    Dim stSql As String
    Dim intOption As Integer

    ' Set the focus to the first button on the form,
    ' and then hide all of the buttons on the form
    ' but the first. You can't hide the field with the focus.
    Me![Option1].SetFocus                                                [2]
    For intOption = 2 To conNumButtons
        Me("Option" & intOption).Visible = False
        Me("OptionLabel" & intOption).Visible = False
    Next intOption

    ' Open the table of Switchboard Items, and find
    ' the first item for this Switchboard Page.
    Set con = Application.CurrentProject.Connection                      [3]
    stSql = "SELECT * FROM [Switchboard Items]"
    stSql = stSql & " WHERE [ItemNumber] > 0 AND [SwitchboardID]=" & Me![SwitchboardID]
    stSql = stSql & " ORDER BY [ItemNumber];"
    Set rs = CreateObject("ADODB.Recordset")
    rs.Open stSql, con, 1   ' 1 = adOpenKeyset

    ' If there are no options for this Switchboard Page,
    ' display a message. Otherwise, fill the page with the items.
    If (rs.EOF) Then                                                     [4]
        Me![OptionLabel1].Caption = "There are no items for this switchboard page"
    Else
        While (Not (rs.EOF))                                 [5]
            Me("Option" & rs![ItemNumber]).Visible = True
            Me("OptionLabel" & rs![ItemNumber]).Visible = True
            Me("OptionLabel" & rs![ItemNumber]).Caption = rs![ItemText]
            rs.MoveNext
        Wend
    End If

    ' Close the recordset and the database.
    rs.Close
    Set rs = Nothing
    Set con = Nothing

End Sub
```

Figure 12-8 presents the table relationships for a portion of the information system. This is the Access equivalent of the entity-relationship diagram. Access classifies several tables as master files: Client, Survey, Question, and SampleVoter. Access also classifies several tables as transaction files: VoterResponse, SurveyVoter, and SurveyQuestions. The structure of this collection of tables is the result of numerous analysis and design activities. It should provide the data needed to generate many of PRC's information requirements. For example, the analyst can combine data from several of these tables to implement the prototype of the Analysis Softcopy View (from Figure 11-13).

FIGURE 12-8 / *PRC Table Relationships*

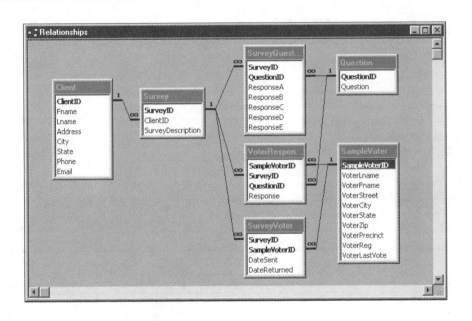

The Survey Summary Form (Figure 12-9) is an example of how a prototype can lead to new information products during the development phase of the project. Having discussed the limited usefulness of the detailed survey results displayed by the Analysis Softcopy View, the analyst and user agreed that a summary form might be better. This new form is composed of objects: some from the original design and some newly constructed during development. The record selectors at the bottom of the form indicate that there are 16 surveys in the database, the first of which is the Property Tax Survey conducted for the Legislative Analyst. The record selectors at the bottom of the lightly shaded subform area tell the user that this particular survey includes 15 questions. The form's underlying query object computes the summary results of the first question. The ability to easily combine and manipulate existing objects into new objects is one of the great advantages of object-oriented languages and objectlike applications.

FIGURE 12-9   / *PRC Survey Summary Form*

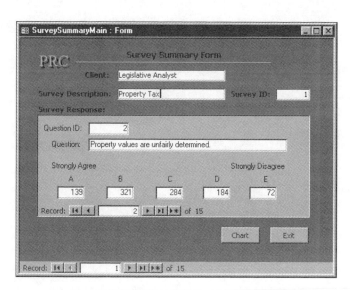

All of the switchboard form objects are viewable through an object browser window (Figure 12-10). In the right panel of the window, the object named cmdChart is selected. The middle panel identifies this object as a member of the Form_SurveySummaryMain class. The bold face type indicates that this is a new class created by the Access user. The left panel shows that this form is part of the PRC project. In essence, the analyst has added a new class to the library.

This series of images (Figures 12-2 through 12-10) brings up several important points. First, there are many types of objects: file-related fields, calculated fields, buttons, and descriptive text. Even the form itself is an object. Second, each object is defined by its attributes and methods: the form has a size attribute and an open method, the buttons have caption attributes and OnClick methods, the descriptive text has font attributes, and so on. Third, some of the methods came from a code library, but others are created to perform a form-specific function. Fourth, some of these objects are dependent on the occurrence of an event. When the user clicks on the Chart command button, a new form appears with the summary displayed in graphic format. These are examples of how *event-driven applications* work. The program waits for the user or some other external object to initiate an event sequence.

For the most part, object-oriented database programming is visual, intuitive, and fast—but not simple. An object-oriented software package alone cannot transform the novice into an effective application developer. Many subtleties surround the use of this powerful tool, and when it comes to hand coding a method, not so subtle programming skills are essential.

FIGURE 12-10   / *PRC Survey Summary Objects*

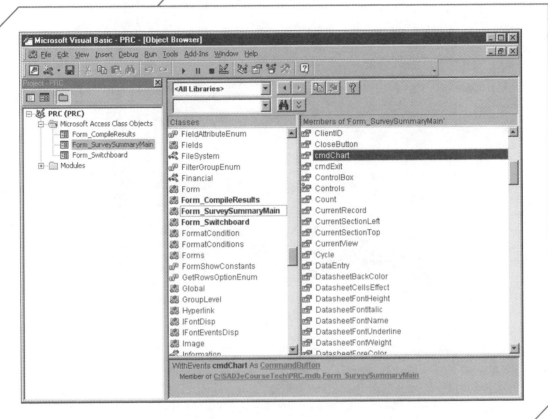

To summarize, a few years ago, 4GL programming described a combination of procedural and nonprocedural programming techniques. Today, 4GL programming incorporates many object-oriented, event-driven features that characterize products such as Microsoft Office.

Objectlike Application Builders

An *application builder* is a software program that automates much of the design and development processes discussed in this textbook. Its purpose is to assist the analyst-programmer in bringing new applications online in a short amount of time. For this reason, application builders are often called *rapid application development (RAD)* tools. The latest versions of application builders include visual tool sets and design interfaces to simplify the design process, along with

TECHNOTE 12-1

HTML AND JAVA PROGRAMMING

The advent of the Internet and World Wide Web inspired the development of new programming languages to develop, deliver, and interpret data in the form of text, graphics, audio, video, animation, and interactivity. **Hypertext Markup Language (HTML)** defines the basic layout of a Web page by using **tags**, or codes, attached to the text. The final appearance of the page depends on the specific characteristics of the client device and browser. HTML documents are simple in format so that they can be delivered quickly across the Internet. The following example illustrates a simple HTML document:

```
<HTML>
<HEAD>
<TITLE>A Simple HTML Example</TITLE>
</HEAD>
<BODY>Hello Virginia.</BODY>
</HTML>
```

When delivered to a browser, this document displays "Hello Virginia." The HTML tags, such as <BODY> and </BODY>, enclose different sections of the document. A quick reference guide for all existing HTML tags fits nicely on two sheets of paper. There are no conditional or looping constructs in HTML. In other words, you will not find IF-THEN-ELSE or DO WHILE tags in an HTML document. Developing pure HTML documents bears little resemblance to computer programming.

Java, on the other hand, is a robust, completely object-oriented programming language comparable in complexity and sophistication to C++. Whereas programmers use C++ to create the vertical, office management software installed on your dentist's computer, for instance, Java is used to develop software products that are deliverable to computers of all types, anywhere in the world, via the Internet. While it is impractical to illustrate the Java source code here, the following HTML document shows how easily you can deliver Java products to a global market.

```
<HTML>
<HEAD>
<TITLE>Java Class Call Example</TITLE>
</HEAD>
<BODY>
<APPLET  CODE="ATMApplet.class"
WIDTH=750  HEIGHT=400>
</APPLET>
</BODY>
</HTML>
```

Imagine that ATMApplet is the name of a sophisticated bank teller application written in Java. An applet is a Java program distributed by a Web page. When a Java-aware browser receives the Java applet, it simply loads and executes the program on the client station. Now imagine that this program is distributed via the Web, rather than sold as a shrink-wrapped application. Instead of buying and installing the entire package on your computer, you simply use parts of the package on an as-needed basis—and only pay for what you use. Someday you may be able to rent your word processor, spreadsheet, or tax preparation programs in a similar fashion.

object-oriented code generators to eliminate most of the design-to-development tasks. Analyst-programmers frequently use these RAD tools to construct medium- and large-scale information systems.

Considering the initial cost of the application builder and the time required to learn how to use the product, the analyst should research carefully these software packages before committing to using one. Just as with research on any other computer product, the Internet, magazine reviews, and references from existing users are good sources for such information. Often you can download a trial version of the software to get a firsthand look before you buy.

Borland Delphi is a popular RAD tool that features a complete set of tools for database programming and Internet application construction. Several case studies reported on Borland's Web site suggest that this is a viable product for large-scale projects. For example, it took a 10-person team just 12 months to create a Delphi-based front-end GUI to a multigigabyte, disease intervention and tracking database that up to a thousand users regularly access.

Microsoft .NET is an integrated Web application development system for professional programmers. It enables Web teams to design, build, debug, and deploy cross-platform Web applications. Visual Studio.NET features an integrated WYSIWYG (what you see is what you get) editor, enhanced database programming tools, and end-to-end debugging facilities for multitier applications.

PowerBuilder from Sybase is another object-oriented application builder. This client/server application development tool provides very broad operating system and database support, a development environment, and its own application server. Several case studies available on Sybase's Web site confirm the successes of this product.

These examples suggest that the systems analyst must stay current with developments in the software industry, especially where it concerns tools that make systems analysis, design, development, and deployment easier.

PROGRAM DEVELOPMENT: STEP-BY-STEP

In keeping with the iterative prototyping strategy presented in the preceding chapter, the next step in the development process is to transform the prototype into the final product. The difficulty of this task varies, depending on the extent to which you prototype the system, the 4GL products you use to develop the program code, and the computer operating environment you use to install the system. Certainly, strong evidence suggests that some combination of automatic code building and hand-coding tools make the job much easier than in the past. Figure 12-11 shows that automatic code building predominates during prototyping activities and hand coding is more evident during product development.

In general, prototyping is accomplished with code-building tools, such as an editor or screen painter, but complete development requires a programming tool as well. Those products that have both types of tools provide a smooth transition from prototype to product. Figure 12-12 transforms this approach into a series of steps.

FIGURE 12-11 / *Code Builders versus Hand Coding*

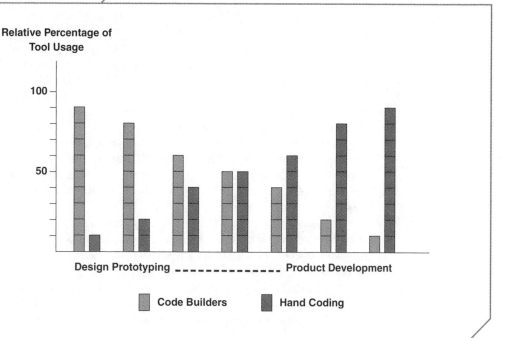

**Relative Percentage of
Tool Usage**

100

50

Design Prototyping _ _ _ _ _ _ _ _ _ _ _ _ _ _ **Product Development**

■ **Code Builders**   ■ **Hand Coding**

FIGURE 12-12 / *Prototype to Product: Step-by-Step*

1. Identify Tasks
   Refer to the Prototype System Flowchart and the Prototype USD (see Chapter 11) to associate the I/O and processing prototypes with the appropriate subsystem.

   Refer to the Detailed Structure Charts (see Chapter 10) to associate the prototypes with specific programs.

   List the master file maintenance programs that are required and the prototypes that must be transformed for their use.

   List the transaction file-processing programs that are required and the prototypes that must be transformed for their use.

   List the overall system control program, which provides user access to the subsystems identified in the Menu Tree and/or System Flowchart (see Chapters 6 and 10).

2. Inspect the Prototype Code
   Eliminate any code that specifically documents the prototype tool and/or duplicates code provided by other modules.

3. Use Code Generators
   Where possible, generate code for those design elements that were not prototyped.

4. Trim and Modify Generated Code
   Eliminate any duplicate code, as described in step 2.

5. Build Hand Code "around" the Generated Code
   Develop code to coordinate and link the various modules created above (e.g., code to implement the menu tree).

Developing Graphical User Interface Dialogs

The graphical user interface dialogs (GUIDs), introduced in Chapter 7 and prototyped in Chapter 11, require careful design and sequencing. Up to this point, the analyst has not needed to put these dialogs into a sequencing pattern associated with any specific controlling processes. However, now that the processes are under development, the analyst must also develop the dialogs.

To illustrate a dialog sequence, imagine using an ATM to withdraw money from your checking account. The opening dialog activates in response to a specific event: placing the access card in the machine. Then a dialog requesting your account access code appears, after which a menu of transaction options appears, and so on. This sequence of dialogs is also closely associated with processing. After you have completed one transaction, the original transaction menu dialog reappears. This replacement is controlled by the transaction processing program.

To help you sequence the information system dialogs and associate them with their controlling programs, we use a variation of an old tool, a *state-transition diagram (STD)*, which shows the dialog-dependent actions of an information system. As the name implies, dialogs cause the system to change (make a transition) from one process (state) to another. Figure 12-13 illustrates a small portion of a dialog for the Political Research Corporation. Descriptions of the events that cause the system to change from one form to another annotate this illustration.

To develop the series of dialogs, the analyst creates a template form based on the basic design, finalized after the Design Review Session, or copied from the form prototype. The analyst copies the template and then customizes it to suit the needs of each form in the system. This makes it easy to implement a consistent look and feel.

The analyst used a mixture of techniques to develop the dialog sequence in Figure 12-13. The two menu forms were created in four steps. First, the analyst used the Switchboard Manager tool to create the switchboard. Second, the analyst copied the switchboard, renamed the copy to Analyze Results, and changed the label and command-button objects to reflect the new menu options. Third, the analyst pasted a copy of the switchboard's OnClick code into the OnClick subprocedure for each button on the new form. Finally, the analyst used the Visual Basic code interface to change the code to open the various forms and reports associated with the new menu options. The analyst used a slightly different technique to develop the two Survey Summary forms. These forms are based on an Access query. Although Access includes a function to create forms automatically based on tables or queries, in this case, the analyst first copied a design template form, and then placed the appropriate query objects onto the form. The chart form was copied from Survey Summary and then customized.

FIGURE 12-13   *PRC Survey Summary GUID*

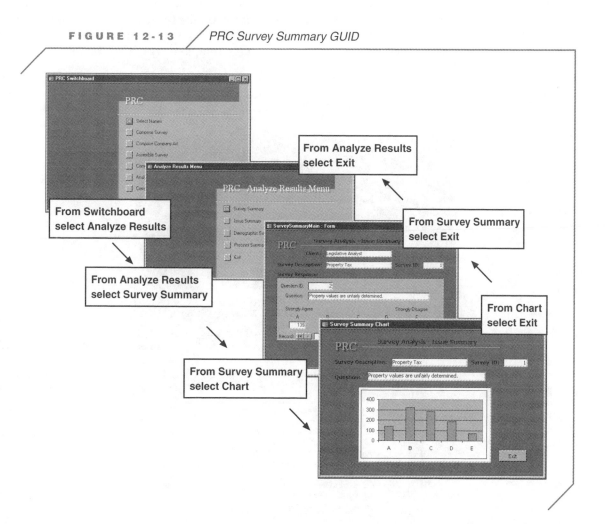

Code Generators

Many CASE tools include features for generating 4GL code. For example, Visible Analyst can automatically generate a database schema from information in the data repository, as well as C and COBOL shell code. Application builder products, such as PowerBuilder from Sybase and Delphi from Borland, offer highly visual design workbenches for design and development activities with automated code generation features. On a smaller scale, Microsoft Office provides automatic Hypertext Markup Language (HTML) code generation as an easy-to-use option on the Save As menu. These 4GL products create source code in response to analyst-programmer specifications entered via a graphical user interface.

The objectlike features of Microsoft Access provide a friendly development tool of this type. Figure 12-14 shows another form, along with a code segment designed to advance the screen values to the next survey. When the analyst placed the Next Survey command button on the form, Access responded with a series of questions about what action to associate with the OnClick event. Access then automatically generated the code to carry out the action.

FIGURE 12-14 / *PRC Survey Questions Form and Code*

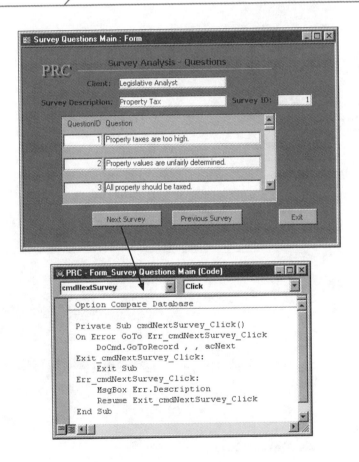

Hand Coding

In spite of the advances in automatic code generation, the analyst-programmer will probably find it necessary to write or revise some code segments to complete the project. For example, the analyst did not have to write a single line of code for the OnClick event associated with the Next Survey button on the screen form in Figure 12-14. In contrast, however, the analyst-programmer had to copy and customize the code segments for the OnClick events associated with each of the buttons on the Analyze Results menu.

Button objects, such as those described above, provide a good example of event-driven programming logic, which is very different from the traditional, linear file-processing logic found in most third- and fourth-generation programs. To the extent that event-driven logic is employed, the user is in charge of the processing sequence in an application. Therefore, the programmer must anticipate the users' needs as they might occur in a real-world setting. This places the focus on determining the way the information product is to be used, rather than on what the information product contains. The changes in programming perspectives promise to be even more dramatic as visual programming products continue to evolve.

Of course, object-oriented languages are available to the analyst-programmer as well. C++ is probably the language of choice for many application developers, but Visual Basic offers an alternative, especially since it is the support language for Access. As described earlier in this chapter, Visual Basic works with objects that have properties and procedures (methods) and is well suited to the event-driven application preferences of today's user.

SystemRecycling

Three months ago you designed, developed, and implemented an information system for Reynolds Auto Supply Store. The automated cash drawer system allows Reynolds to track inventory, reorder stock, and create sales receipts. The system also provides customer and supplier master file maintenance.

Recently, a Reynolds competitor asked you to develop a similar system for their business. You suspect that the new system will be a virtual duplicate of the one you built for Reynolds, requiring a few simple processing routine changes and some user interface modifications.

Should you rework the Reynolds solution for the new contract? Would your answer differ if these were not competing businesses? Who owns the solution designed and developed for Reynolds?

ThinkingCritically

MACROS, SCRIPTS, AND COMMAND FILES

Most users are familiar only with interactive, event-driven computing. They invest many hours in learning the keystrokes, mouse clicks, and menu options of their word processor, spreadsheet, and database management software. Unfortunately, they must reexercise common command patterns each time they choose to perform a particular task. As we know from the preceding discussion, although many 4GL development products use similar interfaces, the analyst-programmer is actually assembling these commands into files for reuse.

The typical small-enterprise computer information system is composed of several subsystems, each of which might rely on a different 4GL product. For the most part, this chapter concentrates on the database processing, but the correspondence and financial analysis subsystems can also be constructed using 4GL development techniques.

Specifically, word processor and spreadsheet macros are really computer programs in disguise. Their *macro languages* include powerful control structures and functions, and they must adhere to a strict syntax, as does any language. Macros are dependable and flexible, because the analyst-programmer can reuse and modify them at runtime. Many products contain a collection of macros. Finally, many third-party macro libraries are available.

To illustrate further how the analyst uses a combination of generated code and analyst-programmer code during the development process, Figure 12-15 presents a spreadsheet with summary data for a PRC survey. The raw data for this spreadsheet is imported from PRC's database into a specific range of cells (B9:F11). The automatic recalculation feature computes the response percentage breakdown, shown in another range of cells (B16:F18). After the

FIGURE 12-15 / *PRC Survey Response Spreadsheet*

Survey Stats.xls							
	A	B	C	D	E	F	G
1		Survey Analysis - Response Data					
2	PRC						
3		Client:	Legislative Analyst				
4		Survey ID:	1				
5		Survey Topic:	Property Tax				
6							
7		Raw Data					
8	Question	A	B	C	D	E	Total
9	Property taxes are too high.	632	117	94	77	68	988
10	Property values are unfairly determined.	139	321	284	184	72	1000
11	All property should be taxed.	11	34	41	517	373	976
12							
13		Percent Breakdown					
14							
15	Question	A	B	C	D	E	
16	Property taxes are too high.	63.20	11.70	9.40	7.70	6.80	
17	Property values are unfairly determined.	13.90	32.10	28.40	18.40	7.20	
18	All property should be taxed.	1.13	3.48	4.20	52.97	38.22	
19							
	Response Data						

import and recalculation, a custom-built macro scans each cell in the response percentage breakdown range. If the value in a cell is greater than 50 percent, the macro changes the font to bold face and highlights the cell.

Figure 12-16 shows the menu selections leading to a window display of the code underlying the macro. The analyst hand coded the ForEachNextOver50 macro, based on a much simpler macro named Over50. First, the analyst used the macro record feature to build the Over50 macro, which highlights a single cell (B16) and changes its font to bold face. Over50 provided a code sample the analyst copied and enhanced for the more complex macro.

The ForEachNextOver50 code demonstrates all three fundamental programming logic structures: sequence, selection, and repetition. In the absence of selection or repetition logic, program instructions execute in a sequence from top to bottom. Selection logic, bounded by "If ...End If," provides condition tests (PercentCell.Value > 50) with two possible outcomes (true or false) and potentially two sets of instructions to execute—one if the test condition evaluates to true, the other if it evaluates to false. Repetition logic, bounded by "For ... Next," provides for the repeated execution of a set of instructions.

The code also demonstrates something about the object-oriented nature of Visual Basic. In the expression "PercentCell.Value > 50," it is the value property of the object named PercentCell that is compared to 50. The fact that Visual Basic supports Access and Excel is no coincidence. Microsoft developed Visual Basic for Applications (VBA) to provide developers with a consistent support language across many of its software products.

FIGURE 12-16 / *PRC Survey Response Macro*

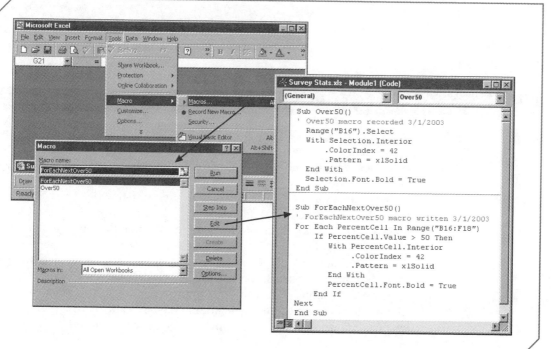

TECHNOTE 12-2

APPLICATIONS DEVELOPMENT WITHOUT PROGRAMMING

An important difference distinguishes an application developed around a horizontal software product from a computer information system. Commonly, a word processor, spreadsheet, or database management product serves as the basis for a **customized application**, which generally refers to an information system product that is adapted from a single horizontal software package. For example, financial specialists often create complex spreadsheet applications to implement budget and forecasting models. A series of small spreadsheet macros allows you to create menus that offer customized options to activate different parts of these models. This, however, is not how we define an information system.

A computer information system is broader in scope and function than a customized application. This text envisions such a system as a well-coordinated collection of resources that can serve the information needs of the finance department as well as the needs of the production, marketing, and sales departments. The system development life cycle is offered as a process through which these needs can be identified, coordinated, and satisfied. However, this does not rule out the SDLC as an appropriate process for the customized application project. Indeed, depending on the size and complexity of the project, an informal SDLC may help the applications developer avoid some of the pitfalls that plague systems work, without burdening the effort with unnecessary project overhead.

Several 4GL products are available that make both application and information system building much easier. Object-oriented horizontal software provides an environment in which object-oriented programming is accessible to analysts, without the normal rigor of a language such as C++. In addition, specialized front-end or back-end software, which helps to design and build database forms, reports, and queries, further reduces the analyst-programmer's hand-coding responsibilities.

Although these products make development work easier, in general, developing a computer information system without programming is still not possible.

4GL PROGRAMMING WITH CASE TOOLS

By definition, a full CASE tool includes product development and automatic code generation. The newest 4GL products, such as those illustrated previously, increase the likelihood of the small-enterprise analyst finding an appropriate CASE tool.

In some ways, the 4GL products, with their systemwide object orientation, already behave like a back-end CASE tool. Certainly, the analyst-programmer can take advantage of these products no matter how they are classified in the CASE tool lexicon.

One commonality of the full CASE tools and the object-oriented 4GL products is that their code generators usually rely on detailed screen form, report, or process specifications. This is a great improvement over the older method of writing line after line of third-generation language code. Furthermore, another significant advance in the development process seems promising: as software companies develop full CASE tools that transform the design models (data flow diagrams, entity-relationship diagrams, etc.) directly into code, the analyst may be able to avoid programming altogether.

PROJECT DELIVERABLE: THE PROTOTYPE REVIEW SESSION

Prototypes are often the first tangible, computer-based products that emerge from the design-development activities. If the analyst and user have collaborated as described in our discussions on joint application design, the user has already seen pieces of the new system design in the context of a computer environment and is well aware of the project's status. The prototype review session provides an opportunity for the analyst to bring the design and development work together in a well-coordinated and documented oral and written presentation. Figure 12-17 provides general guidelines for the report and oral presentation.

FIGURE 12-17 / *Prototype Review Report Content*

Written Material
1. System design update
   Revised DFD, ERD, USD
   System flowchart, menus, and GUID
   Macros, queries, and subprocedure code specifications
2. USD and/or menu tree with prototyped segments highlighted
3. Samples of prototyped user interfaces
4. Project budget and status updates

Oral Presentation and Demonstration
1. Describe prototyping methodology
2. Visuals for items 2 and 3 above
3. Hands-on demonstration
4. Question-and-answer period

The analyst should schedule this session to occur early in the development phase. This gives the analyst sufficient time to incorporate suggestions that may result from the review. However, the analyst should be very careful about agreeing to make fundamental design changes at this point in the project. A quick look at the project PERT chart introduced in Chapter 6 (Figure 6-18) provides a reminder that any delay in the completion of the development activities will result in commensurate delays in the product delivery.

THE CORNUCOPIA CASE

The analysts decide to use Microsoft Access to develop the majority of this project. Along with similar mainstream, relational database products such as Corel Paradox and Lotus Notes, Access has matured over the years. It now includes powerful object-oriented features, automatic code-generating options, a visual development workbench, a query-based structured query language (SQL), a useful set of predefined macro commands, and well-articulated associations with an object-oriented programming language (Visual Basic). Furthermore, Access is compatible with its Microsoft Office Suite companion products Word, Excel, and PowerPoint, and it runs well under Microsoft Windows. As you explore the Cornucopia project, be aware that the completed product includes very little analyst-supplied programming code.

Prototype Review Session

The prototype review session begins with a description of prototyping and how this technique is used during the development phase of the project. The analysts distribute a copy of their User's Guide to Prototyping introduced in Chapter 11 (Figure 11-21). Using the Prototype Menu Tree, illustrated in Figure 11-22, the analysts describe the prototyped elements. Many of the prototypes are presented as a series of PowerPoint slides displaying the sequence of screen images associated with the prototyped element. Each slide contains an image copied from development software, such as Access, Excel, and Word. For example, the CD Sales element requires a two-slide sequence, first displaying the Switchboard form and then, when the analysts click on the CD Sales menu option, the CD Sales form. One advantage of this presentation technique is that the analysts do not need to worry about any last-minute problems with the application software. In some cases, the prototype is presented using the actual implementing software. For example, it might be easier to present a Web site prototype in a browser rather than a slide show. Figure 12-18 illustrates one of the slides from the prototype slide show.

Prototype Conversion

The menu, master file maintenance screen forms, and sales transaction processing output prototypes serve as the basis for developing the working product. In fact, all but the transaction processing products are already close to final form. This is not surprising because the prototypes were created with Access, the same software used to develop the final product. As application development software matures, the time and effort required to transform prototypes into final products will diminish.

Implementing the features that are promised by transaction processing prototypes is usually a complicated task. Transaction data usually stretches over several files. Cornucopia's sales transaction processing involves five tables: CD Transaction, CD Transaction Detail, CD Master, Supplier Master, and Customer Master.

FIGURE 12-18 / *Cornucopia Prototype Review Session Slide Show*

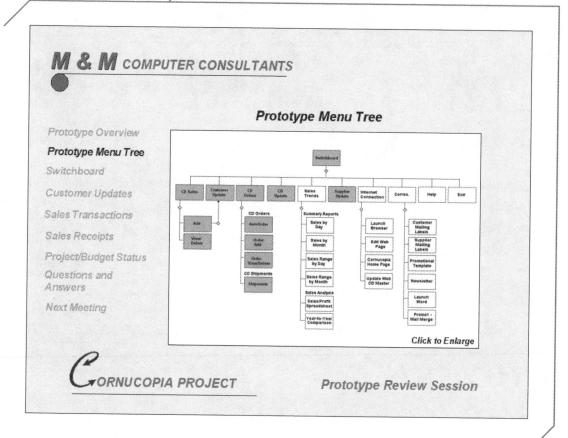

Finally, some elements of the system were never prototyped. The Sales Trends, Internet, and Correspondence subsystems must be constructed either from scratch or from very general design specifications. Certainly, the array of macro operations, queries, filters, and Wizards make information system development easier than ever before. Programming is kept to a minimum, and even that is made easier with languages such as Visual Basic.

System Flowchart

Figure 12-19 presents an annotated system flowchart. The shaded elements represent the subsystems, with CD Sales broken down into four parts: Sales, AutoOrder, Order, and Shipment. A screen form presents each element to the user, which serves as the focus for the discussion on how the system works.

FIGURE 12-19 / *Cornucopia System Flowchart*

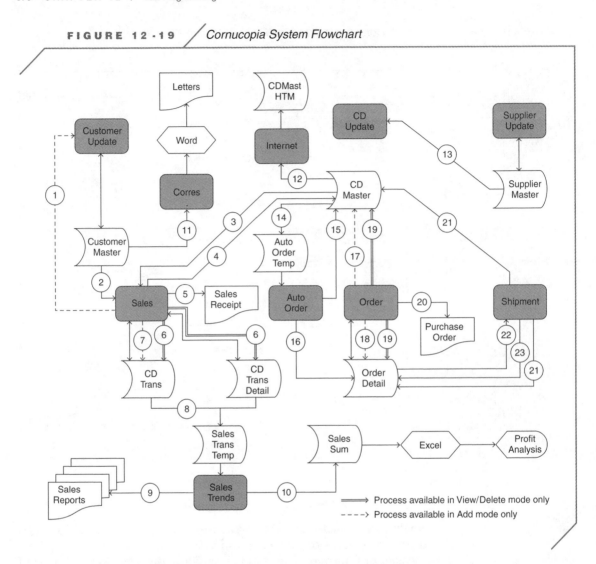

Figure 12-20 provides the narrative for each circled number on the system flowchart. Item number 1 identifies a button on the CD Sales Form (Add New Customer) that permits the user to jump directly to the Customer Update Form, while item number 2 describes how the active customer list is supplied to a combo box, and so on.

FIGURE 12-20 / *Cornucopia System Flowchart Annotations*

Items 1–7 reference: CD Sales Form (see Figure 8-17)

1. The "Add New Customer" command button only appears on CD Sales Form when that form is opened in Add mode. When this button is clicked, the "Open Customer Update" macro executes, which opens Customer Update Form.
2. The cboCustomerID control's Row Source property selects active customers from the Customer Master table.
3. The cboCDNumber control's Row Source property selects active CDs from the CD Master table. This object is located on a subform that displays the detailed transactions associated with the Transaction ID appearing at the top of the form.
4. The "Close" command button's click event invokes the "Close Sales Form" macro. This macro opens three queries and then closes the form. The "NoTransDetail Query" marks CD Transaction table records that have no corresponding detail records. "Update Inventory w Sales" and "Update Inventory w Returns" adjust the UnitsInStock value on the CD Master table.
5. When clicked, the Print icon button opens the "CD Transaction Sales Receipt" report. This report's Record Source property invokes the "Sales Receipt Query" in order to print the sales and/or returns associated with the Transaction ID (see Figure 9-12).
6. The Delete icon button is enabled when CD Sales Form is opened in View/Delete mode. When this button is clicked, the "Delete Trans" macro executes, which opens the "DeleteTrans Update" query. This query marks the appropriate CD Transactions and CD Transaction Detail records as deleted and reverses the quantity amounts so that when the form is closed, the CD Master inventory is also adjusted.
7. The New Record icon button is enabled when the form is opened in Add mode. When this button is clicked, a new record is added to the CD Transaction table.

Items 8–10 reference: Select Sales Trend Option (see Figure 10-16)

8. When this form opens, it invokes the "Make SalesTrans Temp" query. This query categorizes each sales transaction by its price range (< $11, $11-$20, >$20) and adds a record to the SalesTrans Temp table.
9. When either of the daily summary report options is selected, the corresponding report is opened. These reports use the Filter property to select the appropriate records from SalesTrans Temp. When either of the range summary report options is selected, the corresponding report is opened. These reports use the Record Source property to invoke the "Sales Range Summary Query" to summarize the SalesTrans Temp records.
10. When the "Sales/Profit Spreadsheet" option is selected, a relatively complex event procedure is executed. First, the SalesSum table is deleted. Second, a sequence of three SQL commands adds sales range summary records to SalesSum. Third, a TransferSpreadsheet command exports SalesSum as an Excel file named SaleProfit.xls. Finally, the Excel application program is launched and SaleProfit.xls is opened (see Figure 9-17).

Item 11 reference: Select Correspondence Option (see Figure 10-16)

11. When the "Promo1-Mail Merge" option is selected, the Word application program is launched and a preformatted merge document named Promo1.doc is opened. When the user selects "Promo1Macro" from the Tools menu in Word, the Customer Master table is used to create customized letters (see Figure 12-31).

Item 12 reference: Select Internet Option (see Figure 10-16)

12. When the "Update Web CD Master" option is selected, the "CD Web Macro" executes. This macro opens "CD Web Query," which selects the active CDs from CD master and then formats the data into HTML using a predefined template named CD Master Template.htm. This action permits the user to update the Web page titled "CD Inventory Table" (see Figure 9-20).

Item 13 reference: CD Update Form (see Figure 8-16)

13. The cboSupplierID control's Row Source property selects active customers from the Supplier Master table.

Items 14–16 reference: CD AutoOrder Form (see Figure 10-19)

14. When this form opens, the "Create AutoOrders" macro executes, which opens the "Make AutoOrder Temp" query. This query adds a record to the AutoOrder Temp table for every CD Master table record in which UnitsInStock + UnitsOnOrder is less than ReorderLevel.
15. The event procedure associated with clicking the "Confirm" button opens the "Update CD UnitsOnOrder from AutoOrder" query. This query adjusts the UnitsOnOrder value on the CD Master table.
16. The click event procedure associated with the "Close" button opens the "Append Order Detail" query. This query adds the confirmed orders to the Order Detail table.

Items 17–20 reference: CD Order Form (see Figure 9-14)

17. The "Confirm" command button only appears on the CD Order Form when that form is opened in Add mode. The click event procedure associated with this button opens the "Update CD UnitsOnOrder from Order" query. This query adjusts the UnitsOnOrder value on the CD Master table.
18. The New Record icon button is enabled when the form is opened in Add mode. When this button is clicked, a new record is added to the Order Detail table.
19. The Delete icon button is enabled when the CD Order Form is opened in View/Delete mode. The click event procedure associated with this button opens the "Cancel Orders Update" query. This query marks the appropriate CD Order Detail records as cancelled and adjusts the UnitsOnOrder value on the CD Master table.
20. When clicked, the Print icon button opens the "CD Purchase Order" report. This report's Record Source property invokes the "Purchase Order Query" in order to print the orders associated with the SupplierID (see Figure 9-13).

FIGURE 12-20 / *Cornucopia System Flowchart Annotations (continued)*

Items 21–23 reference: CD Shipment Form (see Figure 12-26)

21. The "Received" button's click event procedure opens the "Received Order Update" query. This query moves the UnitsOnOrder quantity to the UnitsInStock quantity on the CD Master table. This event also sets BackOrdered to false, if previously true, and posts the received date to the Order Detail record.
22. When this form opens, it invokes the "Outstanding Orders" query. This query selects those Order Detail records that show no date received or date cancelled.
23. The "Back Ordered" button's click event procedure toggles the BackOrdered value between true and false.

As you explore the Cornucopia files on your Data Disk, you should note that several areas are "under construction." One area of particular interest is the Cornucopia logo that appears on each form. The analysts intend to use the pad as a link to user documentation for each form and its attendant subsystem processing. This demonstrates an important developmental strategy in which each component is developed and tested separately. The menuing system was initially prototyped with such "under construction" notations and later modified as working subsystems were incorporated into the product.

What follows is a systematic discussion of how the Cornucopia system works. You can use this information to help you develop your Portfolio Project. You will learn about the Internet connection subsystem in the next chapter.

Object Properties

Access embodies many of the object-oriented concepts discussed in this and earlier chapters. Recall that objects are derived from a class, which provides the blueprint for the object's data attributes and methods. The form object is an excellent example. When the analysts open a new form in design mode, Access creates an instance of the form class, with an entire set of properties (data attributes) and behaviors (methods).

In the case of the CD Sales Form, the analysts change several of the default property values. The Caption property is changed to "Cornucopia," Scroll Bars is changed to "neither," Navigation Buttons is changed to "no," and so on. The Record Source property links the form to the specified table or query. Here, the analysts imbed a SELECT statement, rather than a predefined table or query, which illustrates the flexibility afforded the developer. This single SELECT statement creates an ordered data file of matching sales transaction and customer master file records. By comparison, more than two dozen lines of procedural programming code are required to accomplish the same thing.

Event Procedures

As the name implies, event procedures contain program instructions that execute when a particular event occurs. Event procedures are associated with objects. For example, the custom tool pad on the CD Sales Form introduced in

Chapter 11 (Figure 11-25) contains a button to advance to the next record. When the user clicks on this button, Access executes an event procedure named "cmdNextRec_Click()," to remind users that new sales records require a valid customer ID. Figure 12-21 shows the Visual Basic code, much of which the Access Wizard created. The analysts added the conditional statement highlighted in the shaded box. This is one of only a handful of custom program lines in the entire project.

FIGURE 12-21 / *Cornucopia Event Procedures*

```
Sub cmdNextRec_Click()
    On Error GoTo Err_cmdNextRec_Click

    DoCmd.GoToRecord , , acNext

    If IsNull(cboCustomerIID) Then
        MsgBox "If you want to add a new sales record you must
            select a valid CustomerID before selecting a CD"
        TransDetailSubform.Locked = True
        cboCustomerID.SetFocus
    End If

    Exit_cmdNextRec_Click:
        Exit Sub
    Err_cmdNextRec_Click:
        MsgBox Err.Description
        Resume Exit_cmdNextRec_Click

End Sub
```

Macros

A macro is a collection of predefined commands that are customized to perform specific tasks. Access has almost 50 actions from which to select. Most actions have arguments that take on specific values. For example, the first argument for the OpenForm action is Form Name. To open a particular form, all the analysts need supply is a valid form name for this argument. Figure 12-22 illustrates the Macro Builder screen for the macro that closes the sales form. This macro consists of four actions, each with its own argument values. The macro is associated with the click event of the Close button on the sales form. The full detail of all the project macros appears in Figure 12-23.

FIGURE 12-22 / *Cornucopia Close Sales Form Macro*

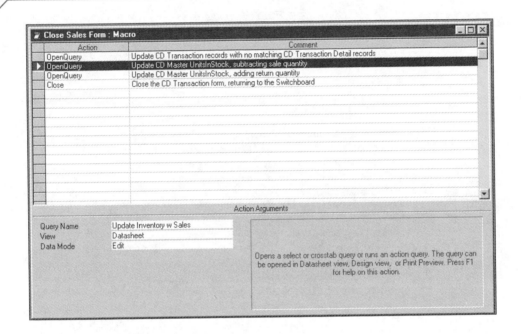

FIGURE 12-23 */Cornucopia Macro Instructions/*

Macros 1–9 are associated with CD Sales Form events. Macros 2–6 are used in the Add mode. Macros 7–8 are used in the View/Delete mode. Macro 9 is used in both modes.

1. Open Sales Options

Action	Argument	Value
OpenForm	Form Name:	Sales Form Options
	Data Mode:	Edit

2. Open Sales Add

Action	Argument	Value
Close	Object Type:	Form
	Object Name:	Sales Form Options
OpenForm	Form Name:	CD Transaction
	Data Mode:	Add
SetValue	Item:	[Forms]![CD Transaction]![lblMode].[Caption]
	Expression:	Add Mode
SetValue	Item:	[Forms]![CD Transaction]![cmdNew].[Enabled]
	Expression:	Yes
SetValue	Item:	[Forms]![CD Transaction]![cmdDelete].[Enabled]
	Expression:	No
GoToRecord	Record:	New
SetValue	Item:	[Forms]![CD Transaction]![TransDetailSubform].[Locked]
	Expression:	Yes
GoToControl	Control Name:	cboCustomerID

3. Add New Sales

Action	Argument	Value
GoToRecord	Record:	New
SetValue	Item:	[Forms]![CD Transaction]![TransDetailSubform].[Locked]
	Expression:	Yes
GoToControl	Control Name:	cboCustomerID

4. Set CustomerID to AANON

Action	Argument	Value
GoToControl	Control Name:	cboCustomerID
SetValue	Item:	[Forms]![CD Transaction]![cboCustomerID]
	Expression:	AANON

5. Unlock Sales Subform

Action	Argument	Value
SetValue	Item:	[Forms]![CD Transaction]![TransDetailSubform].[Locked]
	Expression:	No
GoToControl	Control Name:	TransDetailSubform

6. Set ListPrice Default

Action	Argument	Value
SetValue	Item:	[TransactionPrice]
	Expression:	[ListPrice]

FIGURE 12-23 / *Cornucopia Macro Instructions (continued)*

SetValue	Item:	[ListPriceTOT]
	Expression:	[ListPrice]
Requery		

7. Open Sales View

Action	*Argument*	*Value*
Close	Object Type:	Form
	Object Name:	Sales Form Options
OpenForm	Form Name:	CD Transaction
	Data Mode:	Read Only
SetValue	Item:	[Forms]![CD Transaction]![lblMode].[Caption]
	Expression:	View/Delete Mode
SetValue	Item:	[Forms]![CD Transaction]![cmdNew].[Enabled]
	Expression:	No
SetValue	Item:	[Forms]![CD Transaction]![lblAddNewCustomer].[Visible]
	Expression:	No
SetValue	Item:	[Forms]![CD Transaction]![lblAnonymousCustomer].[Visible]
	Expression:	No
GoToControl	Control Name:	cmdNextRec

8. Delete Trans

Action	*Argument*	*Value*
OpenQuery	Query Name:	Delete Trans Update
Requery		

9. Close Sales Form

Action	*Argument*	*Value*
OpenQuery	Query Name:	NoTransDetail Update
OpenQuery	Query Name:	Update Inventory w Sales
OpenQuery	Query Name:	Update Inventory w Returns
Close	Object Type:	Form
	Object Name:	CD Transaction

Macros 10–14 are associated with CD order and shipment forms.

10. Open Order Options

Action	*Argument*	*Value*
OpenForm	Form Name:	Order Options
	Data Mode:	Edit

11. Create AutoOrders

Action	*Argument*	*Value*
Close	Object Type:	Form
	Object Name:	Order Options
OpenQuery	Query Name:	Make AutoOrder Temp
OpenForm	Form Name:	AutoOrder Selections
	Data Mode:	Edit

FIGURE 12-23 / *Cornucopia Macro Instructions (continued)*

12. Open Order Add

Action	Argument	Value
Close	Object Type:	Form
	Object Name:	Order Options
OpenForm	Form Name:	CD Order
	Data Mode:	Edit
SetValue	Item:	[Forms]![CD Order]![lblMode].[Caption]
	Expression:	Add Mode
SetValue	Item:	[Forms]![CD Order]![cmdNew].[Enabled]
	Expression:	Yes
SetValue	Item:	[Forms]![CD Order]![cmdDelete].[Enabled]
	Expression	No
GoToRecord	Record:	New
GoToControl	Control Name:	cboCDNumber

13. Open Order View

Action	Argument	Value
Close	Object Type:	Form
	Object Name:	Order Options
OpenForm	Form Name:	CD Order
	Data Mode:	Read Only
SetValue	Item:	[Forms]![CD Order]![lblMode].[Caption]
	Expression:	View/Delete Mode
SetValue	Item:	[Forms]![CD Order]![cmdNew].[Enabled]
	Expression:	No
SetValue	Item:	[Forms]![CD Order]![cmdConfirm].[Visible]
	Expression"	No
SetValue	Item:	[Forms]![CD Order]![cboCDNumber].[BackColor]
	Expression:	12632256
SetValue	Item:	[Forms]![CD Order]![cboSupplierID].[BackColor]
	Expression:	12632256
SetValue	Item:	[Forms]![CD Order]![ReorderQuantity].[BackColor]
	Expression:	12632256
SetValue	Item:	[Forms]![CD Order]![cboCustomerID].[BackColor]
	Expression:	12632256
GoToControl	Control Name:	cmdNextRec

14. Open Shipment

Action	Argument	Value
Close	Object Type:	Form
	Object Name:	Order Options
OpenForm	Form name:	CD Shipment
	Data Mode:	Edit

FIGURE 12-23 / *Cornucopia Macro Instructions (continued)*

Macros 15–19 are associated with their named subsystems.

15. Open Sales Trend Options

Action	*Argument*	*Value*
OpenForm	Form Name:	Sales Trend Options
	Data Mode:	Edit

16. Open Customer Update Form

Action	*Argument*	*Value*
Open Form	Form Name:	Customer Update
	Data Mode:	Edit

17. Open CD Update Form

Action	*Argument*	*Value*
OpenForm	Form Name:	CD Update
	Data Mode:	Edit

18. Open Supplier Update Form

Action	*Argument*	*Value*
OpenForm	Form Name:	Supplier Update
	Data Mode:	Edit

19. Open Correspondence Options

Action	*Argument*	*Value*
OpenForm	Form Name:	Correspondence Options
	Data Mode:	Edit

Macros 20–21 are associated with Internet forms.

20. Open Internet Options

Action	*Argument*	*Value*
OpenForm	Form Name:	Internet Options
	Data Mode:	Edit

21. CD Web Macro

Action	*Argument*	*Value*
Output To	Object Type:	Query
	Object Name:	CD Web Query
	Output Format:	HTML(*.html)
	Output File:	c:\...\CD Master.htm
	Auto Start:	No
	Template File:	c:\...\CD Master Template.htm

Macros 22–23 are invoked from numerous forms.

22. Requery

Action	*Argument*	*Value*
Requery		

23. Under Construction Message

Action	*Argument*	*Value*
MsgBox	Message:	Under Construction

Queries

A query is a specification for selecting data from one or more tables. For example, the Outstanding Orders query selects those order records from the Order Detail table where both the Received Date and the Cancelled Date are blank. Figure 12-24 shows a query-by-design screen for the DeleteTrans Update query. This query selects and updates data from two tables, matching the TransactionID values in the CD Transaction and CD Transaction Detail tables. This query is associated with the click event of the Delete button on the sales form. Access creates SQL code for each query, as illustrated in Figure 12-25.

As a precaution, Access generates a series of dialog boxes associated with certain action queries. To disable these messages, click Tools on the menu bar, and select Options. Click the Edit/Find tab, and select or clear the appropriate check boxes in the Confirm section.

FIGURE 12-24 / *Cornucopia DeleteTrans Update Query*

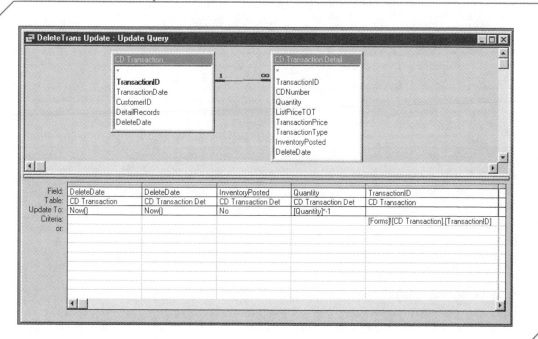

FIGURE 12-25 / *Cornucopia Query SQL Statements*

Queries 1–6 are associated with CD Sales Form events.

1. CD Transaction Detail Query (Select Query)

SELECT DISTINCTROW [CD Transaction Detail].TransactionID, [CD Transaction Detail].CDNumber, [CD Transaction Detail].DeleteDate, [CD Master].Title, [CD Master].ListPrice, [CD Transaction Detail].ListpriceTOT, [CD Transaction Detail].TransactionPrice, [CD Transaction Detail].Quantity, [CD Transaction Detail].TransactionType, CCur([CD Transaction Detail].[TransactionPrice]*[Quantity]) AS TransactionAmount
FROM [CD Master] INNER JOIN [CD Transaction Detail] ON [CD Master].CDNumber = [CD Transaction Detail].CDNumber
WHERE ((([CD Transaction Detail].DeleteDate) Is Null))
ORDER BY [CD Transaction Detail].TransactionID;

2. DeleteTrans Update (Update Query)

UPDATE DISTINCTROW [CD Transaction] INNER JOIN [CD Transaction Detail] ON [CD Transaction].TransactionID = [CD Transaction Detail].TransactionID SET [CD Transaction].DeleteDate = Now(), [CD Transaction Detail].DeleteDate = Now(), [CD Transaction Detail].InventoryPosted = No, [CD Transaction Detail].Quantity = [Quantity]*-1
WHERE ((([CD Transaction].TransactionID)=[Forms]![CD Transaction].[TransactionID]));

3. Sales Receipt Query (Select Query)

SELECT DISTINCTROW [CD Transaction].TransactionID, [CD Transaction].TransactionDate, [Customer Master].Firstname, [Customer Master].LastName, [Customer Master].StreetAddress, [Customer Master].City, [Customer Master].State, [Customer Master].ZIPCode, [CD Transaction Detail].CDNumber, [CD Master].Title, [CD Transaction Detail].Quantity, [CD Transaction Detail].TransactionPrice, [CD Transaction Detail].TransactionType
FROM [Customer Master] INNER JOIN ([CD Transaction] INNER JOIN ([CD Master] INNER JOIN [CD Transaction Detail] ON [CD Master].CDNumber = [CD Transaction Detail].CDNumber) ON [CD Transaction].TransactionID = [CD Transaction Detail].TransactionID) ON [Customer Master].CustomerID = [CD Transaction].CustomerID
WHERE ((([CD Transaction].TransactionID)=[Forms]![CD Transaction]![TransactionID]));

4. NoTransDetail Update (Update Query)

UPDATE DISTINCTROW [CD Transaction] LEFT JOIN [CD Transaction Detail] ON [CD Transaction].TransactionID = [CD Transaction Detail].TransactionID SET [CD Transaction].DetailRecords = No
WHERE ((([CD Transaction].DetailRecords)=Yes) AND (([Cd Transaction Detail].TransactionID) Is Null));

5. Update Inventory w Sales

UPDATE DISTINCTROW [CD Master] INNER JOIN [CD Transaction Detail] ON [CD Master].CDNumber = [CD Transaction Detail].CDNumber SET [CD Master].UnitsInStock = [UnitsInStock]-[Quantity], [CD Transaction Detail].InventoryPosted = Yes
WHERE ((([CD Transaction Detail].InventoryPosted)=No) AND (([CD Transaction Detail].TransactionType)="Sale"));

6. Update Inventory w Returns

UPDATE DISTINCTROW [CD Master] INNER JOIN [CD Transaction Detail] ON [CD Master].CDNumber = [CD Transaction Detail].CDNumber SET [CD Master].UnitsInStock = [UnitsInStock]+[Quantity], [CD Transaction Detail].InventoryPosted = Yes
WHERE ((([CD Transaction Detail].InventoryPosted)=No) AND (([CD Transaction Detail].TransactionType)="Return"));

FIGURE 12-25 / *Cornucopia Query SQL Statements (continued)*

Queries 7–12 are associated with CD Order Form events.

7. **Make AutoOrder Temp (Make Table Query)**

   SELECT DISTINCTROW [CD Master].CDNumber, [CD Master].Title, [CD Master].UnitsInStock, [CD Master].UnitsOnOrder, [CD Master].WhlsalePrice, [CD Master].ReorderLevel, [CD Master].Discontinued, [CD Master].SupplierID, 0 AS ReorderQuantity, [SupplierID] & " " & Date() & " " & Time() AS OrderItemID, " " AS CustomerID, Date() AS OrderDate, "No" AS OrderPosted INTO [AutoOrder Temp]
   FROM [CD Master]
   WHERE ((([CD Master].UnitsInStock)<[Reorder Level]-[UnitsOnOrder]) AND (([CD Master].Discontinued)="No"));

8. **Update CD UnitsOnOrder from AutoOrder (Update Query)**

   UPDATE DISTINCTROW [AutoOrder Temp] INNER JOIN [CD Master] ON [AutoOrder Temp].CDNumber = [CD Master].CDNumber SET [CD Master].UnitsOnOrder = [CD Master]![UnitsOnOrder]+[AutoOrder Temp]![ReorderQuantity], [AutoOrder Temp].OrderPosted = "Yes"
   WHERE ((([AutoOrder Temp].OrderPosted)="No"));

9. **Append Order Detail (Append Query)**

   INSERT INTO [Order Detail] (CDNumber, SupplierID, ReorderQuantity, OrderItemID, CustomerID, OrderDate, OrderPosted)
   SELECT DISTINCTROW [AutoOrder Temp].CDNumber, [AutoOrder Temp].SupplierID, [AutoOrder Temp].ReorderQuantity, [AutoOrder Temp].OrderItemID, [AutoOrder Temp].CustomerID, [AutoOrder Temp].OrderDate, [AutoOrder Temp].OrderPosted
   FROM [AutoOrder Temp]
   WHERE ((([AutoOrder Temp].ReorderQuantity)>0));

10. **Cancel Order Update (Update Query)**

    UPDATE DISTINCTROW [CD Master] INNER JOIN [Order Detail] ON [CD Master].CDNumber = [Order Detail].CDNumber SET [CD Master].UnitsOnOrder = [CD Master]![UnitsOnOrder]-[Order Detail]![Reorder Quantity], [Order Detail].CancelledDate = Now()
    WHERE ((([CD Master].CDNUmber)=[Forms]![CD Order]![cboCDNumber]) AND (([Order Detail].OrderItemID)=[Forms]![CD Order]![OrderItemID]));

11. **Update CD UnitsOnOrder from Order (Update Query)**

    UPDATE DISTINCTROW [CD Master] INNER JOIN [Order Detail] ON [CD Master].CDNumber = [Order Detail].CDNumber SET [CD Master].UnitsOnOrder = [CD Master]![UnitsOnOrder]+[Order Detail]![ReorderQuantity], [Order Detail].OrderPosted = "Yes"
    WHERE ((([Order Detail].OrderPosted)="No"));

12. **Purchase Order Query (Select Query)**

    SELECT DISTINCTROW [Supplier Master].SupplierID, [Supplier Master].SupplierName, [Supplier Master].StreetAddress, [Supplier Master].City, [Supplier Master].ZIPCode, [Supplier Master].State, [Supplier Master].PhoneNumber, [Supplier Master].FaxNumber, [Supplier Master].EmailAddress, [Order Detail].CDNumber, [Order Detail].CancelledDate, [CD Master].Title, [Order Detail].ReorderQuantity, [Order Detail].OrderItemID, [Order Detail].OrderDate
    FROM ([CD Master] INNER JOIN [Order Detail] ON [CD Master].CDNumber = [Order Detail]. CDNumber) INNER JOIN [Supplier Master] ON ([Supplier Master].SupplierID = [CD Master].SupplierID) AND ([Order Detail].SupplierID = [Supplier Master].SupplierID)

FIGURE 12·25 / *Cornucopia Query SQL Statements (continued)*

WHERE ((([Supplier Master].SupplierID)=[Forms]![CD Order]![cboSupplierID]) AND (([Order Detail].
CancelledDate) Is Null) AND (([Order Detail].OrderDate)=[Forms]![CD Order]![OrderDate]))
ORDER BY [CD Master].Title;

Queries 13–14 are associated with CD Shipment Form events.

13. Outstanding Orders (Select Query)

SELECT DISTINCTROW [Order Detail].CDNumber, [Order Detail].SupplierID, [Order Detail].ReorderQuantity,
[Order Detail].CustomerID, [Order Detail].OrderDate, [Order Detail].OrderItemID, [Order Detail].ReceivedDate,
[Order Detail].BackOrdered, [Order Detail].CancelledDate, [Order Detail].OrderPosted
FROM [Order Detail]
WHERE ((([Order Detail].ReceivedDate) Is Null) AND (([Order Detail].CancelledDate) Is Null));

14. Received Order Update (Update Query)

UPDATE DISTINCTROW [CD Master] INNER JOIN [Outstanding Orders] ON [CD Master].CDNumber =
[Outstanding Orders].CDNumber SET [CD Master].UnitsInStock = [CD Master]![UnitsInStock]+[Outstanding
Orders]![ReorderQuantity], [CD Master].UnitsOnOrder = [CD Master]![UnitsOnOrder]-[Outstanding
orders]![ReorderQuantity]
WHERE ((([CD Master].CDNumber)=[Forms]![CD Shipment]![CDNumber]) AND (([Outstanding Orders].
OrderItemID)=[Forms]![CD Shipment]![OrderItemID]));

Queries 15–16 are associated with Sales Trend Form events.

15. Make SalesTrans Temp (Make Table Query)

SELECT DISTINCTROW [CD Transaction].TransactionDate, [CD Transaction Detail].Quantity, [CD Transaction
Detail].TransactionPrice, [Quantity]*[TransactionPrice] AS TransAmount,
IIf([TransactionPrice]<11,1,IIf([TransactionPrice]>20,3,2)) AS Range INTO [SalesTrans Temp]
FROM [CD Transaction] INNER JOIN [CD Transaction Detail] ON [CD Transaction].TransactionID = [CD
Transaction Detail.TransactionID
WHERE ((([CD Transaction].DetailRecords)=Yes) AND (([CD Transaction].DeleteDate) Is Null) AND (([CD
Transaction Detail].TransactionType)="Sale"));

16. Sales Range Summary Query (Select Query)

SELECT DISTINCTROW [SalesTrans Temp].TransactionDate, IIf([TransactionPrice]<11,[Quantity],0) AS
Units1, IIf([TransactionPrice]<11,[TransAmount],0) AS Amount1, IIf(([TransactionPrice]>=11 And
[TransactionPrice]<20),[Quantity],0) AS Units2, IIf(([TransactionPrice]>=11 And
[TransactionPrice]<20),[TransAmount],0) AS Amount2, IIf([TransactionPrice]>20,[Quantity],0) AS Units3,
IIf([TransactionPrice]>20,[TransAmount],0) AS Amount3
FROM [SalesTrans Temp];

Query 17 is associated with an Internet Options Form event.

17. CD Web Query (Select Query)

SELECT DISTINCTROW [CD Master].CDNumber, [CD Master].Title, [CD Master].Composer, [CD Master].
Artist, [CD Master].Label, [CD Master].Description, [CD Master].ListPrice
FROM [CD Master]
WHERE ((([CD Master].Discontinued)="No"));

Master File Update Subsystem

All three master files are serviced in the same way. Each is associated with an update form of similar design. The Customer Update and Supplier Update forms follow the same pattern as the CD Update Form introduced in Chapter 8 (Figure 8-16), with a custom tool pad and drop-down lists to make data entry easier for the user. The Supplier ID drop-down list on the CD Update Form is populated from entries in the Supplier table via a SELECT statement in the Row Source property value.

Sales Transaction Processing Subsystem

Figures 8-17, 10-18, and 11-25 present the CD Sales Form in various ways, each revealing a little of how this pivotal form captures sales transactions. The system flowchart and processing annotations (Figures 12-19 and 12-20) provide detail on seven critical operations. The form's primary association is implemented through the Record Source property with a complex SELECT statement that draws data from the Customer and CD Transaction tables to complete the upper half of the form. The lower half of the form contains the details of the transaction in a subform, which is associated with the CD Transaction Detail table through another SELECT statement in the Record Source property value.

The user can view existing sales data (View/Delete Mode) or record new sales data (Add Mode). The View/Delete Mode opens the form in edit mode, disables the Add Record icon button, and enables the Delete icon button. The Add Mode opens the form in add mode, enables the Add Record Icon button, and disables the Delete icon button. The Print icon button is available in both modes. These two modes may seem redundant, but they are required in order to synchronize properly and control updates to the transaction tables and the inventory fields in the CD Master table.

Order Processing and Shipments Subsystem

There are two ways to create an order. First, the CD Order Form introduced in Chapter 8 (Figure 8-18) allows the user to order any CD on the CD Master table. Second, the AutoOrder Form, introduced in Chapter 10 (Figures 10-19 and 10-20), allows the user to view a temporary table of CDs that have low inventory levels. The user can create an order by simply clicking the command button captioned Confirm.

When CD shipments arrive, the user can view all of the outstanding orders by activating the CD Shipment Form (Figure 12-26). By clicking the Received button, the user activates a query to update the CD Orders table and the inventory fields in the Customer Master table.

FIGURE 12-26   Cornucopia CD Shipment Form

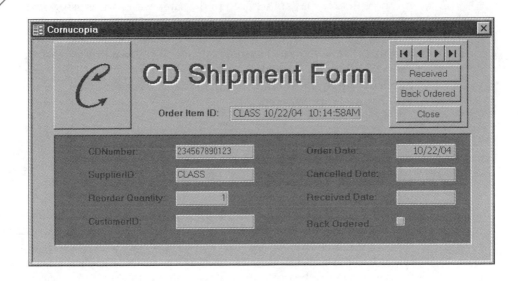

Sales Trends Subsystem

When the user selects the Sales Trends subsystem, the Make SalesTrans Temp query creates a summary table of all valid sales transactions. This table is a source for the information that appears in the various summary reports and the spreadsheet application.

The Sales/Profit spreadsheet introduced in Chapter 9 (Figure 9-16) provides the user with a quick look at the relative profitability of different price ranges. When the user selects this option, an event procedure (Figure 12-27) evaluates records in the SalesTrans Temp table and creates a small summary table named SalesSum. The event procedure includes five DoCmd statements and one Call statement. Four of the DoCmd statements execute SQL code to build the summary records; the fifth transfers the data to Excel. Finally, the Call statement launches Excel and opens the Sales/Profit spreadsheet. Figure 12-28 shows the spreadsheet formulas, with the destination cell range highlighted.

FIGURE 12-27 / *Cornucopia Sales/Profit Event Procedure*

```
Private Sub cmdSaleProfit_Click()

    DoCmd.RunSQL "DELETE * FROM SalesSum;"

    DoCmd.RunSQL "INSERT INTO SalesSum" &_
    "SELECT DISTINCTROW Sum([SalesTrans Temp].Quantity) AS Units," &_
    "Sum([SalesTrans Temp].TransAmount) AS Amount" &_
    "FROM [SalesTrans Temp]" &_
    "WHERE ((([SalesTrans Temp]![TransactionPrice])<11));"

    DoCmd.RunSQL "INSERT INTO SalesSum" &_
    "SELECT DISTINCTROW Sum([SalesTrans Temp].Quantity) AS Units," &_
    "Sum([SalesTrans Temp].TransAmount) AS Amount" &_
    "FROM [SalesTrans Temp]" &_
    "WHERE ((([SalesTrans Temp]![TransactionPrice])>11) AND" &_
    "((([SalesTrans Temp]![TransactionPrice])<20));"

    DoCmd.RunSQL "INSERT INTO SalesSum" &_
    "SELECT DISTINCTROW Sum([SalesTrans Temp].Quantity) AS Units," &_
    "Sum([SalesTrans Temp].TransAmount) AS Amount" &_
    "FROM [SalesTrans Temp]" &_
    "WHERE ((([SalesTrans Temp]![TransactionPrice])>20));"

    DoCmd.TransferSpreadsheet acExport, 5, "SalesSum", "c:\SaleProfit.xls",0, "B4:C7"

    Call Shell("c:\MSOffice\Excel\Excel.exe c:\SaleProfit.xls", 1)

End Sub
```

FIGURE 12-28 / *Cornucopia Sales/Profit Spreadsheet Formulas*

	A	B	C	D	E	F	G	H
1	$\mathcal{C}$		Sales/Profit Spreadsheet					
2	=NOW()							
3	Price	Sales	Sales	Profit	Profit	Average	Percent	Percent
4	Range	Units	Amount	Margin	Amount	Profit/Unit	Sales	Profit
5	< $11			0.3	=C5*D5	=E5/B5	=C5/C$8	=E5/E$8
6	$11 - $20			0.4	=C6*D6	=E6/B6	=C6/C$8	=E6/E$8
7	> $20			0.5	=C7*D7	=E7/B7	=C7/C$8	=E7/E$8
8	Total	=SUM(B5:B7)	=SUM(C5:C7)		=SUM(E5:E7)	=E8/B8		

Range: B4:C7 imported from SalesSum table

Several reports are available through the Sales Trends Option Form. Figure 12-29 shows the data properties associated with the Sales Summary by Day report introduced in Chapter 9 (Figure 9-14). The Filter property value directs Access first to prompt the user to enter the number of days to summarize and then to select only those records from the SalesTrans Temp table where the transaction date is within the number of days specified.

FIGURE 12-29   Cornucopia "Sales Summary by Day" Report Procedures

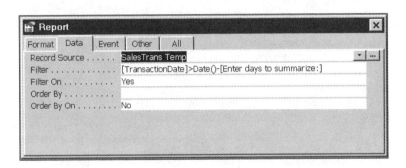

Correspondence Subsystem

The Correspondence subsystem was identified at the beginning of the project, but this is the first serious mention of exactly how the analysts intend to satisfy this information requirement. As depicted in the system flowchart, the Customer Update subsystem plays a critical role in the Correspondence subsystem. The other major component is the file of form letters developed with a word-processing program. In combination, these two elements form the basis for a mail-merge operation, in which form letters are personalized for selected records in the Customer Master table.

A standard mail-merge operation requires a main document containing text and merging specifications, plus a data source from which to draw the merged data. Figure 12-30 presents the main document with the field codes toggled on. The IF statement in the body of this form letter evaluates the status of each data record in the Customer Master table. Active customers receive one message, while inactive customers receive another. The output from this operation is a file of letters (Figure 12-31), each personally tailored to a different customer. This operation is called from the cmdPromo1_Click() subprocedure by the following statement:

Call Shell ("c:\MSOffice\Winword\Winword.exe c:\Promo1.doc", 1)

Several other correspondence products are listed on the options form, none of which has been developed. Because this subsystem is very important to the user, the analysts agree to build skeletons of the other products, but they inform the user that she will need to create the content. In addition, they point out that because the sales transaction subsystem captures customer information for each sale, correspondence can be further customized based on detailed analysis of customer buying patterns and preferences.

FIGURE 12-30 / *Cornucopia Mail-Merge Field Codes*

{TIME\@"MMMM d, yyyy"}

{MERGEFIELD FirstName} {MERGEFIELD LastName}
{MERGEFIELD StreetAddress}
{MERGEFIELD City}, {MERGEFIELD State} {MERGEFIELD ZIPCode}

Dear {MERGEFIELD FirstName}:

We want to thank you for your valued patronage in the past. {IF Active = "Active" "As you may know, there are several new, limited edition recordings available to you as a Preferred Customer. Please be sure to inquire about this plan the next time you are in the store." "We are sorry that we have not been able to serve you recently, but we hope you will come in to browse through our new CD selections."}

Enclosed is a discount coupon that you may redeem at any time.

Sincerely,

Margaret Height
—

FIGURE 12-31 / *Cornucopia Mail-Merge Letter*

"Without music, life would be a mistake."

November 9, 2004

Rocky Racoon
123 Creek Road
Eureka, CA 95501

Dear Rocky:

We want to thank you for your valued patronage in the past. As you may know, there are several new, limited edition recordings available to you as a Preferred Customer. Please be sure to inquire about this plan the next time you are in the store.

Enclosed is a discount coupon that you may redeem at any time.

Sincerely,

Margaret Height

Documentation

Even though the project is in the early stages of the development phase, the analysts begin work on some important implementation phase activities to develop training materials and system documentation. The detailed task list and PERT chart introduced in Chapter 6 (Figures 6-16 and 6-18) show that one analyst can work on implementation activities, while another analyst works on development activities.

Time and Money

Figure 12-32 shows the Status Report for Week 11. The analysts charged 20 hours to the development phase to cover the work performed during Week 10, considerably more than the estimate. During Week 11, the analysts charged four hours to development and one hour to implementation. As of Week 11, actual labor hours exceed estimated labor hours by almost 20 percent. Still, the hardware and software savings cover the overage; taken together, the budget is okay. Development is estimated to be 75 percent complete, which means the schedule is satisfactory as well. It is common to experience overages in one area and savings in another.

FIGURE 12-32 *Cornucopia Status Report – Week 11*

Status11.xls

	A	B	C	D	E	F	G	H	I	J	K	L	M	N	O	P	Q	R	S	T
1	Date:	Cornucopia Project Status							As of: Week 11											
2																				
3	Activity	% Comp.	Status	1	2	3	4	5	6	7	8	9	10	11	12	13	14	15	16	Total
4	Analysis - Estimate	100%		4	5	5	4													18
5	Actual	100%	ok	2	4	2	3													11
6	Design - Estimate	100%					4	4	5	5	4									22
7	Actual	95%	ok			1	3	8	5	7	6									30
8	Develop - Estimate	80%									4	4	5	5	4					22
9	Actual	75%	ok								1	8	20	4						33
10	Impl. - Estimate	0%													4	5	5	5	5	24
11	Actual	5%	ok											1						1
12	Total - Estimate			4	5	5	8	4	5	5	8	4	5	5	8	5	5	5	5	86
13	Actual			2	4	3	6	8	5	7	7	8	20	5	0	0	0	0	0	75
14																				
15	Contract	100%	ok	C																
16	Prelim. Present.	100%	ok			C														
17	Design Review	100%	ok								C									
18	Prototype Review	100%	ok											C						
19	Training Session	5%	ok															S		
20	Final Report	5%	ok																S	
21																				

Status / Sheet2 / Sheet3

Cornucopia with Visible Analyst

Figure 12-33 illustrates the state-transition diagram for the AutoOrder Form. This model defines each stable state of an object and then adds the triggers or events that cause the object to change to another state. It is an effective technique for mapping the graphical user interface dialogs that accompany a product. In this case, the analysts see the transitions that are triggered by selecting different options on the various menu screens and the AutoOrder Form.

FIGURE 12-33 / *Cornucopia State–Transition Diagram with Visible Analyst*

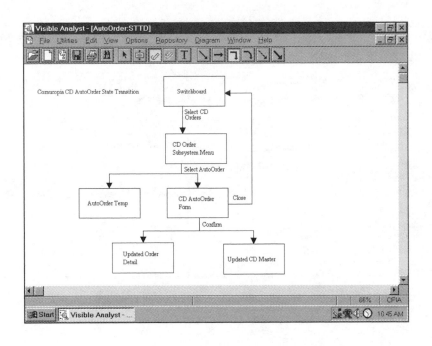

SUMMARY

This chapter holds out the promise of an analyst's toolbox that fully automates the transformation of a prototype into a finished product. The direct programming responsibilities of the analyst-programmer are important, and the newer 4GL development tools require a substantial amount of indirect knowledge of programming fundamentals.

An object-oriented future awaits every analyst, no matter whether the work is on small-, medium-, or large-enterprise projects. The challenge is to select one of these products and devote the time necessary to become a skilled user—they all have the same fundamental programming underpinnings. This chapter illustrates how to use a very small set of 4GL products to generate, customize, and integrate multiple subsystems into one cohesive information system.

TEST YOURSELF

Chapter Learning Objectives

1. How do you transform a prototype into an operational product?

2. What tools are available to the analyst during the development phase?

3. How do you develop graphical user interface dialogs (GUIDs)?

4. How do you blend 4GL-generated code with analyst-programmer code segments?

5. How do you develop effective file sharing between applications?

Chapter Content

1. In the small-enterprise environment, low-level programming languages, rather than high-level languages, offer the best tools for developing the system design into a real product. True or False?

2. Very high-level languages usually embody elements of both procedural and nonprocedural languages. True or False?

3. Given the accessibility of very high-level languages and powerful application builders, the analyst does not need to know anything about programming. True or False?

4. Object-oriented programming (OOP) incorporates several significantly different concepts into computer programming. True or False?

5. In order to effectively use an objectlike application builder, the analyst must possess some appreciation for computer programming concepts and terminology. True or False?

6. The combination of automatic code generation and goal-directed programming commands is commonly called 4GL programming. True or False?

7. A graphical user interface dialog (GUID) serves the same purpose as a state-transition diagram. True or False?

8. The analyst should expect to hand code a significant part of the information system design. True or False?

9. Program development is made much easier by the use of 4GL products, but the task still demands great attention to detail and strict adherence to proven systems development methodologies. True or False?

10. 4GL programming is a catchall term used to describe activities associated with a collection of very high-level languages. True or False?

ACTIVITIES

Activity 1

Review the table relationships for the Political Research Corporation (Figure 12-8).

a. Carefully inspect the fields of SurveyQuestion, VoterResponse, and SurveyVoter. The analyst, feeling SurveyQuestion is misnamed, changes the table name to SurveySummary. Explain the reasoning behind this name change.

b. If a particular survey includes 20 questions, and 50 voters provide a response to each question, how many records with the same SurveyID and SampleVoterID are there in VoterResponse?

c. Design a form to display the survey response from a sample voter.

Activity 2

Review the Cornucopia event procedure named cmdNextRec_Click (Figure 12-21). In the statement "cboCustomerID.SetFocus," cboCustomerID is the name of a combo box object. Use the Microsoft Knowledgebase at *www.microsoft.com* to research the Visual Basic command SetFocus.

Explain how this command exemplifies object-oriented programming.

Activity 3

Use your favorite Internet search engine to locate information on 4GL programming. Prepare a summary of your findings from two different sites.

DISCUSSION QUESTIONS

1. Describe how you would apply the concept of functional obsolescence to evaluate an existing information system that is based on a third-generation language.

2. In what way is a new VCR, with audio programming features, similar to nonprocedural programming?

3. Object-oriented 4GL products are described by some as the perfect tool for application developers. Considering what you have experienced with programming languages, explain why you think this description is justified or not justified.

4. Assume that your system analysis and design work, which is virtually complete, calls for the use of a well-established 4GL product such as Access. How would you respond to the suggestion that you switch to a newly announced version of the product?

5. Given the similarities between GUIDs and UML statechart diagrams, which technique should you invest time in learning?

6. What is the difference between object linking, as described in Chapter 10, and the sales/profit event procedure used in the Cornucopia case (Figure 12-27)?

7. How would you respond to the want ad, "Systems analyst wanted—only programmers need apply"?

8. After evaluating the Time and Money discussion in this chapter and Figures 11-27 and 12-32, develop a response to the user's question, "Will the savings associated with software purchases result in lowered project costs?"

PortfolioProject

Team Assignment 12: 4GL Programming

This assignment is designed to document your development methods. Understanding that your efforts are concentrated on development activities, which can be all consuming, you are required to submit brief summaries, annotations, or examples of the following:

1. Describe your implementing software (Access, Excel, Word, Visual Basic, FrontPage, etc.). Submit a copy of your description to your instructor.
2. Summarize your use of customizing features of your implementing software (event procedures, macros, SQL, etc.). Include at least one example. Submit a copy of your summary and example to your instructor.
3. Describe the ways in which you integrate different implementing software (OLE, import, export, etc.). Submit a copy of your description to your instructor.
4. Annotate your detailed system flowchart with descriptions of the processing performed within each subsystem. Submit a copy of your detailed system flowchart and annotations to your instructor.
5. File a copy of your summaries, annotations, and examples in the Portfolio Binder behind the tab labeled "Assignment 12."

Project Deliverable: Prototype Review Session

1. Submit a prototype review report containing the following:
   a. Cover letter
   b. USD and menu tree with prototyped segments highlighted
   c. Summary of any major design changes since the design review report, including revised DFD and ERD
   d. Prototype GUID screen images, reports, and Web pages
   e. Project status report
   f. Project budget
   g. Oral presentation slide show handout (three slides per page)
2. Prepare a slide show and appropriate handouts to support a 20–30 minute oral presentation covering items 1b through 1f.
3. File a copy of your report in the Portfolio Binder behind the tab labeled "Prototype Review Session."

networking

When you complete this chapter, you will be able to:

OVERVIEW

Small-enterprise computer information systems are designed and developed for installation on a specific computer system. Chapter 11 discusses the specifications of the required system resources. This chapter assumes that those resources are now in place and that the analyst-programmer team is almost finished with product development. The next challenge is to install the product on a specific platform and environment.

The variety of hardware and software options available and the pace with which the industry introduces new options is staggering. Furthermore, small-enterprise information system analysts normally do not have the benefit of a specialized, in-house computer operations staff. In fact, they may find the existing staff to be very apprehensive about the operating complexity of their new system. This chapter first addresses the major issues concerning installation on PC platforms and then introduces networking as a potential solution to cross-platform computing and information system dispersal. This chapter also introduces system resource performance and management principles.

- Describe the key hardware issues relating to the system under development.

- Describe the key software issues relating to the system under development.

- Install an information system that complies with the computer's hardware and system software environment.

- Describe how networking technology influences small-enterprise information systems.

- Describe how Internet and intranet technologies influence small-enterprise information systems.

THE PC SOLUTION

Inherent in all the previous chapters is the assumption that the small-enterprise information system could function in an independent desktop computer environment. This assumption has been valid only since the early 1990s, and it will certainly not remain so for long. In the noncorporate marketplace, networking and mobile computing will soon become the norm, introducing powerful computing options. Nevertheless, this brief period in which the small-enterprise system remains relatively independent affords the analyst the opportunity to see clearly the relationship between the hardware and specific software adaptations and industry-prescribed system software.

Over the years, the term PC has come to describe those systems based on the Intel microprocessor, characterized by an *open architecture*, in which individual corporate proprietary interests are kept to a minimum. PCs dominate the market, in part, because of the abundance of software and ancillary hardware developed due to this open architecture.

Any of dozens of company names might appear on the front panel of a PC. Regardless of a well-publicized record of accomplishment for performance, reliability, and affordability of each machine, all PCs have one thing in common—the basic microprocessor design. This lends a sense of standardization to an otherwise confusing industry. Figure 13-1 shows the history of the PC microprocessor.

FIGURE 13-1 / *Evolution of the PC Microprocessor*

Processor	Introduced	Speed	Word Size	Bus Width
8086	1978	4.77 – 8 MHz	16 bits	16 bits
80286	1982	8 – 20 MHz	16 bits	16 bits
80386	1985	16 – 66 MHz	32 bits	32 bits
80486	1989	33 – 100 MHz	32 bits	32 bits
Pentium	1993	60 – 200 MHz	32 bits	64 bits
Pentium Pro	1995	133 – 200 MHz	32 bits	64 bits
Pentium II	1997	266 – 333 MHz	32 bits	64 bits
Pentium III	1999	.45 – 1.13 GHz	64 bits	64–128 bits
Pentium 4	2000	1.4 – .... GHz	64 bits	128 bits

PC Hardware Issues

Although an exhaustive treatment of hardware technology is beyond the scope of this text, several critical hardware-related features influence the small-enterprise analyst's work. To establish a common frame of reference, Figure 13-2 presents a gross simplification of the hardware architecture for a generic PC desktop computer.

FIGURE 13-2 / *PC Hardware Diagram*

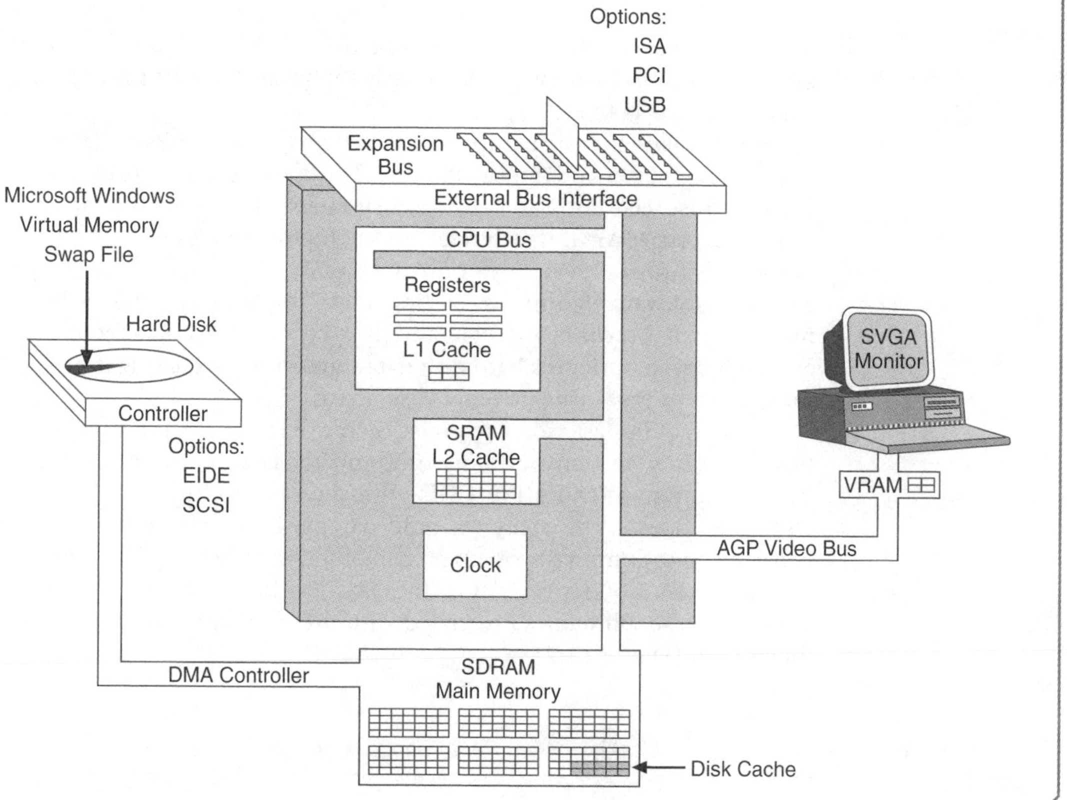

Options:
ISA
PCI
USB

Expansion Bus

External Bus Interface

CPU Bus

Registers

L1 Cache

Microsoft Windows
Virtual Memory
Swap File

Hard Disk

Controller

Options:
EIDE
SCSI

SRAM
L2 Cache

Clock

SVGA
Monitor

VRAM

AGP Video Bus

DMA Controller

SDRAM
Main Memory

Disk Cache

The *register width* (measured in bits) determines the instruction formats, data manipulation parameters, and addressing capabilities of the microprocessor. The CPU *bus width* (measured in bits) determines the amount of data or instruction that can be moved from one place to another at one time. The speed (measured in megahertz or gigahertz) sets the rhythmic pace for microprocessor operations. The *memory cache* serves to supercharge memory access. All these features are carefully matched to complement one another, with the overall goal of maximizing the number of instructions that can be performed in a given period. The analyst must also match the software-processing requirements of the information system to the processing capacity of the hardware.

If the information system requires specialized peripherals, such as a modem or scanner, the expansion bus becomes a concern to the analyst. The *expansion bus* is specialized circuitry that permits peripheral devices to connect to the microprocessor. The speed differential between the expansion bus and the CPU bus, along with the width and architecture options, make this a particularly puzzling problem. Although the microprocessor may work at speeds measured in gigahertz, the expansion bus is much slower in order to avoid timing conflicts between the speedy CPUs and the slower peripherals. Fortunately, there are alternatives to the expansion bus when the information system is critically dependent upon high-speed I/O. A *local bus,* which is a specialized communication link between components, provides a remedy to the I/O bottleneck. A local bus speeds up transfer rates considerably by moving data directly to the CPU, unlike the expansion bus, which moves data to RAM.

Most small-enterprise information systems require substantial disk access to maintain master and transaction files. Fortunately, hardware designs accommodate this need fairly well using customized disk control logic and *direct memory access (DMA),* which refers to the high-speed transfer of data directly between peripheral devices and memory, without going through the CPU. In addition, software algorithms can be used to implement a disk cache in main memory. A *disk cache* is a special random access memory area reserved to serve as a high-speed buffer for hard disk data transfer. Disk caching is probably the most efficient way to supercharge disk access.

GUIs require fast video response. Because the speed-inhibited expansion bus is too slow to transport large amounts of data in a timely fashion, *video bus* technology provides the GUIs with direct access to the CPU bus's full width and speed. The analyst should pay particular attention to this detail, considering the emphasis on graphical, object-oriented applications.

Figure 13-3 presents a list of these critical hardware performance issues, along with the full names associated with the acronyms used in the hardware illustration (Figure 13-2).

FIGURE 13-3 / *Critical Hardware Performance Issues and Acronyms*

Hardware Performance Issues:
1. The processor (word size, speed, CPU bus, cache)
2. Expansion bus
3. Disk caching
4. Local-bus video

Acronyms:
1. ISA—Industry Standard Architecture
2. PCI—Peripheral Component Interconnect
3. USB—Universal Serial Bus
4. AGP—Accelerated Graphics Port
5. SVGA—Super Video Graphics Display
6. DMA—Direct Memory Access
7. EIDE—Enhanced Integrated Drive Electronics
8. SCSI—Small Computer System Interface
9. SRAM—Static Random Access Memory
10. SDRAM—Synchronous Dynamic Random Access Memory
11. VRAM—Video Random Access Memory

PC System Software Issues

Most information system computers need an operating system (OS) to manage system resources and make the computer easier to use. Although at a slower pace, desktop computer operating systems have evolved right along with the microprocessors they serve. Many of the features of these operating systems were replicated from the more sophisticated mini and mainframe computer systems. As with hardware, the small-enterprise systems analyst must address several critical system software issues. Figure 13-4 presents a generic PC operating system diagram.

FIGURE 13-4   *PC Operating System Diagram*

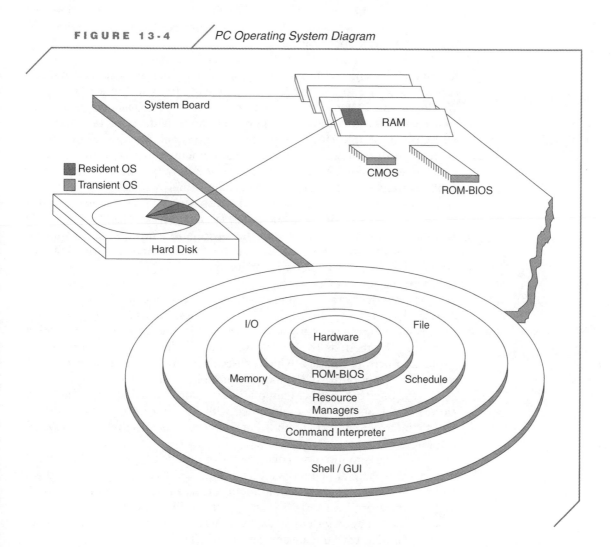

The upper portion of this illustration shows how the operating system is implemented on the hardware. As you can see, *ROM-BIOS* and *CMOS* are implemented on a chip, which explains their hybrid nickname, "firmware."

ROM-BIOS contains a small set of instructions to direct the CPU to load the *resident operating system* instructions from a disk drive. The resident operating system consists of the most frequently used instructions and, as the name implies, remains in memory. In contrast, the infrequently used *transient operating system* instructions are loaded into memory only when needed. CMOS contains changeable information concerning the status of peripheral devices. Some operating systems have utilities that can automatically update CMOS to accommodate plug-and-play features. The rest of the operating system is loaded onto the hard disk, with only the essential resident commands loaded into main memory. The principal concern here is the size of the operating system, with substantial disk requirements and RAM requirements. Obviously, the analyst must accommodate these specifications in conjunction with the hardware specifications.

The lower portion of Figure 13-4 shows the operating system as a series of functional elements, within a concentric circle model. Imagine that the user issues a disk directory command to open a file from within a GUI such as Windows. The GUI passes the user command to the *command interpreter*, which validates and associates user commands with a set of microcode. *Microcode* are simple, single-purpose instructions tailored to work on specific microprocessing circuitry. Next, three components of the resource manager work on the request. The *schedule manager* must evaluate the request in terms of the current system resource activities and other requests pending action. Once the user's file directory command is dispatched, the *I/O manager* secures access to the disk device, and then the *file manager* executes the microcode. The ROM-BIOS provides help with the device and hardware specific instructions. In the case of an open command, the *memory manager* must secure sufficient memory to load the appropriate application program, as well as the requested file.

The preceding example illustrates the way the operating system coordinates and sequences complex hardware activities. *Interrupt management* is a strategy whereby the currently active operation is interrupted to serve another, higher-priority request. The user's initial mouse click or keyboard operation creates an interrupt message, which causes the operating system to halt its current action temporarily so that it can respond in some way. Depending on priorities, the interrupt is either temporarily set aside or acted upon immediately. Normally, the analyst is not concerned with this, but in some cases, device-specific interrupt conflicts require some adjustments in setting up the system software.

A final background topic concerns the multitude of system software products; currently several major operating system choices are available: Windows, Linux, Unix, and Mac OS. In addition, of course, a staggering number of desktop, file, memory, and system management utilities are available. To make wise decisions about system software resources, the small-enterprise analyst must be familiar with the advantages of each. This text addresses several critical system software issues, rather than discussing these products individually. Figure 13-5 details these issues.

FIGURE 13-5   / *Critical System Software Issues*

1. Speed and user interface
2. Application software compatibility
3. Resource requirements (disk and RAM)
4. Resource management features
5. Multitasking and multithreading capabilities
6. Multiuser and multiprocessor capabilities
7. Portability

Fast, responsive, user-friendly operating systems are the norm rather than the exception, so long as the system is matched properly to the hardware. Microprocessor speed, bus width, disk and memory cache, and disk and ROM capacity are the critical hardware issues here. As mentioned in previous chapters, overbuying on the hardware platform and accessories can prolong the useful life of an information system because it leaves room for hardware and software upgrades. However, this is not necessarily the way to approach operating system selection.

Application software compatibility is critical if the information system is to function properly. Purchasing the very latest, high-end operating system may actually impede the overall information system performance because of the time-consuming system maintenance and management functions that are required in a complex operating system environment. Furthermore, the introduction of operating system innovations and upgrades is accelerating, so your advanced operating system may become obsolete before you ever use the advanced features. Finally, operating system upgrades and conversions are relatively inexpensive, so delaying your purchase does not cause major financial problems during the system development life cycle.

In addition to the need for substantial disk and RAM space for modern operating systems, you should also examine the way in which the operating system manages the system resources. The swap file, virtual memory, and disk cache features illustrated in Figure 13-2 are examples of essential resource management techniques. A *swap file* is a disk area the operating system uses to store data as if the data were in RAM, thus virtually increasing the amount of memory available to the system. *Virtual memory* allows your system to work with extremely large application programs and makes it possible for the operating system to implement multitasking. Disk caching dramatically improves the performance of database intensive applications.

Multitasking allows more than one program to run on the computer at the same time by giving each task a "slice" of CPU time. A single foreground program is available for interactive user interfacing, while background programs are processing out of sight, so to speak. This capability can be particularly useful in small-enterprise settings with only one computer. For example, the information system may primarily perform online transaction processing, but when the need arises for a large master file backup or data communications file transfer, the multitasking operating system can launch the appropriate subsystem in the background.

Multithreading allows the CPU to work simultaneously on multiple processing threads, or instruction segments, of a program. The theoretical advantages to this operating system technique are obvious, but the practical application for small-enterprise situations is not. Multithreading is really best suited for processing intensive applications rather than a modest file-processing information system.

Multiuser, multiprocessor, and portability issues are on the cutting edge of operating system features. Although it is unlikely that any of these would be the deciding factor in the selection of an operating system for a Stage-I (see Figure 7-8) small-enterprise information system, the analyst should carefully consider the discussion on networking issues later in this chapter.

PC Information System Installation

Hardware, system software, 4GL software, and customized software combine to create the *system environment*. The task is to make all of these components correspond in conjunction with the other system components (people, procedures, networks, and data), into the seamless information system discussed in Chapter 1. Figure 13-6 illustrates the layered relationships between hardware and users.

As discussed in the past several chapters, the development phase of the system development life cycle (SDLC) is itself a process. With your hardware, system software, and 4GL products in place, the work to transform the prototype and full system design into a working product becomes very specific. Subsystems, although developed and tested (as discussed in Chapter 14) separately under very controlled, workbench-like conditions, must be easily launched via a common user interface. It is impractical to assume that users will be able to navigate operating system and 4GL interfaces in order to locate and launch their information system. The analyst must install the finished product on the hardware-software platform so that the startup procedure is obvious to the user.

Turnkey systems offer the ultimate information system packaging in that the computer is configured to boot directly into the information system interface. In such cases, the user may not even be aware of the operating system or the 4GL products that are parts of the system.

FIGURE 13-6   *System Environment*

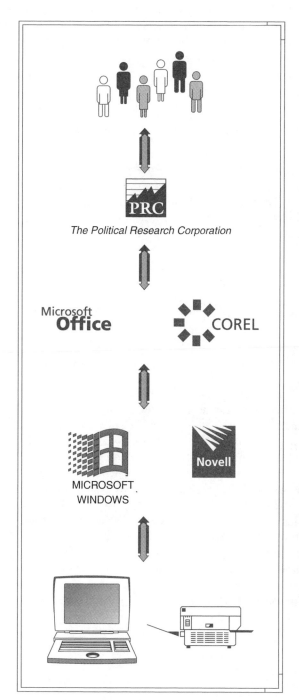

Users

Information Systems

4GL Products

System Software

System Hardware

Another approach is to place the information system interface into the operating system interface, thus allowing the user to choose from any number of applications, with your information system assuming the same access level as system utilities and 4GL packages. Users of different experience levels can then select the tool appropriate for their task. It should be no surprise that users who were once mortified by the computer come to embrace the technology—if they can do so on their own terms. The Cornucopia case study employs this same approach.

Information System Installation: Step-by-Step

Figures 13-7 and 13-8 briefly illustrate how analysts might install the PRC information system on a PC running Windows XP. The first step is to add new user accounts. To begin, click the Start menu on your Windows taskbar, select Settings, and click Control Panel. Double-click User Accounts to open the first User Accounts dialog box (Figure 13-7). Select Create a new account to open the second dialog box in Figure 13-7. After entering the new user name, click the Next button to open the third dialog box (Figure 13-7). Select the Limited account type, and click the Create Account button. To add a password to the new account, return to the first dialog box (Figure 13-7), and select Change an account.

The administrator account type is reserved for the person responsible for managing the computer environment. The administrator may establish groups, to which users can be added, and Group Policies, which control user access to system functions. For example, a Group Policy can customize the desktop view and Start menu for each user. These features permit the administrator to establish minimal system startup security.

The analyst can allow easy access to the information system by adding a reference on the Start menu, and placing a shortcut icon on the desktop (Figure 13-8). First, right-click on the Windows taskbar and click Properties. The Taskbar and Start Menu Properties dialog box is displayed. From the Start Menu tab, click the Customize button to open the Customize Classic Start Menu dialog box shown in Figure 13-8. Use this dialog box to add or remove items from the Start menu.

To create a desktop shortcut icon, right-click the file to which you want to make a shortcut and drag it from Windows Explorer to the desktop. Release the right mouse button, and click Create Shortcut. The Shortcut to PRC Information System Properties dialog box shown in Figure 13-8 is displayed when you right-click the new shortcut and click Properties. You can proceed to set the target path, desktop icon, and execution parameters.

FIGURE 13-7 / *Windows XP User Accounts Dialog Boxes*

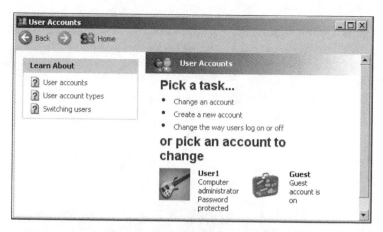

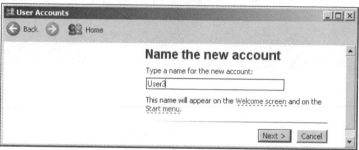

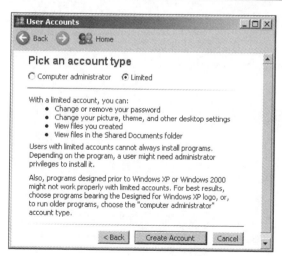

FIGURE 13-8   /   *Adding PRC to the Start Menu and as a Shortcut Icon*

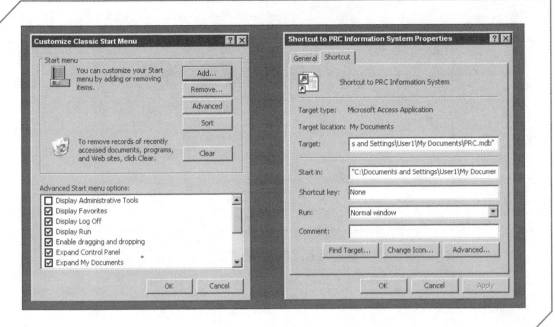

Figure 13-9 presents the Security menu option associated with a Microsoft Access database. Security begins with a high-level database password and continues down into specific form and database object protection for individuals, as well as groups of individuals. Because passwords inhibit free flow for developers, such protection is usually reserved for the final stages of development, testing, and installation.

FIGURE 13-9 / *Information System Security*

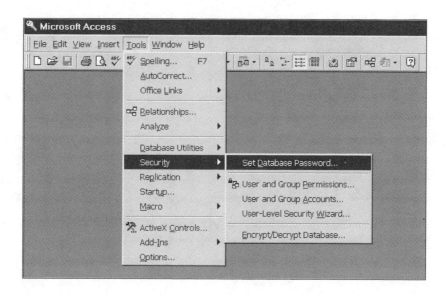

NETWORKING SOLUTIONS

Networking is fast becoming the accepted baseline standard for information systems of all sizes. The small enterprise can begin with a simple modem connection to the Internet and a peer-to-peer network, and then progress to file server and intranet. The hardware and software products necessary to establish these connections are readily available, affordable, and reliable. In fact, this may accelerate the migration of the small enterprise from a Stage-I to a Stage-II information system (Figure 7-3). Accordingly, this section of the text serves two purposes. First, it complements the discussions on hardware and software issues by presenting a very broad introduction to network technology. Second, it alerts the analyst to the need to consider the present and future networking requirements of the enterprise.

Figure 13-10 presents a network diagram that combines several communications technologies. First, an internal LAN connects all enterprise users to a single file server through a switching device called a hub. The file server runs a *network operating system (NOS)*, such as Windows NT or Novell NetWare, that provides network management services, application distribution, file sharing, and peripheral sharing. Second, a Web server employs *Web server software* to store and transmit the enterprise's Web pages over the Internet. *Firewall software* is installed on the Web server to protect against unauthorized Internet access to the LAN server. Third, an enterprise intranet allows internal and external users access to special enterprise Web pages. You will learn more about Internet and intranet technologies later in this chapter.

FIGURE 13-10 / *Network Diagram*

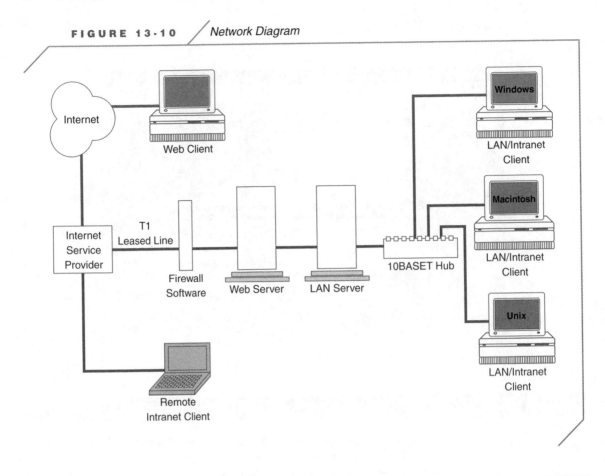

Networking Hardware Issues

A network connects computers in both a physical and logical relationship. The physical relationship allows the machines to pass instructions and data from one to another. The logical relationship establishes one computer as a *server*, or provider of services, and another computer as a *client*, or user of services.

Sometimes the physical connection is established with radio, microwave, or satellite transmissions, but the most common connector remains cabling of some sort. Different cabling types (coaxial, twisted-wire pairs, fiber-optic) provide different transmission capabilities. The connecting medium's transmission capacity is its *bandwidth*. In general, narrowband media (twisted-wire pairs) transmit data much more slowly than do broadband (coaxial cable) or wideband (fiber-optic) media. For example, Ethernet is a popular local area network protocol that uses cable to obtain transmission speeds of up to 10 Mbps (10 million bits per second). A newer, faster Ethernet standard delivers 100 Mbps transmission speeds. A communications *protocol* is a set of processing and transmission instructions that govern such things as the speed at which data moves across the network and the way data errors are detected and corrected. Naturally, each of these cabling systems has different costs, which must be justified by the network's transmission performance requirements.

At each *node*, or client station, of the network, the cabling system connects to a *network interface card (NIC)*, which is usually tied into the computer's expansion bus. Such an arrangement, as discussed earlier in this chapter, can be subject to speed and width restrictions on the data path. One solution to this potential bottleneck is to connect the network card directly into a local expansion bus to benefit from the same performance advantage associated with the video bus.

Beyond the cabling, network interface card, and expansion bus connection, several other hardware issues are associated with the logical relationship between the server and client computers. In particular, the analyst should be concerned about the differing needs of the server and client with respect to microprocessor speed, memory capacity, disk storage capacity, and monitor quality. In the typical setting, the server needs a fast microprocessor, a lot of fast-access memory, an easily expandable disk storage capacity, and a low-cost monitor. The client needs a microprocessor with sufficient speed to handle the applications it is served, a memory capacity that meets the application recommendations, a small amount of disk storage for temporary files, and a color monitor. The client station so equipped is sometimes called a *thin client*, deriving its name from the fact that it is considerably less powerful than the server is.

Finally, the logical relationship extends to the manner in which the computers serve or are served by one another. The typical local area network establishes one computer as a *dedicated server* and all others as clients. This restricts the server to the sole function of providing network services. The small enterprise, however, may be interested in a simpler *peer-to-peer network*, which allows each computer to take on either the client or the server role, depending upon the changing resource-sharing needs of the enterprise.

The Wireless LAN

A wireless LAN makes use of a wireless transmission medium to network client stations. The server, hub, and client stations are equipped with a transceiver to send and receive signals. Although this frees clients from stationary, hardwired configurations, it still imposes a restricted service area of 300–1000 feet from the hub.

There are several advantages to a wireless LAN. Hard-to-wire work areas are no longer off limits to network services. For example, relatively inexpensive wireless PC expansion cards, tailored to fit into a laptop's *PCMCIA slot*, make it possible to quickly assemble ad-hoc network communities in conference rooms, classrooms, and lecture halls.

There are some potential, but solvable, problems with wireless LANs. They require special equipment, are subject to the vagaries of battery-operated client stations, and transmissions are vulnerable to interruption, interception, and interference.

One very common application of this technology is to connect a wireless LAN to an existing wired LAN. Figure 13-11 presents a configuration in which a control module provides the same service as a hub, in this case coordinating signals to and from wireless clients.

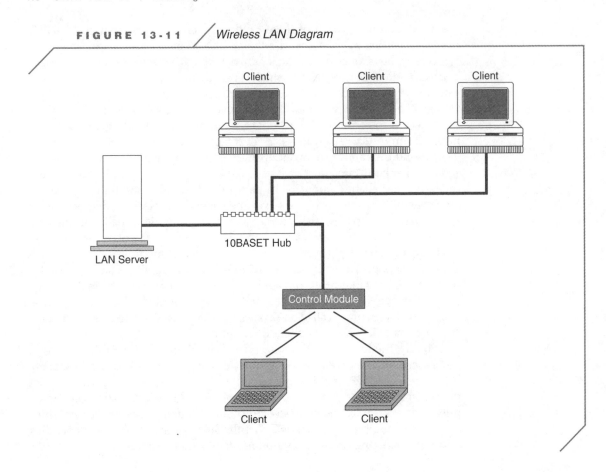

FIGURE 13-11 / *Wireless LAN Diagram*

Networking Software Issues

In a dedicated server environment, the network operating system (NOS) is resident on the server, while a small amount of network software is loaded on each node. A NOS is, in fact, an operating system with many of the same functions as described earlier in this chapter. Of course, a NOS must also provide network services, such as file transfers, application software access, and peripheral sharing. One significant difference between a NOS and a single-system operating system is that a network specialist, appropriately referred to as a *network administrator*, uses the NOS interface. NOS interfaces generally are not designed for the everyday computer user. Instead, the NOS interface presents the network administrator with a collection of tools to establish authorized user accounts and access privileges, to control network-friendly system and application software distribution, to monitor network utilization, and to maintain network backup and security.

The peer-to-peer environment requires the analyst to install the networking software on each computer system, thereby emphasizing the accessibility of the network interface because it is available to every user.

TECHNOTE 13-1
TCO AND THE NET PC

Total cost of ownership (TCO) is a term that describes the growing concern for the maintenance of information system resources. Analysts realize that the need to upgrade continually often compounds the initial investment in hardware, software, and training. The search for the perfectly configured computer confronts the analyst with questions about how much resource (speed, memory, disk space, etc.) is required for network client stations. In theory, a **thin client** is an inexpensive (less than $500) computer with just enough resource to communicate with the server, leaving the "heavy lifting" to more expensive and robust workstations that are employed as LAN or Web servers. In practice, the thin client with its promised reduced TCO has been hard to achieve.

The **NetPC** is a standardized, sealed-case computer that is centrally configured and managed.

It is similar to your television set or radio in that it comes to the user in a preset configuration with well-defined functionality and very little expandability. A consortium of high-profile computer companies developed the NetPC specifications. Briefly, the NetPC is designed to be a stable thin client, in that it is set up at a central location, placed at a client site, and left alone. In the basic configuration, there are no expansion slots, no floppy disk drives, and no CD-ROM drives. The idea is to reduce the maintenance costs of client stations, thus reducing the TCO of the overall information system.

The persistent advances in hardware performance, increases in software resource requirements, developments of efficient application languages such as Java, and lowering desktop computer costs have made analysts reconsider their initial fascination with thin client computing. As with many other good ideas, accelerating technological innovation often supersedes the implementation of the idea.

After any concerns about the network interface are resolved, the analyst must consider the system and application software needs of the client stations. In a dedicated server environment, each client station must have access to a distributed or native operating system. Therefore, either the local operating system resides on the client's hard disk or the NOS needs to serve it to the client. Either way, the analyst must make some special provisions for client operating system services. With regard to application software, the analyst must acquire network editions of horizontal software, which software developers design for installation on the server system and distribution by the NOS.

In the largest sense, network software is supposed to facilitate the user's access to software and hardware resources across the entire network without compromising client system performance. To do this, the analyst must be sure that the network server distributes only horizontal software that is compatible with the client system's operating system.

Client/Server Computing

Client/server computing defines a particular logical and physical environment of computing resources, including hardware and software. As you learned earlier in this chapter, servers come in many varieties. File servers and Web servers are two examples. The basic idea is that servers connect client stations together and provide services that encourage the sharing of software, hardware, data, and processing workload. Consider the many possibilities of such an arrangement with respect to database applications. Where do the database tables reside? The server is the logical answer if you expect all clients to share the data. If so, does the server transmit entire tables to clients, or just the data requested by client inquires? Furthermore, which hardware platform, client or server, is responsible for performing the actual table searches and computation?

In a *file server* environment (Figure 13-12), the database is stored on the server, but each client is expected to perform the database operations, which requires the distribution of the entire database from the server to the client. The amount of network traffic generated by this arrangement depends on the size of the database and the frequency of client requests. Thus, the analyst must evaluate the merits of network implementation versus stand-alone implementation based on databases such as Microsoft Access and Borland dBase.

FIGURE 13-12 / *File Server Diagram*

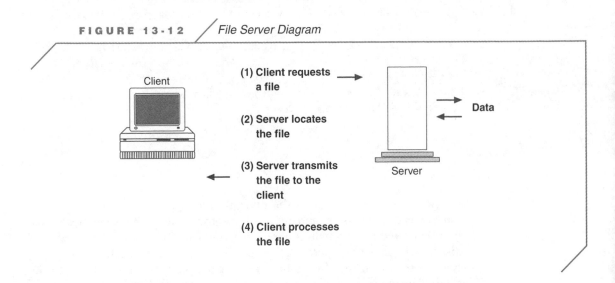

A *database server* environment (Figure 13-13) provides a remedy for high-traffic situations by not only storing the database on the server, but also processing much of the database logic on the server. Oracle and Microsoft provide database products for this market.

Frequently, the client/server model is applied to the Internet, making portions of an enterprise database available via Web pages. You will learn more about this important e-commerce application later in this chapter.

Networking-Induced CIS Changes

These added network hardware and software concerns obviously complicate the analyst's work and increase expenses. However, networking offers the potential for substantial long-term savings and increases in productivity. Understanding

FIGURE 13-13 / *Database Server Diagram*

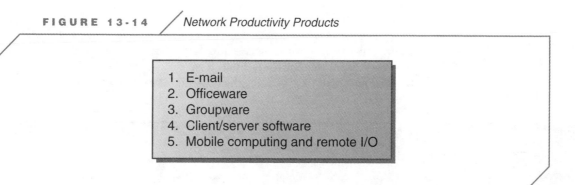

that even the most elementary computer information system influences the way we work, we can say with some degree of confidence that several relatively new, network-based software products have brought dramatic changes to the workplace. Figure 13-14 lists these product categories, most of which are appropriate for a small enterprise that has migrated to a networked system.

FIGURE 13-14 / *Network Productivity Products*

1. E-mail
2. Officeware
3. Groupware
4. Client/server software
5. Mobile computing and remote I/O

The first four items on this list enable the enterprise to implement management principles that rely on the natural tendencies of workers to collaborate. Electronic mail, or e-mail, is usually the first product workers use. The term *officeware* describes a collection, or suite, of horizontal software products that are assembled and presented to users as a single office package. Such a product might include e-mail with calendar and scheduling maintenance, word processing, spreadsheets, and presentation graphics. *Groupware* goes further by allowing workers to access and update the same documents, with the resulting changes available to the entire group. Finally, *client/server software* streamlines the delivery of specific information to clients (network nodes) from the server.

Mobile computing and *remote I/O* applications free network users from their wired confines. Such distribution of data collection and reporting activities usually improves the quality of information by reducing input errors and encouraging end-user feedback as workers throughout the enterprise become active participants in the computer information system.

At the outset, the analyst must exercise extreme caution in adopting any but the most basic of these products. The people and procedures elements of the information system can easily become difficult to manage when too much innovation is applied. However, if it seems likely that these sophisticated solutions will be required in the near future, your hardware platform and software operating system specifications should include sufficient resources to accommodate such a migration.

The SDLC and Networking

Networking solutions are subject to the same cyclical model discussed throughout this book. The analysis, design, development, implementation, and maintenance phases operate regardless of the hardware and software mix of the system. Perhaps even more important, this also holds true for the recurring SDLC concept introduced in Chapter 1 (Figure 1-6). It is reasonable to assume that as the initial information system reaches functional obsolescence, the replacement design will incorporate networking fundamentals. The analyst can then gradually implement networking updates during the system maintenance phase.

To illustrate such a situation, Figures 13-15 and 13-16 present a synopsis of the hardware and software upgrade costs associated with the Political Research Corporation's migration to a simple peer-to-peer network. One advantage to this peer-to-peer upgrade is that the PRC staff will be able to organize into an electronic workgroup, with e-mail and shared access to data files. In addition, the entire staff will have easy access to the laser printer and Internet. Figure 13-17 illustrates the specifics of a file server upgrade.

ThinkingCritically

E-mailPrivacy

Rocky Mountain Cycles wholesales high-performance touring motorcycles throughout the United States. On a daily basis, its remote sales force teleprocesses orders to the national headquarters in Denver, Colorado. In addition, e-mail software installed on the network server services the entire company.

Recently, the network administrator noticed that e-mail traffic increased more than 30 percent each month over the last six months. Suspecting that much of this increase was due to an informal sports betting pool, the network administrator initiated a message sampling procedure. Each morning she looks for system abusers by reading about a dozen e-mail messages, selected at random from the previous day's history file. Although no evidence of misuse has surfaced, the sampling procedure persists after several months.

To what degree is the analyst responsible for advising network users and administrators about unauthorized use of network resources? What course of action, if any, should the analyst recommend to the network administrator? How might you react if you discovered that the network administrator had read your e-mail?

FIGURE 13-15 / *PRC's Peer-to-Peer LAN Upgrade Specifications*

Existing System:
Three stand-alone 650 MHz Pentium desktop computers, each with 128 MB of RAM, 10 GB of hard disk storage, 24X CD-ROM, Windows, and Microsoft Office Pro.

One 15 ppm, 300 dpi, 8 MB RAM laser printer, with an electronic switch connected to the three microcomputers.

System Upgrade:
Client Hardware:

Ethernet network interface cards (3 @ $35)	$105
Cat 5 twisted pair (1 box)	70
Cable connectors	10
256 MB RAM (3 @ 50)	150
4-port Cable/DSL Web Safe Router	130
Printer-server interface card	350
8 MB printer memory	50

Client Software:

Microsoft Office Pro upgrade to the latest version (3 client license)	300

Subtotal	$1,165
Tax and Shipping	120
Installation (3 hours @ $80)	240

Grand Total	$1,525

FIGURE 13-16 / *PRC's Peer-to-Peer LAN Upgrade Diagram*

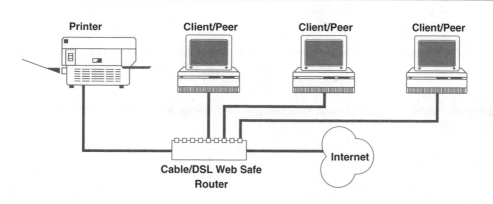

FIGURE 13-17 / *PRC's File Server LAN Upgrade Specifications*

Existing System:
   Three stand-alone 650 MHz Pentium desktop computers, each with 128 MB of RAM, 10 GB of hard disk storage, 24X CD-ROM, Windows, and Microsoft Office Pro.

   One 15 ppm, 300 dpi, 8 MB RAM laser printer, with an electronic switch connected to the three microcomputers.

System Upgrade:
   Client Hardware:

Ethernet network interface cards (3 @ $35)...$105
Cat 5 twisted pair (1 box)...70
Cable connectors..10
256 MB RAM (3 @ 50)...150
4-port Cable/DSL Web Safe Router...130
Printer-server interface card..350
8 MB printer memory..50

   Client Software:

Microsoft Office Pro
upgrade to the latest version (3 client license)...300

   Server Bundle:

Dell PowerEdge 1400SC (Small-Business Server)...................................4,744
   Intel Pentium III 1.4 GHz w/512K cache
   512 MB SDRAM
   15 inch monitor
   36 GB hard drive
   3.5 inch disk drive
   48X CD-ROM
   Operating system: Microsoft Small Business Server 2000
   UPS: APC SmartUPS 700, 700VA
   Tape backup unit: PowerVault 100T 20/40 G Internal
   Tape backup software: CA Arcserve Server

Subtotal...$5,909
Tax and Shipping..600
Installation (8 hours @ $80)...640

Grand Total...$7,149

INTERNET, INTRANET, AND EXTRANET SOLUTIONS

The Internet is a network of networks, permitting users to access millions of Web sites worldwide. An intranet is an enterprise-specific Internet, with special protections built in to prevent unauthorized access from outside the enterprise. Intranets normally require the commitment of some enterprise resource to lease private communication lines, manage usage, and secure the network. An *extranet* extends this concept to include enterprise partnerships. The extranet, illustrated in Figure 13-18, connects an enterprise LAN to other enterprise LANs via the Internet. Such business-to-business (B2B) electronic partnerships are a very common form of e-commerce.

The *virtual private network (VPN)* is an alternative to an enterprise-specific intranet. VPNs combine the technology of the Internet and intranet by establishing a private network via a public network. Through VPN technology, the enterprise transforms its private LAN into a virtual, worldwide private network, without needing to lease lines and otherwise be concerned about the network beyond their own LAN. This concept can be extended to create a *virtual intranet* (connecting several enterprise LANs) or a *virtual extranet* (connecting enterprise LANs to partner LANs). VPN technology, which includes specialized hardware and software, is available from communications equipment manufacturers, such as Cisco Systems, Inc.

FIGURE 13-18 / *Extranet Diagram*

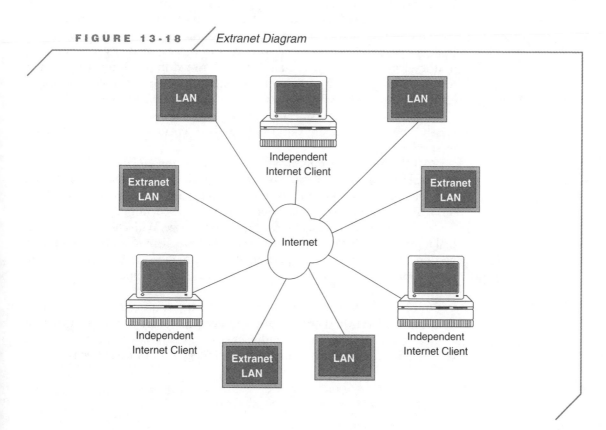

Historically, intranets, extranets, and the Internet do not provide computer information system processing in the sense that we have defined it in the enhanced SDLC. Fundamentally, these technologies have been a transporting medium for information products. Thus, they have supplied collection and distribution services for information systems, but they have not updated master files, sorted tables, prepared formatted text documents, or the like.

More recently, however, these technologies have expanded to include powerful tools and languages that can implement information system processing across the Web. Therefore, the analyst must consider the Internet and intranet as a means to distribute processing, as well as simply an alternative or complementary communication system. For example, Microsoft Office products are Web enabled, in that they provide a platform from which to interact with the Internet. Users can embed a Web address anywhere in a Word document to link to other Office products or any site on the Web. More sophisticated Internet-distributed processing requires use of a programming language such as Java, but a growing number of simplified scripting languages are available to make this task easier. If you have searched the Web recently, no doubt you have found that most pages have come alive with interactive options.

Web Page Interactivity

The hyperlinks common to Web pages make the Internet inherently interactive. More recent software, such as Macromedia Flash, enables developers to include sophisticated animated sequences, or movies, in Web pages. However, from an information system perspective, the systems analyst's concern is the next level of interactivity. The analyst is interested in providing a convenient way for users to send information to an enterprise Web site and receive specifically requested information from an enterprise's Web site. The analyst might also be interested in transmitting custom applications to visitors of an enterprise's Web site. There are various methods to create and implement highly interactive pages.

HTML, the most common Web page construction language, was originally designed to drop the connection between the Web server and the visitor after the transmission was complete, thus making interactivity, other than through hyperlinks, impossible. Newer versions of HTML include client pull and server push features, with which the analyst can animate the page. With *client pull*, the original transmission includes instructions that tell the client to reestablish the connection to the server and request update information. With *server push*, the connection is never broken, so the server can initiate update transmissions whenever necessary.

In order to establish a data-driven Web site, you need something more than standard HTML. Furthermore, you need to consider how to divide the CPU processing tasks between the client and server, where the database resides, and exactly which implementing strategy is most cost effective.

Data-Driven Web Sites

Given the central role of databases in information systems, it is not surprising that specialized tools have been developed to make it possible for users to interact with these databases via the Internet. E*TRADE and eBay are two successful Internet companies that rely on such technology to conduct e-commerce.

In response to the processing workload issue associated with client/server computing, as well as the Internet's transmission-speed bottleneck, Web page developers employ different programming strategies when creating *data-driven Web sites*. In addition to the traditional HTML programming, developers use *client-side programming* to offload processing to the client's microprocessor and *server-side programming* to access enterprise databases. This spreads the processing workload, reduces transmission times, and addresses some of the security and privacy issues surrounding database information. Figure 13-19 illustrates a scenario in which both strategies might be used in a single application.

FIGURE 13-19   *Data-Driven Web Page Scenario*

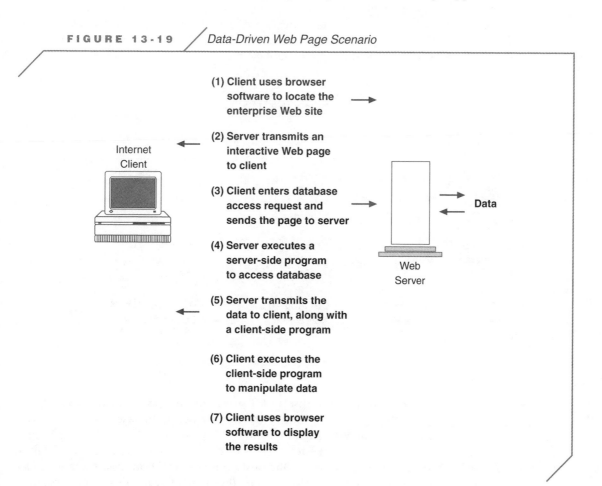

(1) Client uses browser software to locate the enterprise Web site

(2) Server transmits an interactive Web page to client

(3) Client enters database access request and sends the page to server

(4) Server executes a server-side program to access database

(5) Server transmits the data to client, along with a client-side program

(6) Client executes the client-side program to manipulate data

(7) Client uses browser software to display the results

Internet Client

Web Server

Data

Client-side programming is usually implemented in JavaScript or VBScript in a very straightforward manner, as illustrated in Cornucopia's JavaPot.htm program (Figure 13-25). In this example, the JavaScript is used to present a short music history quiz to the Web page visitor.

Server-side programming is more complex because of the need to access a database that resides on the server. Remember that a protocol is a set of rules that governs the way computers talk to each other. There are several such server-side protocols for Internet data communications. The *common gateway interface (CGI)* was the first protocol to specify how a client's browser can pass requests to and receive responses from a Web-based database. CGI scripts are programs written in languages such as Visual Basic, C/C++, Perl, and Java. *Active Server Pages (ASP)*, a popular protocol from Microsoft, employs JavaScript and VBScript to implement server-side programming. Both of these approaches fit the *client/server database model* (Figure 13-20), wherein the server does most of the processing work, the database stays on the server, relatively small amounts of data are transmitted across the Internet, and the client can use its standard browser software to access the database. However, because both approaches require a fair amount of programming skill to implement, there are three other solutions to consider for small-enterprise applications.

FIGURE 13-20 / *Client/Server Database Model*

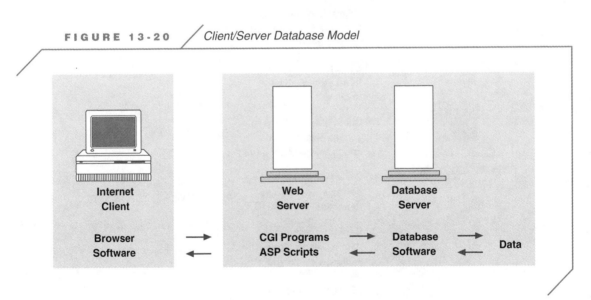

| Internet Client | Web Server | Database Server | |
| Browser Software | CGI Programs ASP Scripts | Database Software | Data |

One alternative is to use an HTML editor, such as Microsoft FrontPage, to create Web pages that interact with an underlying database. A second alternative is to use the Web-friendly features of the database itself. The Page object in Microsoft Access creates Web-enabled pages and links to the database through a series of wizards. Obviously, both of these options are relatively easy to use and inexpensive. However, neither may offer the degree of sophistication required by the small enterprise. The third option, *Web hosting*, is perhaps the most efficient. For a small monthly fee, a Web-hosting service provides all the hardware, software, and support required to implement and maintain a database-driven Web site. With little effort, you may find a Web-hosted Internet mall that presents a sizable collection of small-enterprise e-commerce sites. On a larger scale, an *application service provider (ASP)*—not to be confused with an active server page (ASP) discussed in the preceding paragraph—offers a variety of Internet-deliverable information system solutions. ASPs engage in everything from simple charge-for-use generic applications, such as payroll management, to full-scale custom information systems.

Figure 13-21 presents a summary of the options the analyst should consider in the pursuit of making enterprise databases available to Internet users.

FIGURE 13-21 / *Web-Accessible Database Options*

Option	Ease of Use	Time to Implement	Cost to Implement	Quality
Use database software tools, such as MS Access Page objects, to develop interactive Web pages	Very easy	Very fast	Very low	Very limited
Use an HTML editor, such as MS FrontPage, to develop ASP scripts	Easy	Fast	Low	Limited
Use a text editor to create ASP scripts	Moderate	Moderate	Moderate	Moderate
Develop CGI programs	Difficult	Substantial	High	High
Use a Web-hosting service	Easy	Moderate	Moderate	High

XML Web Services

With each new implementation of Internet technology, there is increasing demand to share data on an even broader scale than that provided by data-driven

Web sites. Standard HTML allows users to format and display information on the Web. Now, *eXtensible Markup Language (XML)* allows users to structure and define the information in their documents. XML permits applications to share data across computer platforms and operating systems regardless of the programming language used to create the application. In a very practical sense, a collection of universal data format and communication standards, known as XML Web services, makes it possible to transport data files more easily between Web sites.

Briefly, XML "extends" HTML by allowing users to create their own collection of tags to define the data structure of a document. As described in Chapter 8, data structures provide the skeletal framework for data file definition. The core building blocks of an XML document are tags, elements, and attributes. As with HTML documents, tags delineate starting and stopping points in the document, while elements and attributes are features that actually specify data names and characteristics. Data elements may be nested within one another, thus providing the flexibility needed to define complex data structures.

Improved networking technologies, such as XML Web services, increase the opportunities for the systems analyst to include connectivity in information system design and implementation. Microsoft delivers XML Web services as its primary integrated-application development environment through its *.NET platform*.

Web Security Issues

There are two major security issues surrounding Internet activity. The first concerns the need to protect valuable computing resources from unauthorized access. Although physical intrusion is always a concern, computers are most vulnerable to electronic invasion. The Internet provides an obvious access route, especially when the enterprise extends a well-publicized invitation to visitors through its Web site. *Firewall software* counters this potential hazard by forcing all entry and exit transmissions to a Web site through a security check. In short, firewalls protect internal enterprise computers from the outside world.

The second major concern is for data transmission security. The most frequently used example is the need to secure credit card account numbers that flow from clients to servers. Various methods can be used to encrypt or code an electronic transmission. *Secure hypertext transmission protocol (SHTTP)* and *secure sockets layer (SSL)* are two Internet security protocols supported by Web browsers. Virtually all e-commerce software includes some transmission security features. Obviously, the analyst must investigate the reliability and costs associated with these products before adopting any solution.

Web Site Management and Maintenance

Unlike much of the rest of the information system, an enterprise Web site requires daily monitoring and maintenance. A Webmaster usually handles this task. Either someone within the enterprise has to be trained to become the Webmaster or the enterprise has to contract an outside source to assume these duties. Because Web pages must be revised to maintain their relevancy and appeal and because the Internet is a 24-hour operation, the Webmaster's workload can be substantial, even if the enterprise maintains its Web pages on the Internet service provider's computer. The Webmaster is responsible for the overall design, content, and hyperlink integrity of the site. This means that individual pages should conform to a design standard and all links to other pages must be valid. Because Web page content must meet the same high standards of quality applied to other information system products, as introduced in Chapter 2 (Figure 2-1), content maintenance is probably the most time-consuming task.

Should your information design include an enterprise Web server, the server administration tasks compound the Webmaster's responsibility. These include the creation and maintenance of user accounts, authentication procedures, and logging file access. In addition, a Web administrator must coordinate with telecommunication companies and the Internet service provider (ISP) to establish and maintain the connection to the Internet. Finally, the administrator is responsible for the server hardware and software.

Intranet and extranet users may require support services at all hours of the day or night, every day of the week. This demand would overwhelm even a tireless Webmaster. In such situations it may be wise to contract for every day, around-the-clock support, or so-called "24/7" support services.

Given the rapid rate of change in Internet technologies, maintaining an effective Internet presence is a time-consuming task that demands a number of special skills. The analyst must factor this into any design that includes these technologies and any of their various implementations.

SYSTEM ARCHITECTURE AND CASE TOOLS

The term *system architecture* as used here refers to the combination of hardware and system software that the analyst designs to support the installation and implementation of the information system. The previous discussions concerning system resource requirements point out the need to estimate these requirements as the design development proceeds.

An alternative to this approach is to prepare detailed estimates of the data file size and maintenance activities, processing volumes, output loads, system backup demands, and any other system performance requirements necessary for the successful operation of the information system. Given this data, some CASE tools can suggest a suitable combination of system hardware and software, that is to say, the system architecture. This recommendation may include a system configuration diagram, which might even specify details about expansion bus widths, bandwidths, CPU speeds, and so on.

In the absence of this tool, the small-enterprise analyst can search the Internet to read about product performance tests and comparisons. Internet sites for hardware and software manufacturers and popular industry magazines, such as *PC Magazine*, *InfoWorld*, or *PC Week*, are useful for finding accurate information about current and future products.

THE CORNUCOPIA CASE

Soon after the Prototype Review Session, the user called the analysts with a question about the point-of-sale options to the system design. She was particularly interested in the possibility of mechanizing the inventory control procedures of the enterprise, which did not surprise the analysts because they had noticed the owner's computer confidence level and information system product interest growing with each JAD working session. To the analysts' credit, they did not immediately reject the suggestion. Rather, they commented that such a major change at this point in the project might endanger the delivery date and cause significant budget overages. They agreed to pursue the matter and return with a careful analysis of this suggestion in the near future. In the mean time, they continued with the development activities.

The system resource requirement specifications introduced in Chapter 11 (Figure 11-16) include a mainstream PC desktop computer with an Internet connection. Nothing in the SDLC process suggests that the system requires a network solution at this time, although the possibility of a future upgrade to use telecommunications to help maintain the CD database has been mentioned. The analysts choose to install the information system as a new entry in the Start menu and as a shortcut on the desktop.

System Interface

The system interface appears in Figure 13-22. The entire system takes advantage of the Windows interface. The new Cornucopia program item and the shortcut both point to the database application. When this application launches, the switchboard (Figure 10-16) automatically opens.

FIGURE 13-22 / *Cornucopia Windows Interface*

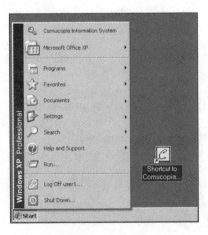

Web Site

Figure 13-23 presents Cornucopia's homepage, with annotations that indicate the linked HTML files. Home, Clogo, and Cpiatoc are HTML files that fill the three frames defined by Cindex. These frames reference other pages within the site. The areas outlined by dashed lines are image maps, within which are more page references. For example, Clogo Map links to Clogo.gif and the entire area links to CpiaInfo. The Marquee Map area is more involved, with five mapped areas. Underlying the map is Marquee.gif, which is a layered image, created in Photoshop.

FIGURE 13-23 / *Cornucopia Web Page Files*

Figure 13-24 shows the HTML code for the four primary files needed to present Cornucopia's homepage. Figure 13-25 shows the JavaPot file referenced through the Marquee image map. This code is included to illustrate the use of JavaScript to add some interactivity to the site. There are two script passages, each bounded by HTML script tags. This tiny application, or applet, is just for fun, but it demonstrates how easily a Web page can deliver functional products.

FIGURE 13-24   */Cornucopia HTML Code*

Cindex.htm

```
<HTML>
<HEAD>
<TITLE>Cornucopia</TITLE>
</HEAD>

<!Frame definition>
<!Divide the screen into two columns>
  <FRAMESET COLS="200,*">
      <!Divide the first column into two rows, define their source files>
        <FRAMESET ROWS="130,*">
            <FRAME NAME="FrLogo" SRC="Clogo.htm"
            MARGINWIDTH = "0" MARGINHEIGHT =
            "0" SCROLLING="none" NORESIZE>
            <FRAME NAME="FrTOC" SRC = "Cpiatoc.htm"
            MARGINWIDTH = "0" MARGINHEIGHT = "0">
        </FRAMESET>
      <!Define the second column's initial source file>
    <FRAME NAME="FrMain" SRC = "Home.htm">
  </FRAMESET>

</HTML>
```

Cpiatoc.htm

```
<HTML>
<HEAD>
<TITLE>Untitled</TITLE>
<BASE TARGET = "FrMain">
</HEAD>

<BODY BACKGROUND="TOCback.gif" TEXT="#C0C0C0" LINK="#C0C0C0" VLINK="#008080">
<H2><CENTER><FONT SIZE=5 COLOR=#C0C0C0>Choices </FONT></I></CENTER>
</H2>
<UL>
<LI><A HREF="CD Master.htm"><I>View CD Inventory</I></A>
<LI><A HREF="CDNotes.htm"><I>View CD Notes</I></A>
<LI><A HREF="MList.htm"><I>Join our Mail List</I></A>
<LI><A HREF="RClasses.htm"><I>Register for Classes</I></A>
<LI><A HREF="SMsg.htm"><I>Send us a Message</I></A>
<LI><A HREF="Home.htm"><I>Home</I></A><FONT COLOR=#C0C0C0> </FONT>
<LI><FONT COLOR=#C0C0C0>Vacant</FONT>
<LI><FONT COLOR=#C0C0C0>Vacant</FONT>
</UL>
</BODY>

</HTML>
```

FIGURE 13-24 / *Cornucopia HTML Code (continued)*

Clogo.htm

```
<HTML>
<HEAD>
<TITLE>Untitled</TITLE>
<BASE TARGET = "FrMain">
</HEAD>

<BODY BACKGROUND="GRAY.JPG>
   <MAP NAME="Clogo Map">
   <AREA SHAPE=RECT COORDS="60,16,140,88" HREF=CpiaInfo.htm>
   </MAP>
   <H2><CENTER><IMG SRC="Clogo.gif" BORDER = "0" USEMAP = "#Clogo Map">
   </CENTER></H2>
</BODY>

</HTML>
```

Home.htm

```
<HTML>
<HEAD>
<TITLE>Untitled</TITLE>
</HEAD>

<BODY BACKGROUND="GRAY.JPG">
<H1><CENTER><FONT SIZE=6 COLOR=#008080>Welcome to</FONT><FONT SIZE=6 COLOR=#000080>
</FONT><FONT SIZE=7 COLOR=#000080>Cornucopia</FONT><FONT SIZE=6>
</FONT></CENTER></H1>
<P>
<CENTER><I><FONT COLOR=#000080>&quot; Without music, life would be a mistake.&quot;</FONT>
<FONT COLOR=#008080> <BR>
<BR>
<MAP NAME="Marquee Map">
<AREA SHAPE=RECT COORDS="119,6,300,53" HREF=MInfo.htm>
<AREA SHAPE=RECT COORDS="36,68,115,161" HREF=MArtists.htm>
<AREA SHAPE=RECT COORDS="134,68,207,163" HREF=MEvents.htm>
<AREA SHAPE=RECT COORDS="222,69,294,162" HREF=MSpecials.htm>
<AREA SHAPE=RECT COORDS="307,69,384,163" HREF=JavaPot.htm>
</MAP><I>
<BR>
</I></FONT></I></CENTER>
<P>
<CENTER><IMG SRC="Marquee.gif" BORDER="2" USEMAP =
"#Marquee Map">
<BR>
</CENTER>
<P>
<CENTER><FONT SIZE=2 COLOR=#000080>122 Gate Street . Eureka.
CA . 95501 . (707) 345-6789 . cpia@stones.com<BR>
Monday through Saturday, 9am to 5pm </FONT></CENTER>
</BODY>

</HTML>
```

FIGURE 13-25 / *Cornucopia JavaPot.htm*

```
<HTML>
<HEAD>
<TITLE>Untitled</TITLE>
<SCRIPT Language = "JavaScript">

    // Define global variables
      var QNumber = 0
      var UserAns = null

    // Define and initialize an array
    function QuizArr(n) {
      this.length = n
            for (var i=1;i<=n;i++){
         this[i]=0
      }
      return this
    }

    // Define a custom object with two properties
    function MusicQuestion(Question,Answer){
      this.Question = Question
      this.Answer = Answer
    }

    // Display the question, clear and set focus to the answer text object
    function ShowQuestion() {
      QNumber += 1
      if (QNumber > MusicQuiz.length) {
         alert("Sorry, there are no more questions.")
         QNumber = 1
      }
      document.GetAns.DisplayQuestion.value =
(MusicQuiz[QNumber].Question)
      document.GetAns.UserInput.value = " "
         document.GetAns.UserInput.focus()
    }

    // Check the user's answer
    function CheckAns() {
      UserAns = document.GetAns.UserInput.value
            if (UserAns == MusicQuiz[QNumber].Answer) {
         alert("Correct")
      }
      else {
         alert("Sorry, the correct answer is "+ MusicQuiz{QNumber].Answer)
         document.GetAns.UserInput.value = " "
      }
    }

</SCRIPT>
</HEAD>
```

FIGURE 13-25 / *Cornucopia JavaPot.htm (continued)*

```
<BODY>
<H2><CENTER><BODY BACKGROUND="GRAY.JPG"><IMG SRC="Mpotpourri.GIF">
<BR>
</CENTER></H2>
<P>
<B><FONT COLOR=#000080>Marquee Potpourri is always a surprise, even to us. The little musical quiz that
follows was inspired by Harold Schonberg's <I><B>The Lives of the Great Composers</B></I><B> (Norton,
1981), which makes a great gift item.</B>
</FONT>
<BR><BR>
<B><FONT COLOR=#000080>

<SCRIPT Language = "JavaScript">
    MyTitle = "Musical Quiz"
    document.write ('<CENTER>')
    document.write (MyTitle.fontsize(5),'<BR>','<BR>')
    document.write ('<CENTER>')

// Create the array and load the quiz questions and answers
    MusicQuiz = new QuizArr(2)
    MusicQuiz[1] = new MusicQuestion('Mozart died in 1791. How old was he?','36')
    MusicQuiz[2] = new Music Question('Beethoven premiered his 9th Symphony in 1824. When did he go
deaf?','1817')

</SCRIPT>

<!-- Present the user interface in a form -->
<FORM Name = "GetAns">
Click to display a new question.
<Input type = "button" value = "NEW QUESTION" onClick = "ShowQuestion()">
<BR>
<BR>
The question:<BR>
<INPUT type = "text" name = "DisplayQuestion" size = 63 maxlength = 100>
<BR>
Enter your answer:<BR>
<INPUT type = "text" name = "UserInput" size = 63 maxlength = 100>
<BR>
Click to check your answer.
<Input type = "button" value = "CHECK ANSWER" onClick = "CheckAns()">
<BR>
<BR>

</FORM>
</B></CENTER>
<BR><BR>
<P>
<CENTER><IMG SRC = "Elogo.gif"></CENTER>
<P>
<CENTER><FONT SIZE=2 COLOR=#000080>Last Update: 11/11/99<BR>
</FONT></CENTER>
</BODY>

</HTML>
```

Time and Money

The analysts report six hours to development activities, bringing the completion up to 90 percent. In addition, they report four hours to the implementation activities, with a 10 percent completion. Figure 13-26 presents the project status as of Week 12.

FIGURE 13-26 / *Cornucopia Project Status – Week 12*

Activity	% Comp.	Status	1	2	3	4	5	6	7	8	9	10	11	12	13	14	15	16	Total
Date: Cornucopia Project Status As of: Week 12																			
Analysis - Estimate	100%		4	5	5	4													18
Actual	100%	ok	2	4	2	3													11
Design - Estimate	100%					4	4	5	5	4									22
Actual	95%	ok			1	3	8	5	7	6									30
Develop - Estimate	100%								4	4	5	5	4						22
Actual	90%	ok							1	8	20	4	6						39
Impl. - Estimate	15%													4	5	5	5	5	24
Actual	10%	ok											1	2					3
Total - Estimate			4	5	5	8	4	5	5	8	4	5	5	8	5	5	5	5	86
Actual			2	4	3	6	8	5	7	7	8	20	5	8	0	0	0	0	83
Contract	100%	ok	C																
Prelim. Present.	100%	ok			C														
Design Review	100%	ok							C										
Prototype Review	100%	ok									C								
Training Session	10%	ok														S			
Final Report	10%	ok															S		

The user's request to investigate the possibility of changing the system design to incorporate point-of-sale hardware and software is answered, in part, by the budget report (Figure 13-27), which shows a modest budget overrun ($250) to date. The status report reminds the user that only four weeks remain before the system's scheduled completion date. Together with the analysts' quick estimate that the changes will require about $1,000 in hardware and software and at least 15 labor hours ($750), the user agrees to wait for the final report, which promises to include a more complete analysis of the proposed upgrade.

FIGURE 13-27 / *Cornucopia Project Budget – Week 12*

	A			1	2	3	4	5	6	7	8	9	10	11	12	13	14	15	16	Total
1	Date:		Cornucopia Project Budget									As of: Week		12						
3				1	2	3	4	5	6	7	8	9	10	11	12	13	14	15	16	Total
4	Estimates	Hardware							2500	500	500	500								4000
5		Software							1000	250	250									1500
6		Labor		200	250	250	400	200	250	250	400	200	250	250	400	250	250	250	250	4300
7		Total		200	250	250	400	200	3750	1000	1150	700	250	250	400	250	250	250	250	9800
8	Actuals	Hardware							4250											4250
9		Software							225		425									650
10		Labor		100	200	150	300	400	250	350	350	400	1000	250	400					4150
11		Total		100	200	150	300	400	4725	350	775	400	1000	250	400	0	0	0	0	9050
12	Weekly +/-	Hardware		0	0	0	0	0	1750	-500	-500	-500	0	0	0	0	0	0	0	250
13		Software		0	0	0	0	0	-775	-250	175	0	0	0	0	0	0	0	0	-850
14		Labor		-100	-50	-100	-100	200	0	100	-50	200	750	0	0	0	0	0	0	850
15		Total		-100	-50	-100	-100	200	975	-650	-375	-300	750	0	0	0	0	0	0	250
16	Cum +/-	Hardware		0	0	0	0	0	1750	1250	750	250	250	250	250	0	0	0	0	
17		Software		0	0	0	0	0	-775	-1025	-850	-850	-850	-850	-850	0	0	0	0	
18		Labor		-100	-150	-250	-350	-150	-150	-50	-100	100	850	850	850	0	0	0	0	
19		Total		-100	-150	-250	-350	-150	825	175	-200	-500	250	250	250	0	0	0	0	

Budget / Sheet2 / Sheet3

Cornucopia with Charting Software

From the outset of the project, the analysts felt that Cornucopia would soon want to enhance the basic design of their stand-alone information system. The first design drafts and prototypes did not include point-of-sale processing or networking, other than the publication of a straightforward Web site. The user recently broached the subject of a point-of-sale enhancement. Feeling that a simple peer-to-peer LAN might be a very modest enhancement to consider, the analysts used Microsoft Visio to begin charting a solution. Figure 13-28 shows the future peer-to-peer LAN upgrade diagram.

FIGURE 13-28 / *Cornucopia Future Network Diagram*

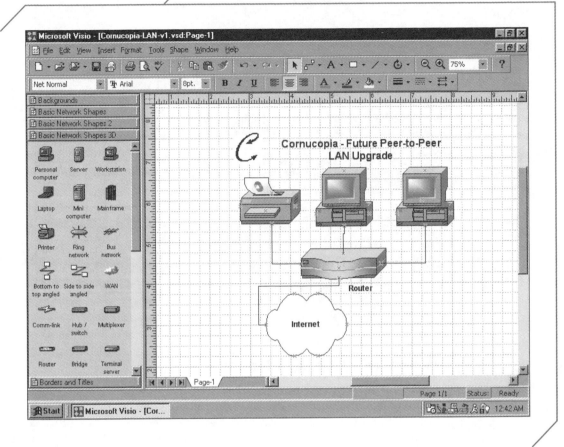

SUMMARY

This chapter demonstrates the diverse hardware and software issues that the small-enterprise analyst must address when preparing to install the information system. In practice, these concerns permeate the SDLC from beginning to end. Speed, memory, and storage dominate hardware decisions, whereas user interfacing, resource management, and resource sharing require special attention during system software selection.

Once the horizontal and vertical software is in place, the actual information system installation begins when the custom programs (scripts,

command files, macros, batch files, etc.) are loaded into hard drive directories. Then the analyst constructs a unified user interface to integrate the information subsystems with the rest of the software on the computer. This activity naturally occurs near the end of the development phase, after which the analyst proceeds with system testing, which you will learn about in the next chapter.

The chapter presents networking hardware and software issues, along with an introduction to client/server computing. Internet, intranet, and extranet solutions are discussed in terms of how a small enterprise can extend its information system reach well beyond its geographic neighborhood. Several types of implementing software, from transmission protocols to programming languages, are introduced.

TEST YOURSELF

Chapter Learning Objectives

1. What key hardware issues relate to the small-enterprise information system under development?

2. What key system software issues relate to the small-enterprise system under development?

3. How do you install an information system that complies with the computer's hardware and system software environment?

4. How does network technology influence small-enterprise information systems?

5. How do Internet, intranet, and extranet technologies influence small-enterprise information systems?

Chapter Content

1. The PC solution applies to enterprises of all sizes. True or False?

2. The small-enterprise analyst need not be concerned with system hardware issues. True or False?

3. While it is difficult to imagine overbuying in everyday hardware categories such as CPU speed, disk capacity, and memory cache, the analyst should be very careful when selecting operating system software. True or False?

4. The process used to install an information system is slightly different from the process used to install application software. True or False?

5. Typically, an operating system provides some security features that the analyst can use to prevent inadvertent, illegal, or improper access to an information system. True or False?

6. Networking technology may accelerate the migration of the small enterprise from a Stage-I to a Stage-II information system. True or False?

7. While a file server and a Web server perform different functions, they require similar hardware features. True or False?

8. There is only one type of logical relationship between a client and a server. True or False?

9. Networking solutions are subject to the same obsolescence pressures faced by stand-alone solutions. True or False?

10. Advances in communications technology provide a means by which processing can be distributed to remote Internet and intranet clients. True or False?

11. Web page designers must be cautious about employing server push technology. True or False?

12. A fully interactive enterprise Web site does not pose any special security risks. True or False?

13. Maintaining an effective Internet presence requires regular Web site management. True or False?

14. When expanding the enterprise information system to interface with the Internet, the analyst must exercise great care to design and develop something that is easy to change because someone will regularly have to maintain all of the site's files and links. True or False?

15. XML Web services loosely describe organized collections of Internet-based companies that provide fast, reliable, and affordable data file transport via the Internet. True or False?

ACTIVITIES

Activity 1

1. Review the passwords and profiles image for PRC (Figure 13-7). Develop some plausible entries for the Program Item Properties required to include all of the subsystems identified in PRC's data flow diagram and system flowchart (Figures 11-11 and 11-12).

2. Review the LAN upgrade specifications for PRC (Figures 13-15 and 11-17). Update the specifications to reflect the most current prices.

3. Develop the upgrade specifications for a dedicated server network for PRC.

Activity 2

Use your favorite Internet search engine to investigate application service providers. Prepare a brief summary of the services offered by three ASPs.

Activity 3

Use your favorite Internet search engine to investigate client/server databases. Prepare a brief summary of three client/server database products.

DISCUSSION QUESTIONS

1. To what extent should the analyst instruct the user about the critical hardware and software issues presented in this chapter?

2. How would you respond to the user who insists that you build the information system around a particular PC platform?

3. What are the advantages and disadvantages of installing the computer information system into the operating system interface?

4. In what ways might network productivity products actually change the way people work?

5. Describe a scenario for migrating to a LAN from a single-user system. Consider all six information system components in your answer.

6. How does the possibility of different hardware platforms (e.g., PC and Macintosh) within the same information system complicate the software compatibility issue?

7. Why might a small enterprise, implementing its first computer information system, be reluctant to engage a Web-hosting service?

8. Why are programming skills important to implementing client/server computing?

9. What are the advantages of a simple peer-to-peer LAN for a small enterprise with only two or three desktop computers?

10. Explain how you might use the Internet as a resource for developing alternative networking solutions for a client.

PortfolioProject

Team Assignment 13: Networking

As described in this chapter, networking includes many different technologies. It is very likely that, at a minimum, your small-enterprise project requires you to develop a Web site. Your project might also include a simple peer-to-peer LAN or one of the more complicated networking environments. As a practical consequence of the academic environment surrounding your project, there are varying degrees to which you can implement your networking design. While this assignment asks only that you summarize your intended use of networking technology, you may be able to elaborate considerably in your descriptions and illustrations, depending upon your circumstances. In order to document your networking efforts you must complete the following tasks:

1. Prepare a slide show illustrating the various pages of the enterprise Web site. Submit a printout of the slides (three slides per page), along with a printed sample of the code for the homepage.
2. Prepare a brief summary of the current and potential use of networking technology in the solution to your project. Submit a copy of your networking summary to your instructor.
3. File a copy of your slide show handouts, sample code, and networking summary in the Project Binder behind the tab labeled "Assignment 13."

Implementation

testing,documentation,
andtraining

When you complete this chapter, you will be able to:

OVERVIEW

The repetitive and cumulative developmental approach to each part of the SDLC is once again evident in the system-testing task. This begins with individual program segments and proceeds, with advancing complexity, to complete programs and subsystems, before full system testing begins. This chapter presents a discussion of error tolerance levels in small-enterprise information systems to remind you that perfection is an expensive illusion. The testing section of this chapter concludes with a discussion on the importance of test procedure documentation.

In this chapter, you will review and summarize the elements of system documentation, which provides a foundation for producing three manuals—training, procedure, and reference.

This chapter will discuss traditional training cycles and materials, with a careful treatment of training methodologies and ideas covering how users can best be motivated to learn about a new system. Finally, the chapter concludes with a brief, but important, look at the analyst's responsibilities to train users in the ethical use of their information system.

- Distinguish between various product testing goals and procedures.

- Document testing plans and results.

- Assemble product documentation by building on SDLC project deliverables.

- Develop training strategies and materials to accommodate different audiences.

- Understand how to include information system ethics in user-training sessions.

TESTING

Testing is designed to make sure the product performs as promised. *Debugging* is the identification and correction of errors in an information system, which comes from the programmer's attempt to eliminate the *bugs*, or errors, in computer programs. This term can be misleading if it is used in the narrow sense of program debugging. Figure 14-1 presents a collection of testing goals that demonstrate broader responsibilities. In short, the analyst plans, conducts, supervises, and evaluates testing activities that concern all six components of the computer information system. These activities occur throughout the SDLC.

FIGURE 14-1 / *Testing Goals*

1. Eliminate syntax errors
2. Identify and screen user input errors
3. Eliminate logic errors
4. Eliminate the potential for runtime errors
5. Evaluate system speed performance
6. Evaluate system security
7. Validate system integrity
8. Evaluate the level of user acceptance
9. Evaluate system documentation

Syntax errors occur when computer instructions conflict with the system, programming language, or application rules. For example, if a user enters the Visual Basic command ADDEM acct_amt TO acct_total, the program generates a syntax error because Visual Basic does not recognize the command ADDEM. When the user substitutes SUM for ADDEM, the command executes. However, this does not mean that the command is logically correct. What if acct_total is supposed to reflect the cumulative account purchases and acct_amt contains only the balance owed on the account? The command executes, correctly adding all the account amounts into a total, but the result is invalid. *Logic errors*, such as this, are the most difficult to test, identify, and correct.

In the preceding example, imagine a screen form for the user to enter customer account payments. At some point, the user is required to type the amount paid into a field named acct_pymt. If the user keys "abc.de" instead of "123.45" you would expect the system to report this mistake immediately, with an error message such as, "This field requires numeric input. Retry." In this case, the programmer provided a test for a common user input error. To make sure that the test works properly, the programmer should purposely enter nonnumeric data during the user interface testing activity.

Another common user input error is capitalization, or *case sensitivity*. If the program code reads IF last_name = "ROBERTS" and the user enters "Roberts" when prompted for a last name, the program determines that "Roberts" does not equal "ROBERTS." To avoid this problem, the programmer could change the command to read IF UpperCase(last_name) = "ROBERTS", which employs the Visual Basic function UpperCase to transform all the data field letters into uppercase. Once again, the programmer should test this code to be sure that it works.

Runtime errors occur when the computer can't process an instruction. Discrepancies in data definition and data input, illegal machine operations such as dividing a number by zero, and a missing or incorrect file path are examples of situations that cause runtime errors. Such errors usually cause the application program to abort, which understandably creates a great deal of user anxiety.

To evaluate system speed, the analyst must prepare sufficiently complex and voluminous processing and data access tests. For example, assume that a sales transaction system must access a database to verify the customer's account balance before permitting the customer to make a charge against the account. This operation may take a fraction of a second when the database contains only a few, sample entries. What happens when the database contains 10,000 entries? Alternatively, assume that a system inquiry asks for a statistical analysis (average sale, number, and percentage of sales above and below the average, etc.) on all transactions during the past six months. The analyst must test such a request with data volumes that approximate those anticipated when the system is operational.

System security testing presents special challenges to the analyst, because it is often impossible to imagine the sequence of either unintended or intended actions users will introduce into the system. For the small-enterprise system, the analyst should begin by testing the security features that come packaged with the horizontal and vertical software used in the system. For example, Microsoft Access provides multiple levels of security (passwords, table, form, and query access privileges, and so on). Another effective and relatively inexpensive security testing method is to ask a user to try to break into the information system, delete or alter files, change a critical processing parameter (such as a reasonableness check on paycheck amounts), and generally try to compromise the system's integrity. A more thorough test is to give this same task to a computer professional.

To validate *system integrity* is to prove that the system performs its functions accurately and consistently across all subsystems. For example, assume that an information system produces several reports, all of which are based on the same sales transaction data. The analyst should conduct a test to make sure that the daily and monthly reports show the same total figures. Furthermore, system integrity tests should prove that any future corrections to the daily reports are also processed against the monthly reports.

Another important system integrity test concerns the *referential integrity* of the database files, which refers to the ability of the information system to prevent inconsistent data manipulations across various subsystems. The database maintenance process should prohibit two inconsistent database file activities: the deletion of a master file record that still has at least one associated record in a related transaction file, and the addition of a transaction file record that contains a reference to a nonexistent master file record.

User-acceptance testing occurs throughout the project. Every meeting between the user and analyst provides an opportunity to determine the user's level of satisfaction with the new information system. However, this does not preclude a more formal evaluation of the nearly finished product. In some cases, the analyst may ask the user to sign a document declaring that the enterprise accepts the product, with or without restrictions or conditions. Regardless of the degree of formality, this process can also help the analyst evaluate system documentation, especially if the user-acceptance testing procedures direct users to the documentation to help resolve error messages or generate test reports.

Incremental Testing

In addition to a set of realistic goals, you should develop a clear and systematic testing plan and some precise measure by which you can judge the test results. As with most complex systems, you can usually achieve complete system testing by breaking the task into smaller, more manageable parts. At the lowest level, you are concerned with syntax errors; at the highest level, you are concerned with user-acceptance testing and documentation. Figure 14-2 presents a generic testing plan that incorporates this notion into our familiar SDLC.

FIGURE 14-2 / *Testing and the SDLC*

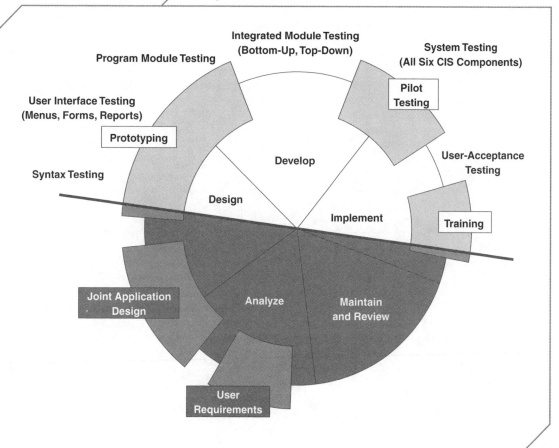

Testing begins early in the SDLC and continues well into the implementation phase. This represents a significant cost to the project in terms of analyst-programmer time and machine time, but no computer products are initially error free. Experienced computer professionals also understand that subtle, but crucial, benefits accompany early testing. Figure 14-3 demonstrates the benefits of early testing by plotting the time required to correct errors, depending on when they are discovered. Although the unit of time is undefined, the exponential shape of the curve shows that error correction costs increase dramatically as the analyst proceeds from one testing activity to another.

FIGURE 14-3 / *Cost of Error Detection and Correction*

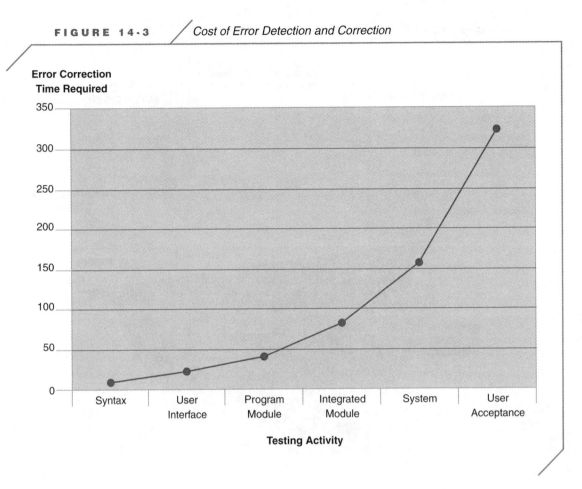

Error Correction Time Required

Testing Activity: Syntax, User Interface, Program Module, Integrated Module, System, User Acceptance

Naturally, the resources that can be devoted to testing are limited; at some point, the potential benefits simply do not justify the expense. However, in the face of a system-critical error, users are generally unsympathetic to any analyst's excuse based on the argument that the system was not tested because of a resource-constrained testing strategy. Analysts may occasionally feel that there is never enough time to do the job right the first time, but always time to do it over. An effective strategy is to identify and test those elements of the system that present the greatest risk to system performance before you run out of resources.

Even if you had unlimited resources to devote to testing, it is unlikely that you would be able to test all the possible situations a moderately complex system will encounter. A program with three branches requires eight tests to check all the different possible branching combinations (two raised to the third power). A program with six branches requires 64 tests, and so on. The next section presents some techniques that help reduce the number of tests required.

Testing Scope

The 4GL word processor, spreadsheet, and database management products that form the foundation for small-enterprise information systems are relatively error free. Thus, the analyst-programmer rarely conducts the type of testing and debugging required for a 4GL program. Rather, the emphasis is on testing the customized application of the various horizontal products, from a stand-alone macro or script to the entire information system.

Although some testing can be performed in a parallel fashion, testing during each phase of the SDLC follows a definable progression of increasingly broader *testing scope*, which are the boundaries or limits of a testing procedure. Figure 14-4 summarizes the progression suggested by Figure 14-2. Remember, testing may not always proceed in this linear fashion because the blurred distinction between the phases of the SDLC and the continual process of review and revision influence all aspects of the project.

FIGURE 14-4   */ Testing Progression*

A. **Design Phase Testing**
  1. Syntax Testing
  2. User Interface Testing

B. **Development Phase Testing**
  1. Program Module Testing
  2. Integrated Module Testing
  3. System Testing

C. **Implementation Phase Testing**
  1. User-Acceptance Testing

During design phase activities, the analyst-programmer performs syntax testing as a natural adjunct to working with horizontal software. The quality of product documentation, help tools, and debugging utilities can make this task easy or difficult. Figure 14-5 presents an *integrated development environment (IDE)*. An IDE is characterized by the familiar Windows-like workspace, with menu options, close boxes, and scroll bars. The Java program in this illustration has failed to compile, as noted in the lower window. Sometimes analysts refer to an IDE as a *programmer's workbench*.

FIGURE 14-5 / *Integrated Development Environment*

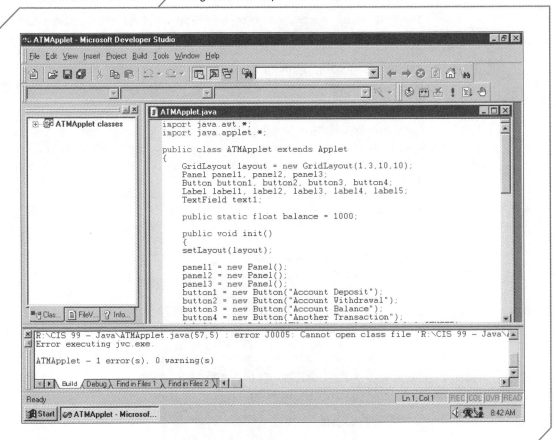

The degree to which the project incorporates object-oriented technologies greatly influences this element of the overall testing plan. Indeed, one of the most significant advantages of object-oriented design and development is the availability of an ample library of error-free objects and methods.

User-interface testing is a natural part of prototyping. Although the sample menus, files, forms, and reports are primarily created to facilitate user involvement in the early design stages, they also provide an ongoing test of the user interface. Invariably the user enters invalid, incorrect, or inconsistent data during prototype design and review sessions, thus providing an inadvertent test of some of the input-error detection provisions. To the extent that the prototype does not include these procedures, the analyst-programmer must test the interface during the development phase.

During development phase activities, prototype testing evolves into program module testing. For example, with a database management application, the prototyped master file maintenance forms and reports eventually require some command file or object-property specification, which in turn must be tested. An integrated module is similar to a subsystem. The name implies that the subsystem contains more than one module. The two common approaches to testing these modules are the top-down and bottom-up methods. Testing individual program modules, prototype files, forms, and reports describes the beginning of the bottom-up approach; testing prototype menus describes the top-down approach. Normally, development phase testing is conducted in the analyst-programmer's development environment, which allows for more controlled testing conditions.

System testing involves all six CIS components: hardware, software, data, networking, people, and procedures. The SDLC, with its emphasis on user involvement, calls for pilot testing soon after the analyst-programmer completes the integrated module testing. A *pilot test* allows the user to interact with the new system, without the analyst-programmer standing nearby to answer every user question or system message. If possible, the pilot test should take place in the user's work setting, which makes it more likely that all six components of the information system undergo some testing.

User-acceptance testing provides the final opportunity to evaluate the entire system before it is implemented. In a 4GL environment such as the one described in this book, few surprises should remain at this point in the project, making this type of testing more or less a formality. You will learn more about the process of user-acceptance testing in Chapter 15.

Test Data

With the exception of some of the highest-level system software components, such as operating system interfaces, the analyst-programmer must test all portions of the information system software with data. At the beginning of the testing process, the analyst-programmer should use a small amount of data, with a moderately representative sample of the anticipated data values. This makes it much easier to determine exactly which results the test should produce. This technique also helps to establish baseline performance parameters against which the analyst-programmer can compare future tests. Real-world data volumes and diversity are used during the later testing stages to make sure that the system performs as specified in the contract, preliminary presentation, and design documents.

The analyst-programmer very likely will use a small set of fictitious data to test any new or unfamiliar horizontal product features. The analyst-programmer usually discards this test data once the syntax is error free. Conversely, it is wise to save the sample files created during product design and prototyping. At first, these files should be small and very simple. Unusual situations, such as extreme or improbable numeric values, are added to the test data as you build more into the prototype and the individual subsystem modules. As different features of the software are exercised in successive tests, these workbench files should be retained for future testing.

Capturing live data for testing is sometimes required for subsystem or higher testing, because with even moderately complex information systems, developing mock data to represent all of the circumstances presented by real environments is impossible. In fact, to some extent you are testing for unpredictable situations. This testing method is used in conjunction with pilot testing and other forms of phased conversion techniques that you will learn about in Chapter 15. This can be an effective bridge between testing, training, and conversion.

Testing Procedures

Testing procedures become more formal as project development moves farther along. When more people, procedures, data, hardware, and software are involved, a carefully prepared testing plan can help avoid duplicate and conflicting testing efforts. This is especially true for pilot testing that involves the user. Remembering the operating constraints that affect small enterprises, the analyst should not introduce unnecessary confusion. A *testing plan* should provide well-defined schedules, user and analyst-programmer responsibilities, and target goals. Figure 14-6 illustrates how the analyst for the Political Research Corporation documents the testing activities for an upgrade project. Notice that these are informal presentations of the required tasks, schedules, and so on. This is typical for a small-enterprise project that involves only a few individuals. In larger settings, the testing plan is more formal. Nevertheless, brief notes, such as those in Figure 14-6, can be invaluable to the analysts managing the upgrade.

A completely error-free system is very difficult to achieve. As discussed earlier in this chapter, this is due to three main factors: complexity, cost, and uncertainty. However, distinguishing among different types of errors is important. A prudent analyst concentrates on identifying and resolving *system-critical errors* before giving attention to other error types. For example, if the subsystems do not communicate properly, the full system is of little use. On the other hand, if an input module does not adequately inform the user about error correction procedures, the system can still operate while a fix is undertaken. Figure 14-7 illustrates a common error-rate pattern experienced by new information systems. Notice that effective system training and documentation can significantly reduce error rates, even if the analyst-programmer never eliminates all the errors.

In theory, the system should be tested after any change to the hardware, software, or data structures, and sometimes even after changes in procedures and personnel. In practice, ongoing information systems are rarely tested as completely as they are during their initial development and implementation. Regardless of the degree to which they are tested, you can significantly reduce the testing effort if future system-maintenance analysts can refer to a well-documented testing procedure that includes sample test data and expected results.

FIGURE 14-6 / *Testing Plan Documentation*

Product Testing Control Sheet

Project | Client Billing Upgrade

Category and Product Description	Comment
Syntax	
mail-merge macro	completed
User Interface	
billing screen form	completed
payment screen form	completed
Program Module	
accounts receivable database maintenance:	
blank form, account balance form	completed
Integrated Module	
accounts receivable database maintenance	completed
mail-merge with accounts receivable database	completed
System	
new PRC mainmenu options:	
mail-merge, accounts receivable maintenance	in progress
Pilot	
PRC upgradable	
User Acceptance	
PRC upgradable	

Product Testing Procedure

Project | Client Billing Upgrade   **Product** | payform.prg

Test Description | user interface - client payment screen form

Test Data	Expected Results	Comment
invalid client id	message: not on database	completed
nonnumeric pay	audio error signal	completed
overpayment	message: credit balance warning	completed
abort transaction	clear screen, restore blank form	completed
update transaction	message: display new balance	
	message: continue?	completed
exit	clear screen, return to main menu	completed

FIGURE 14-7   /   *Error Rates over Time*

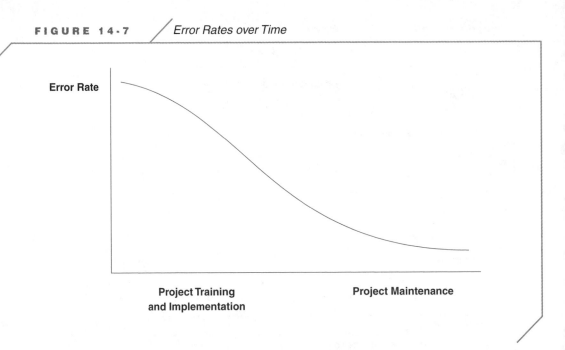

Error Rate

Project Training
and Implementation

Project Maintenance

CASE Tools and Debug Tools for System Testing

This chapter has already discussed several computer-assisted program development, testing, and debugging tools. Some of these tools are embedded in the application (such as options to permit *single stepping* through a macro) and programming language environment (programmer workbenches). Others are separate software utilities, and several are included in the lower CASE products. Almost all of them are visual and allow the user to put a program on an electronic dissecting table of sorts. However, system, pilot, and user-acceptance testing continue to be analyst-controlled and analyst-directed

ThinkingCritically

Dataas**Private**Property

Austin Linen Supply's new information system testing plan calls for a volume test to determine whether the customer database can perform account lookups in a reasonable amount of time. Rather than create mock data, you request and receive a copy of the enterprise's customer accounts, which you load onto your test system. After the test, you realize that many of these customers are also potential clients of your consulting services.

Is it ethical for you to use this data to your advantage?

activities. The CASE and alternative technology tools do not yet extend to these areas of testing.

In 1945, the first computer bug, literally a tiny moth, was physically removed from the Mark-1 computer. Since that time, programs or computers that do not work properly are said to have "bugs" in them. Sometimes the bugs are obvious, as is the case when you misspell a macro or program command. Sometimes the bugs are almost impossible to isolate. Recognizing that users, analysts, and programmers can be desperate when confronted with one of these intractable problems, the software industry developed the *debug tool*. Generally, learning how to use a debug tool is difficult and time consuming because it requires much more than a casual understanding of the underlying software product.

Figure 14-8 shows the debug screen display for Visual Basic. With the debug tool, a programmer can examine a program as each instruction executes. This may help the programmer to isolate the source of the problem. Of course, a thorough understanding of programming fundamentals is required. For example, the breakpoint feature allows the programmer to temporarily enter a logical expression in the program to command the program to halt when certain conditions occur. This small example should demonstrate that debug tools can be complicated.

FIGURE 14-8   *Visual Basic Debug Window*

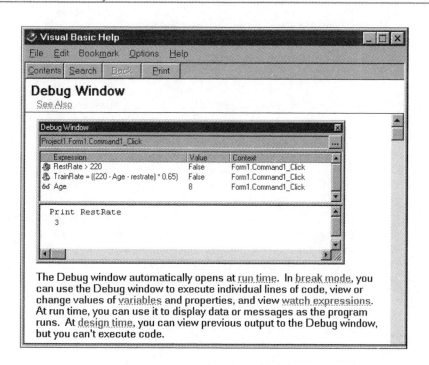

SYSTEM DOCUMENTATION

System documentation is a collection of information that provides a detailed history of the SDLC. Like most histories, it is most useful when it includes sufficient detail and insightful commentary to help the reader understand what happened to make things as they are. For an information system, this begins with the initial project documents, the request for services, and ends with the most current request for system maintenance. As you now know, a lot happens between these events, all of which is part of the project's history.

Documentation serves as a reference and source of diagrams, illustrations, and detailed explanations for the training materials discussed later in this chapter. Analysts also use documentation to develop procedure manuals and reference manuals for the user. In a sense, the complete system documentation, or *project binder*, serves as the source document for several other, more tailored products.

Project By-Products as System Documentation

The PERT chart and detailed task lists introduced in Chapter 6 specify that formal system documentation activities are to commence after the design review session and continue until the project is delivered. However, *all* of the project documents are considered part of system documentation. Accordingly, the analyst should retain and file several early project items in a project binder. In addition, some kinds of documentation could be by-products of the project. For example, the preliminary user's system diagram (USD) is prepared to facilitate communication between the analyst and the user, but you should also file it in the project binder for later reference. Figure 14-9 summarizes the contents of the project binders. Although this illustration suggests a paper-based collection of documents, much of this material actually exists in electronic form. Our CASE tool discussions refer to this as a *central repository*.

Project Manuals as System Documentation

Another, more formal, set of system documentation is designed to develop the collection of electronic materials and manuals that users are accustomed to seeing shrink-wrapped with off-the-shelf application software. The individual items in these packages often have generic names, such as Help or User Guide. In contrast, we define system documentation in more precise terms. For example, Microsoft Office ships on several CDs that include different installation options. The typical installation requires considerable disk space, while the custom option can require almost twice as much disk space. Included in either option is an elaborate system of electronic documentation that consumed a half-dozen printed documentation manuals in a previous version. Generally, the small-enterprise user does not expect and is not willing to pay for professionally packaged documentation such as this. Nevertheless, the purpose of the product documentation is the same, which is to help users and future analysts understand how the information system works, how it can be used, and how it can be modified.

FIGURE 14-9   / *Project Binders*

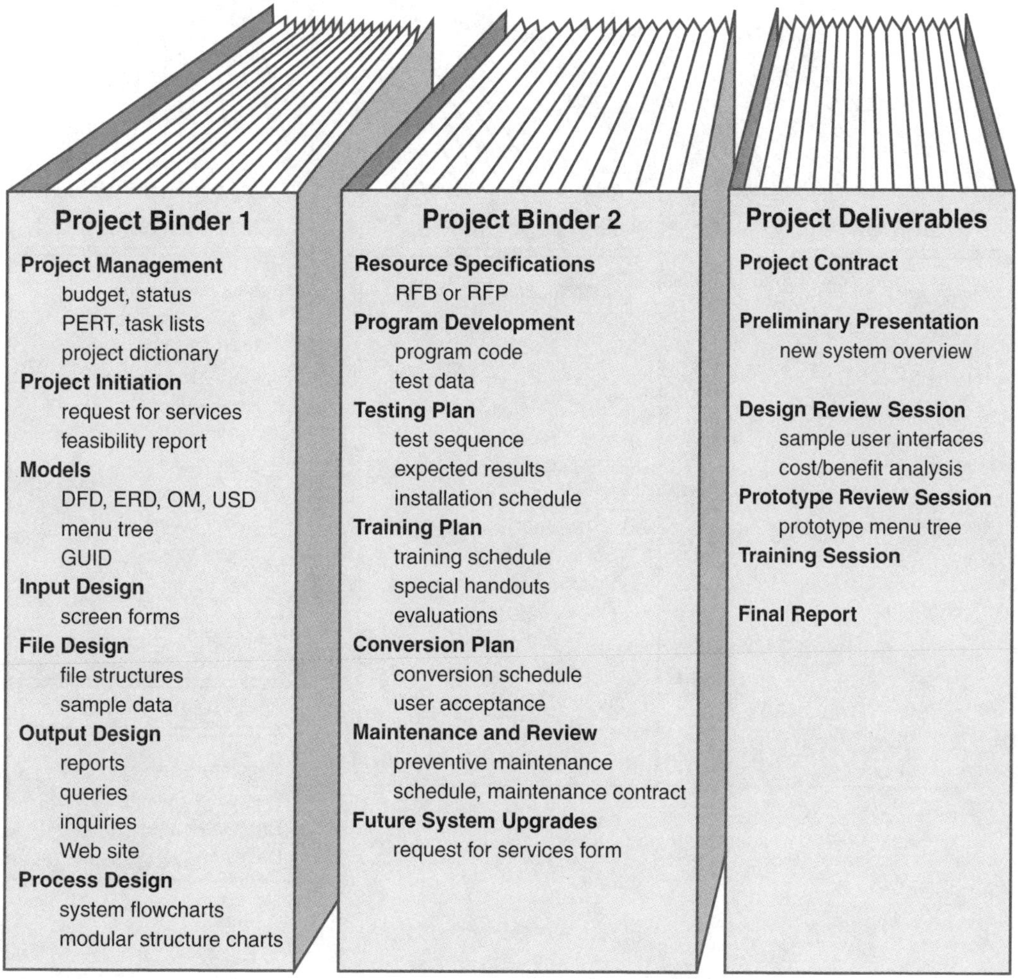

Project Binder 1

Project Management
  budget, status
  PERT, task lists
  project dictionary
Project Initiation
  request for services
  feasibility report
Models
  DFD, ERD, OM, USD
  menu tree
  GUID
Input Design
  screen forms
File Design
  file structures
  sample data
Output Design
  reports
  queries
  inquiries
  Web site
Process Design
  system flowcharts
  modular structure charts

Project Binder 2

Resource Specifications
  RFB or RFP
Program Development
  program code
  test data
Testing Plan
  test sequence
  expected results
  installation schedule
Training Plan
  training schedule
  special handouts
  evaluations
Conversion Plan
  conversion schedule
  user acceptance
Maintenance and Review
  preventive maintenance
  schedule, maintenance contract
Future System Upgrades
  request for services form

Project Deliverables

Project Contract

Preliminary Presentation
  new system overview

Design Review Session
  sample user interfaces
  cost/benefit analysis
Prototype Review Session
  prototype menu tree
Training Session

Final Report

This text identifies three manuals that the product documentation package comprises: the training manual, the procedures manual, and the reference manual. You will learn more about each of these manuals later in this chapter. Figure 14-10 shows the subjects each manual should address; this can help you as you work through the intervening sections on training methodologies and cycles.

FIGURE 14-10 / *System Documentation Manuals*

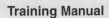

Training Manual

Objectives
Schedule
System Overview
  Narrative
  User's System Diagram
  Menu Options
Demonstration Topics
Hands-On Exercises
Quick Reference Guide
Notes

Reference Manual

System Models
  DFD
  ERD
  OM
  System Flowchart
  Menu Trees
File Structures

Process Specifications
  Structure Charts
  Special Formulas
  Queries/Macros
  Internal Table Values
  Program Code
  Testing Procedures
Web Site

Procedures Manual

System Description
Operating Instructions
  Start-Up
  Menu Descriptions
  Maintenance
  Back-Up
  Shut-Down
User Interfaces
  Switchboard
  Input Forms
  Output Reports
System Security
Emergency Instructions

Appendices

Software Specifications
Hardware Specifications
Error Messages
Definitions

TRAINING

The typical computer professional may not view teaching as a job responsibility, but a teacher is exactly what the analyst becomes during the implementation phase of a system project. Without the benefit of a course in cognition or effective classroom techniques, user-training sessions can be a frustrating experience for everyone involved. Although a discussion of the many theories that attempt to explain how people learn is beyond the scope of this text, we can touch on some of the more important considerations to keep in mind.

First, most small-enterprise users value a straightforward teaching technique. Put simply, they need to know how to get started, what to do when they are ready to learn more about system features, and where to turn for help when they need it. The analyst must develop a coherent strategy to provide training that meets these goals within a reasonable period.

Second, if users perceive that system benefits can directly improve their job performance, they become highly motivated learners. On the other hand, if they view training sessions as time lost from production, they will resist even your best effort. A well-planned, time-efficient, relevant, and rewarding training strategy begins with some understanding about the student-teacher relationship and an appreciation for the learning process itself.

Instructor-Directed Learning

In a traditional instructional setting, the teacher assumes the role of the purveyor of knowledge, whereas the student is expected to be absorbingly attentive. On occasion, the student may ask for clarification on a particular point, but mostly the teacher asks the questions that determine how well the student has absorbed the material. This is *instructor-directed learning* in its purest form. To be effective, this style of instruction requires the skill of a very experienced teacher, whose own knowledge of the subject and experience with students correctly anticipate and adjust to the subtle dynamics of the learning process. How much do the students already know? How do students respond to abstract concepts? What are the questions that arise from a given lesson? What is the limit to the amount of new material that students can digest at one sitting?

Although instructors may still use this approach in some circumstances, most computer training involves much more student-initiated, hands-on experimentation. This is certainly the case with packaged software. Users commonly look to electronic learning features, or tutorials, and context-sensitive Help utilities to learn how to use the software.

Resource constraints do not permit the analyst to include sophisticated electronic learning features in small-enterprise systems. Nor can the analyst devote the time required to become a very experienced teacher. Instead, the analyst should initiate formal training sessions, and then set out a series of activities for the user to follow independently. This strategy capitalizes on the natural tendency of computer users to teach themselves.

User-Directed Learning

Self-teaching, or *user-directed learning*, is the strategy adopted by most major software publishers for introducing their customers to new products. According to the "80/20 rule," users should be able to accomplish 80 percent of what they want to do with the product after learning only 20 percent of the product features. Thus, typical Microsoft Word users might go for years without knowing about language codes, but they must know the basic formatting tools before they can produce even modestly complex documents. All of the product documentation is built on the premise that users "know when they need to learn" about different features. In other words, user expertise creates greater expectations, which can be met only by learning more about the system. At that time, the information must be accessible in a series of increasingly detailed explanations.

This approach can be adapted to the training needs of small-enterprise users as well. The users' active participation in the project from the very beginning establishes a good foundation for the formal training sessions of the implementation phase. The analyst should then provide sufficient time for the users to explore the system on their own, with help nearby.

Figure 14-11 presents several CIS training techniques, some of which apply to either instructor-directed learning or user-directed learning. In practice, training is a perpetual activity that includes many items from both categories. The particular style you adopt may even vary from system to system, depending on the system complexity and the users' familiarity with technology.

FIGURE 14-11 / *Information System Training Techniques*

1. Lectures
   This can be anything from a brief overview to a detailed "by-the-number" marathon. Normally, such sessions are supplemented with overheads and handouts.

2. Demonstrations
   Because users are generally passive observers during product demonstrations, this is one of the least effective training techniques.

3. Hands-On Exercises
   These sessions require the user to participate actively in the training in a controlled fashion. There is an expected outcome to each exercise, so each user has a similar experience.

4. Tutorials
   Self-paced instruction that is well structured can provide the highest return on the invested training hour. It combines the best of both learning formats.

5. Problem Solving
   The user is presented with a problem scenario and asked to develop a solution. This real-world simulation is an excellent way to build user confidence in the product.

6. Unstructured Exploration
   This technique can be very effective if the user is not intimidated by the product. Beginning computer users don't know how to engage in "productive play."

Training Content

The *basic training cycle*, as illustrated in Figure 14-12, begins with a presentation or experience centered on one or more of the techniques described in the preceding sections. Feedback from the student to the teacher, which may prompt corrective or elaborative instruction, follows. Finally, a summary presentation or experience reinforces the lesson. This cycle may be repeated periodically or continuously, depending on the complexity of the subject matter.

FIGURE 14-12   /   *Information System Training Cycle*

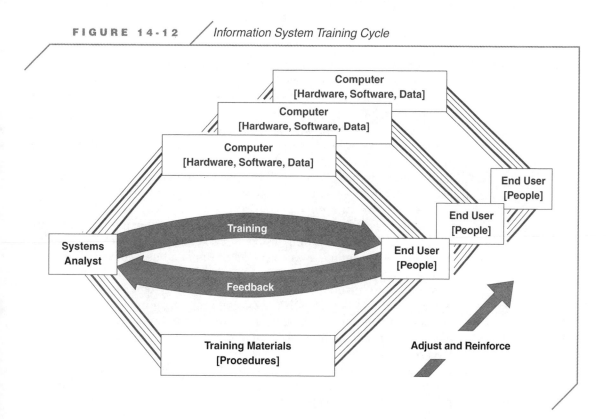

In some cases, training may lead to certification or licensing, although this seems unlikely in the small-enterprise setting. Nevertheless, testing the trainee's comprehension and competence regarding system functions and procedures is always important. If the analyst, the user, or some regulatory agency determines that an informal training cycle cannot satisfy this requirement, the analyst should design a formal testing procedure.

You must consider several factors when you plan your training program for the people who will use, operate, and manage the small-enterprise information system:

- Who are your students?
- What do they need to know?
- Under what conditions are they best instructed?

In a medium-size or large enterprise, the end users, computer operators, and managers are likely to be separated by distance, interest, and attitude. The training needs vary for each of these groups. For example, end users should know how to request a file query, operators should know how to access files to satisfy a query, and managers should know that files are secure and accurate. The techniques used to train different audiences must be appropriate to each audience. You might select the instructor-directed learning technique for the manager, or the user-directed learning technique for the end user.

In the small enterprise, these personnel distinctions may disappear or be significantly reduced. However, this does not necessarily simplify your approach to the training task. Under some circumstances, you may need to train a single individual about file security, file maintenance, and file query. Recognizing that users can view the same information system at different levels of detail is very important.

Figure 14-13 incorporates an integrated computer information system with this notion of providing different levels of detailed training to different audiences. For example, in a corporate setting, end-user training begins with hands-on exercises and tutorials concerning the information system and then may expand to include 4GL products and, finally, system software. Computer operator training begins with the 4GL products, continues with system software, and finally may include system hardware training. Manager training is often a lecture and demonstration and rarely extends beyond the 4GL product level of detail.

Though unusual, it is possible that the analyst applies the full range of detailed training to the same set of people in the small-enterprise environment. In such cases, the training should progress from level to level in a series of cycles.

FIGURE 14-13 / *Training Content and the System Environment*

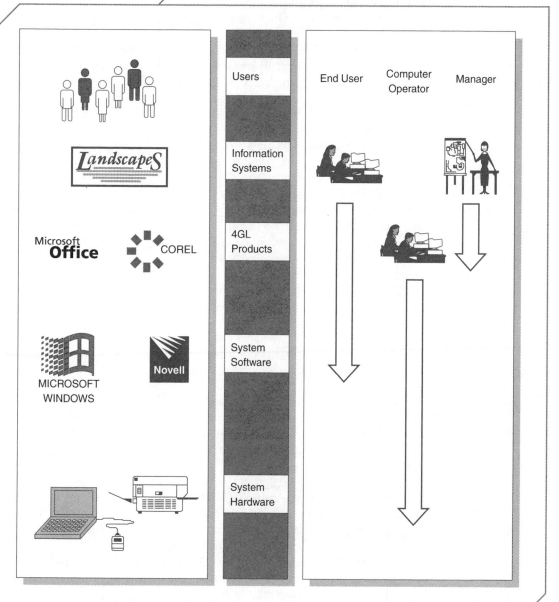

TRAINING MATERIALS

Obviously, the analyst must develop the formal training materials before training can begin. The problem with this is the natural desire to introduce small changes right up to the moment that the system is scheduled to go live. To accommodate both circumstances, the analyst often supplements training materials with corrections soon after implementation. It is important that users have a good foundation of system knowledge so they can easily accept any changes that are sure to come.

Training materials should be designed for the small-enterprise setting, in which the audience is a small group of workers who represent a functional composite of the end user, computer operator, and manager. Vendor documentation, electronic tutorials, and trade books should be available to these groups as their training moves beyond the immediate information system.

The *training manual* should contain everything you need to conduct the formal end-user training sessions. Much of the training material is from the system documentation, but a significant amount of work remains to develop teaching materials, such as handouts, PowerPoint presentations, manuals, tutorials, exercises, and reference guides. Do not wait until the system development is complete to begin this work. Remember, the PERT chart in Chapter 6 identifies this as a task that you can work on while several other tasks are also in progress.

The *procedures manual*, sometimes referred to as a user's guide, is so named to be consistent with our prior references to procedures as one of the six components of a computer information system. This manual should contain detailed instructions about using the information system. This includes system start-up, menu navigation, subsystem operation, and troubleshooting, in addition to the textual and visual descriptions of all inputs, processes, and outputs. Once again, much of the material can be leveraged from the system documentation, but you should remember that the end-user audience is considerably different from the one for which the system documentation is developed.

Control procedures should be part of the routine operations of the system. These procedures require the computer operator to maintain records concerning daily processing, scheduled maintenance, and system malfunctions. The level of formality in record keeping varies from system to system. Simple, single-user systems might require only a brief notation concerning regular system backups; other systems might require detailed statistics on master file additions, changes, and deletions, as well as transaction file record counts.

Regardless of system complexity, system malfunction procedures should always appear in the manual. When system-critical problems occur, users must know how to document the situation, which actions to take either to recover or to preserve the system, and who to contact.

The procedures manual also might contain some short tutorials or lessons that the user can repeat after the formal training session is over. If these exercises are interactive, you might have to create the accompanying sample files. Figure 14-14 shows a brief description of the topics covered in a typical procedures manual.

The *reference manual* contains much of the technical material that is part of the system documentation. Of the three manuals described in this section, this is the least important to the success of the training session. File layouts, special formulas, processing table values, and special codes are some examples of this type of information—often more useful once the system is operational. At this point, the user need only know where to find the reference manual.

FIGURE 14-14 / *Procedures Manual Table of Contents*

1. Information System Description
   This item is slightly more detailed than the overview that appears in the training manual. It includes summarized narrations about each of the subsystems as well as system diagrams and menu trees.

2. System Operating Instructions
   This item includes detailed operating instructions for every facet of the information system, from cold boot to shutdown, from maintenance to backup, and from main menu to submenu.

3. System Input and Output Samples
   This item provides sample input and output forms, screens, and reports, along with descriptions of where they originate in the system.

4. Emergency and Security Instructions
   This item contains special instructions for security problems (e.g., lost passwords) and/or emergencies (e.g., power failures, system software crashes).

Appendix A. 4GL Software Specifications
   This item identifies the specific version of 4GL software used in the information system.

Appendix B. Platform Specifications
   This item identifies the system software and hardware used in the information system.

Appendix C. System Error Messages
   This item contains all the error messages generated by the information system except those normally associated with the 4GL or system software.
   Each message should be explained and recommended actions specified.

Appendix D. Definitions
   This item defines any unusual terms that are used in the information system.

INFORMATION-AGE TRAINING WITH ETHICAL CONTENT

The emergence and maturation of the PC as a viable hardware platform brings a powerful collection of technologies within the economic reach of almost every small enterprise. Furthermore, the evolutionary development of the computing profession, with systems analysis as one of its clearly defined specialties, demands an equally professional standard of conduct. This *code of ethics*, or code of conduct, is a social contract by which information system professionals and users should abide.

The Thinking Critically scenarios in each chapter demonstrate that computer professionals and users frequently confront ethical questions that can be difficult to answer. The analyst has a special responsibility to inform the user of the standards of ethical conduct and to demonstrate commitment to those standards. Although such behavior should be observed throughout the project, the training session is an ideal opportunity to encourage users to consider this issue. The following is a possible approach to educating users about the ethical use of an information system.

An Objective Standard

Several professional organizations offer objective standards for computer professionals as well as public and corporate users. The Association of Information Technology Professionals (AITP) and the Association for Computing Machinery (ACM) offer guidelines for ethical conduct that differ in focus, but agree in intent. Each attempts to provide a set of goals and rules, or objectives, concerning the behavior of the IT community.

The ACM specifies five general principles (canons), professional ideals (ethical considerations), and mandatory rules (disciplinary rules). The general principles include language that either requires or encourages members to "act at all times with integrity," "strive for increased competence," "accept responsibility for his/her work," "act with professional responsibility," and "use his/her special knowledge and skills for the advancement of human welfare."

The ACM offers an ethical consideration that is particularly relevant to your work as an analyst. They suggest that whenever you work with data concerning individuals, you should always consider the principle of the individual's privacy and seek the following:

1. To minimize the data collected
2. To limit authorized access to the data
3. To provide proper security for the data
4. To determine the required retention period of the data
5. To ensure proper disposal of the data

The AITP prescribes a standard of conduct that includes more than 25 specific statements of behavior, such as the following:

1. Do not misrepresent or withhold information concerning the capabilities of equipment, software, or systems.
2. Protect the privacy and confidentiality of all information entrusted to you.

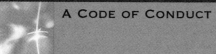

TECHNOTE 14-1

A CODE OF CONDUCT

A formal **code of conduct** is often hard to apply, especially when it contains language that is specific to a particular discipline mixed with legalistic or philosophical phrasing, or both. Richard Mason, writing in the *MIS Quarterly* (March 1986), offers an alternative that is easy to remember and very well suited to many of the situations that confront the modern systems analyst and computer user. Mason argues that "Privacy, accuracy, property, and accessibility are the four major issues of information ethics for the information age." He refers to this collection with the acronym of **PAPA**.

PAPA is easy to remember whenever an ethical question arises. For example, is it ethical to:

- Allow a client to read his or her employees' e-mail?

- Omit the error-checking features of a database maintenance system because your project is behind schedule?
- Install a software product on a local area network if you originally purchased the software for your stand-alone PC?
- Deny someone access to public information because it might undermine your client's competitive advantage?

Although PAPA is a convenient way to focus your attention on ethical concerns, adopting a singular ethical perspective may be risky. The analyst must combine objective standards, the experience of others, and personal judgment to formulate a code of conduct that is consistent and workable. This should provide the basis for the analyst's behavior and serve as the inspiration for the code of conduct presented to users.

---

3. Do not use knowledge of a confidential or personal nature in any unauthorized manner or to achieve personal gain.
4. Do not exploit the weakness of a computer system for personal gain or personal satisfaction.

For Example—LandscapeS

To help illustrate the concepts in this chapter, LandscapeS, which specializes in landscape architectural services, is introduced as an enterprise on the verge of implementing its first computer information system. After three years of manual record keeping, the two owners hired an independent systems analyst to design, develop, and implement a modest computer information system. The original contract called for three information products: customer account maintenance and correspondence, landscape job tracking, and ornamental plant inventory control.

Figure 14-15 presents the user's system diagram. To accommodate the owners' very mobile business operations, the analyst recommended a notebook computer for the enterprise, along with a software suite. The analyst installs the new information system and data files on the notebook's hard disk.

FIGURE 14-15 / *LandscapeS USD*

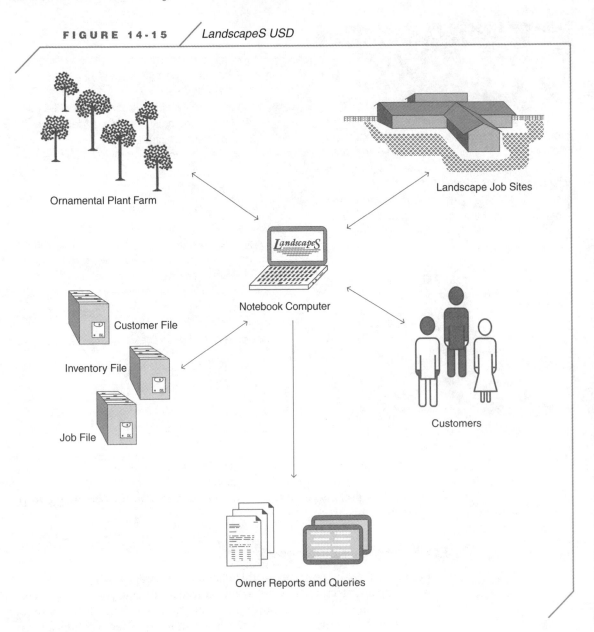

Ornamental Plant Farm

Landscape Job Sites

Notebook Computer

Customer File

Inventory File

Job File

Customers

Owner Reports and Queries

The first training session includes a brief demonstration by the analyst and a one-hour hands-on exercise designed to lead the owners through the product. Figure 14-16 is an excerpt from the training material. Although the exercise becomes less teacher-directed as the user progresses from system startup to system shutdown, this is not an example of user-directed learning. It merely reflects the expectation that the user requires less instruction as familiarity with the product increases. The analyst is still responsible for guiding the user through each exercise and answering questions as they arise.

FIGURE 14-16   *LandscapeS Hands-On Training Exercise*

EXERCISE 6—You have a new customer.

Part A: System Startup
1. Turn on the computer. You have three hours of battery life at this point.
2. Start the LandscapeS information system (double-click on the customized icon).

Part B: Customer Accounts and Correspondence
1. Select the customer option from the main menu.
2. Select the file update option.
3. Add a fictitious customer to the master file.
4. Exit the file update option to the main menu.
5. Select the account update option from the main menu.
6. Enter a $100 charge to the new customer.
7. Exit the account update option to the main menu.

Part C: Landscape Job Tracking
1. Start the landscape job tracking subsystem.
2. Create a new job for the new customer.
3. Exit the subsystem.

Part D: Owner Reports
1. Print a report of the new job.

Part E: System Shutdown
1. Close all applications.
2. Turn off the computer.
3. Recharge the battery.

After the owners complete this part of the training, the analyst gives them the computer to use for two days. This experience is really a combined pilot test and user-directed learning exercise. As you would expect, many questions await the analyst upon return to the enterprise. At this point, the owners are able to describe situations in which the system was confusing or general areas of the system for which they need more training. The analyst can then adjust the training materials accordingly. This illustrates the importance of feedback in the training cycle.

Toward the end of the training activities, the analyst introduces several ethical issues into the discussion. At first, the owners are surprised that this topic is part of user training, claiming that they already understand that it is illegal to pirate software and so on. Nevertheless, the analyst presents them with a fictitious letter (Figure 14-17) that makes them reconsider. Their first question to the analyst concerns the legality of releasing a customer file to some other entity. The next question concerns the ethical standards that might apply in this case.

FIGURE 14-17   /   *LandscapeS Ethical Question*

SOUTH BAY INVESTORS, INC.
222 Century City, Suite 654
Los Angeles, California 90021

April 1, 2004

Mr. C. Dwelley
LandscapeS
Box 789
Alton, CA 95543

Dear Mr. Dwelley:

We represent a mortgage investment group that specializes in real property, secured loans, and financing. As an independent entity, we are interested in locating commercial and private property owners that may be in need of our services.

It occurs to us that your work as a landscape architect puts you in contact with many people who might be interested in a low-cost way to finance all or part of your fees. In fact, some of your customers may be able to expand their plans for custom landscaping because of our services.

If we can work out an arrangement by which you can make your customer file available to us, we can offer you a one percent finder's fee on any subsequent financial contracts negotiated between our group and your customers.

Please contact us at (310) 555-1234 if you would like to offer our help to your valued customers.

Sincerely,

Julie Stein

PROJECT DELIVERABLE: THE TRAINING SESSION

A strong correlation exists between successful systems and well-trained users. Given the consistent user involvement in the SDLC, you should already have a good feel for your users' training needs. Although the SDLC model specifies training as a singular activity bridging the development and implementation phases, we have emphasized that continuous user training is a by-product of sustained user involvement in the project. Beginning with the first prototyping sessions, the user enters a training cycle that focuses on the system interfaces and master file maintenance screen forms and reports. In this case, the feedback portion of the process may be a combination of user confusion and user complaint. The analyst must recognize the difference, retrain on the former, and redesign on the latter.

This indirect type of training usually occurs outside the normal workplace setting, either in another location or during off hours. Enterprise work patterns should not be disturbed during the early part of the SDLC.

Pilot testing, which you will learn more about in Chapter 15, provides an excellent opportunity for further user training. Analysts usually conduct this exercise in the actual work setting, even though it does not involve the entire enterprise. This is a good time to evaluate the dynamics among the six components of the information system. It is the first time that people, data, procedures, networking, hardware, and software come together to form a complete system.

By the time the formal training sessions commence, the user has a good understanding of how the information system works. Figure 14-18 shows the training session report content. In situations involving some users who have not fully participated in the project, the more experienced users can serve as training assistants. As much as possible, the analyst should conduct this training in a real-world setting.

FIGURE 14-18 / *Training Session Report Content*

Written Material
1. Description of training methodology
2. Training schedule
3. Training manual table of contents
4. Samples of training handouts

Oral Presentation, Demonstration, and Training
1. Describe training methodology
2. Visuals for items 2 and 3 above
3. Demonstration
4. Hands-on instruction

THE CORNUCOPIA CASE

Cornucopia's design and development testing sequence is patterned after the plan illustrated in Figure 14-2. At this point in the sample project, testing has progressed through integrated module testing and the operating system interface. The next set of system tests should involve all six components of the information system.

Cornucopia is a good example of a small enterprise that has only two or three people accessing the information system, one of whom is the owner. However, everyone needs some training. At this time, the owner chooses to be the person responsible for the overall maintenance of the computer and the information system. The sales clerks need to know how to work with the master file maintenance and the transaction processing subsystems. All these people participated in at least a portion of the prototyping and testing activities, but the owner has the most experience with the system so far. Fortunately, one of the sales clerks recently completed a computer literacy class at the local community college.

Testing Sequence

Figure 14-19 cross-references the exact sequence of tests with the appropriate illustrations in previous Cornucopia material. The various subsystems are successfully developed and tested in the modular fashion. The object-oriented nature of the development software eliminates much of the programming effort, leaving the analyst-programmer to select the proper macro, query, or Wizard to employ in a particular situation. Errors ensue when the tools are not used properly or applied in the proper sequence; rarely do syntax errors arise.

Training Activities

The training sessions (Figure 14-20) occur during a two-week period, with four of the contracted twelve hours (see the project contract in Chapter 2) reserved for two follow-up visits after the system is fully implemented. All the initial training takes place after the business is closed for the day. The follow-up sessions are scheduled for normal business hours that are not too busy. All training sessions are conducted on site, which means that the computer will be moved back and forth from the development location to the business premises until the system is fully implemented. While most of the training session involves computer demonstrations or hands-on exercises, the analysts package the session in a slide show. Figure 14-21 illustrates one of the slides from the training session slide show.

FIGURE 14-19 / *Cornucopia Testing Sequence*

1. Syntax Testing:
   Continues throughout all testing activities.

2. User Interface Testing:
   Prototype Menus: (Figure 10-16) Several menus were tested with the "Under Construction" message.
   Prototype Master File Update Forms: (Figure 8-16) Sample master files were created before the lower portion of these forms could be built. They were tested to satisfy the visual requirements of the user.
   Prototype CD Sales Transaction Form: (Figure 8-17) The multiple record display on the lower portion of the form required a subform. This subsystem was initially tested with random data. Later, when all of the tables were linked (i.e., had referential integrity), testing was more disciplined.

3. Program Module Testing:
   Master File Update Subsystems: These modules were tested with the same sample master file data used for the prototype. Once the customer master file modules were debugged, they were copied and slightly modified to yield the CD and Supplier master file update subsystems.
   Sales and Order Transaction Processing: (Figures 8-18 and 12-26) These modules were developed and tested in tandem because they both affect the inventory management system. Several different types of transactions come into play (sales, returns, back orders, and cancellations). Each one was tested and debugged separately and together to ensure that they were compatible. This effort consumed the great majority of the development and testing budget.
   Correspondence Mail Merge: (Figure 12-31) Only one form letter was developed and tested. Others will follow the same procedure as the user identifies the need and develops content.
   Sales Trends Spreadsheet: (Figure 9-17) This module only required a little sample sales data, so its development and testing could proceed independent of the others. The implementing subprocedure (Figure 12-27) is the longest and most complicated code in the entire project. Fortunately, most of the code was copied from a general purpose query and replicated to the different sales categories.
   Web Page Construction: (Figures 9-17 through 9-21 and 13-23 through 13-25) Most of this effort was focused on visual design and the experimental JavaScript. Except for the database table, this module was developed and tested independently with two browsers (Netscape and Internet Explorer).

4. Integrated Module Testing:
   Switchboard: (Figure 10-16) The switchboard and secondary menus were continually tested as new subsystems were developed.
   System Links: File paths and program calls were tested periodically, and on different development systems. They need to be retested when the system is installed on the client station.

5. System Testing:
   Cornucopia Desktop Shortcuts: (Figure 13-22) Both the Start menu and desktop shortcut entries were easy to construct and test. Three tasks that remain are (1) turn off the warning messages associated with special queries, (2) remove the native menus and toolbars, and (3) implement system passwords and security protection attributes.

FIGURE 14-20 / *Cornucopia Training Schedule*

Session #1: Tuesday, 5:30–7:30 PM, for all employees.

1. Information System Overview
2. Demonstration of the Windows Interface
3. Demonstration of the Cornucopia Interface
4. Break
5. Hands-On Exercise on the Interfaces
6. Description of the Procedures Manual

Session #2: Wednesday, 5:30–7:30 PM, for all employees.

1. Hands-On Review Exercise
2. Demonstration of the MF Update Subsystems
3. Hands-On Exercise on the Update Subsystems
4. Break
5. Demonstration of the Sales Transaction and Order
   Subsystem
6. Hands-On Exercise on the Sales Transaction Subsystem

Session #3: Tuesday, 5:30–7:30 PM, for all employees.

1. Hands-On Review Exercise
2. Demonstration of the 4GL Software Tutorials
3. Hands-On Problem-Solving Exercise
4. Break
5. Description of Emergency and Security Procedures
6. Review of Procedures Manual

Session #4: Wednesday, 5:30–7:30 PM, for the owner.

1. Demonstration of the Correspondence Subsystem
2. Demonstration of the Sales Trending Subsystem
3. Demonstration of the Internet Connection Subsystem
4. Hands-On Exercise on the above Subsystems
5. Break
6. Overview of the System Hardware

FIGURE 14-21 / *Cornucopia Training Session Slide Show*

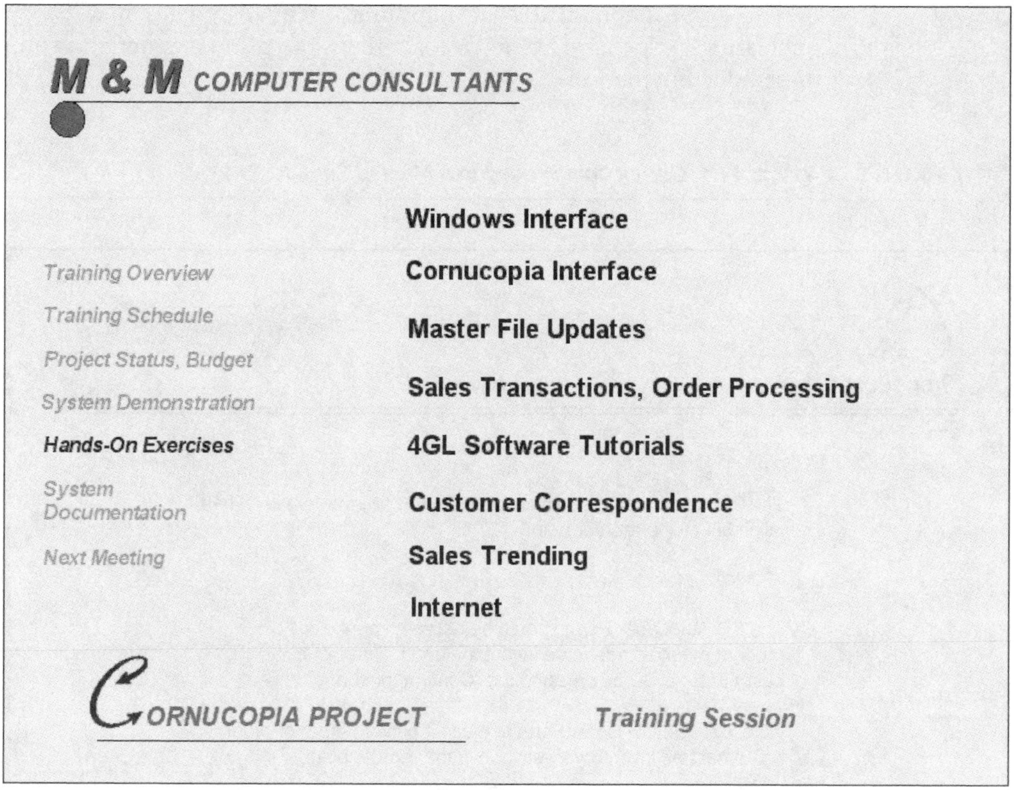

M & M *COMPUTER CONSULTANTS*

Training Overview

Training Schedule

Project Status, Budget

System Demonstration

Hands-On Exercises

System Documentation

Next Meeting

Windows Interface

Cornucopia Interface

Master File Updates

Sales Transactions, Order Processing

4GL Software Tutorials

Customer Correspondence

Sales Trending

Internet

*C*ORNUCOPIA PROJECT   *Training Session*

Product Documentation

Several important documents are prepared for the implementation activities. The training manual and procedures manual are used during the training sessions. The reference manual is primarily used when the system breaks down or for future analysts involved in a system upgrade. The user also receives the hardware and horizontal software materials (CDs, literature, packing materials) that come with those products.

Figure 14-22 shows the procedures manual entry describing the CD AutoOrder Form menu option. The analysts developed a standard format that includes the issue or revision date and the procedures manual section reference, all highlighted by easily recognized content headings. One disadvantage to printed manuals is the amount of human intervention required to print and distribute documentation materials. Assembling product documentation into three-ring binders, with stylized section lettering and dividers, makes it easier to incorporate changes to individual pages to accommodate information system changes. However, any paper-based documentation system is subject to some haphazard page handling. Manual updates may become an increasingly expensive burden. An alternative to printed material is computer-based documentation.

FIGURE 14-22 / *Cornucopia Procedures Manual Sample Page*

Procedures Manual

: AutoOrder Form

Description:

The CD AutoOrder Form presents CDs whose inventory level has fallen below the reorder point.

Use:

The CD AutoOrder Form should be activated periodically in order to maintain adequate inventory levels. The user can easily order a CD by pressing the "Confirm" button.

Note: Use the CD Order Form to place a custom order or order a CD that has inventory levels above the reorder point.

Access:

To activate the CD AutoOrder Form:

1. Select CD Orders from the switchboard.

2. Select AutoOrder from the submenu.

Operating Instructions: Menu Descriptions

M&M   Section 2.2.5   Issue/Revised: 05/19/04

Time and Money

It is not always easy to differentiate the activities performed in one category from another category. This is true for the Cornucopia project team. Over the past three weeks, they reported a total of five hours toward the development activities, increasing the percentage completed to 100 percent, and fifteen hours toward the implementation activities, moving this phase to 75 percent complete. Figure 14-23 presents the status report as of Week 15. Notice that the total actual labor hours for development are double the estimate. Clearly, this would jeopardize the overall budget estimate if it were not for the reduced hardware and software prices that provide an offset. This experience may explain why some analysts prefer to inflate the budget estimates in the beginning. However, this is not a recommended approach. A more productive practice is to evaluate the differences between estimated and actual costs in terms of unexpected project complexity, resource ordering and delivery delays, personnel inexperience, and so on. In other words, the analysts' ability to develop accurate estimates can improve by carefully evaluating past experience and making corrective adjustments.

FIGURE 14-23   *Cornucopia Project Status – Week 15*

Status15.xls

Date:	Cornucopia Project Status							As of: Week 15										
Activity	**% Comp.**	**Status**	**1**	**2**	**3**	**4**	**5**	**6**	**7**	**8**	**9**	**10**	**11**	**12**	**13**	**14**	**15**	**16**	**Total**
Analysis - Estimate	100%		4	5	5	4													18
Actual	100%	ok	2	4	2	3													11
Design - Estimate	100%					4	4	5	5	4									22
Actual	100%	ok			1	3	8	5	7	6									30
Develop - Estimate	100%									4	4	5	5	4					22
Actual	100%	ok								1	8	20	4	6	4	1			44
Impl. - Estimate	80%													4	5	5	5	5	24
Actual	75%	ok												1	2	4	5	6	18
Total - Estimate			4	5	5	8	4	5	5	8	4	5	5	8	5	5	5	5	86
Actual			2	4	3	6	8	5	7	7	8	20	5	8	8	6	6	0	103
Contract	100%	ok	C																
Prelim. Present.	100%	ok		C															
Design Review	100%	ok						C											
Prototype Review	100%	ok										C							
Training Session	95%	ok														C			
Final Report	35%	ok															S		

Status / Sheet2 / Sheet3

Cornucopia with Visible Analyst

Figure 14-24 shows how the analysts can go through the repository to perform syntax checks on the various models of the system. This check is usually performed as the models are developed, but sometimes testing sequences provoke changes that require retesting of specific models.

FIGURE 14-24   /   *Cornucopia Syntax Testing with Visible Analyst*

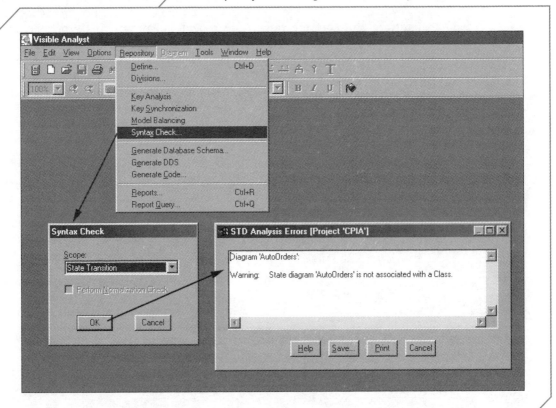

SUMMARY

Testing in a 4GL-based, small-enterprise environment is naturally tailored to fit the customized applications that are developed. The use of object-oriented products, code generators, and graphical application-building utilities allows the analyst to look at the product rather than the detailed program code. The user is also much more involved in testing than ever before. From prototype sessions to final product acceptance, the user is an active participant in testing.

These new testing dynamics still rely on principles developed during the past 40 years of computing. A testing plan is essential. Clearly defined goals must be established. In addition, testing procedures should be as carefully documented as any other element of the project.

System documentation is an ongoing activity throughout the project. As each element of the analysis, design, and development work is completed, the analyst can add a new entry into the project binder. Documentation becomes more formal when the analyst begins to assemble the various manuals required to operate, use, maintain, and update the system. This activity can proceed independent of the final development activities.

Informal end-user training begins with the first JAD working sessions. Prototyping and testing activities provide more training opportunities, as the end user is exposed to more definitive elements of the information system. Given this experience, the formal training session may serve as a summary and review for many end users. Nevertheless, the analyst is responsible for preparing training materials and conducting such sessions in a way that provides a uniform and consistent basis for system implementation and future maintenance. Increased user involvement in the project does not eliminate or diminish the importance of this formal activity.

Training techniques and teaching styles must be tailored to fit the audience, the complexity of the subject, and the training setting. Some standard manuals, however, should accompany any system.

User training should also include a segment devoted to the ethical use of the information system and its resources. In general, users are not well prepared to deal with many of the new ethical issues concerning privacy, accuracy, property, and access as they relate to their computer products.

TEST YOURSELF

Chapter Learning Objectives

1. What are the various stages of product testing and how do their goals and procedures differ from one another?

2. What are the key elements of product-testing documentation?

3. What portions of product documentation are assembled from SDLC project deliverables and what portions remain to be developed by the analyst specifically for documentation purposes?

4. What are the major training methodologies employed by systems analysts?

5. How can you include information system ethics in training sessions?

Chapter Content

1. Only the most trivial systems can be 100 percent tested. True or False?

2. It is best to validate user input as close to the data entry point as possible. True or False?

3. User-acceptance testing occurs at the very end of the project. True or False?

4. It generally costs more to correct design phase errors than it does to correct implementation phase errors. True or False?

5. The project binder is the source for several other detailed documentation manuals. True or False?

6. In the small-enterprise project, printed documentation is preferable to electronic documentation because it is easier to change. True or False?

7. To make sure all the latest system changes are included, training materials should be developed a day or two before the scheduled training session. True or False?

8. What type of error occurs when the syntax is correct, but the code performs the incorrect tasks?
   a. syntax error
   b. logic error
   c. runtime error

9. What type of error occurs when the computer code conflicts with the system, programming language, or application rules?
   a. syntax error
   b. logic error
   c. runtime error

10. What type of error occurs when the processing or supervising program is confused?
   a. syntax error
   b. logic error
   c. runtime error

11. What type of testing occurs when the sample menus, screen forms, and reports are evaluated during prototype design and review sessions?
   a. pilot testing
   b. system testing
   c. user-interface testing

12. What type of testing occurs when all five of the information system components are exercised to determine how well they work together to meet the performance specifications of the project?
   a. pilot testing
   b. system testing
   c. user-interface testing

13. What type of testing occurs when the user interacts with the system, without the analyst's help?
   a. pilot testing
   b. system testing
   c. user-interface testing

14. What type of teaching methodology is best suited to small-enterprise information system training?
   a. instructor-directed learning
   b. user-directed learning
   c. online Help tutorials

15. What is the appropriate audience when training includes information systems and 4GL products?
   a. end user
   b. computer operator
   c. manager

16. What is the appropriate audience when training includes 4GL products, system software, and system hardware?
   a. end user
   b. computer operator
   c. manager

17. What is the appropriate audience when training includes information systems, 4GL products, and system software?
   a. end user
   b. computer operator
   c. manager

18. A testing plan should provide all but the following:
    a. well-defined schedules
    b. user and analyst responsibilities
    c. target goals
    d. detailed documentation on the system design

19. The complete system documentation is assembled in the:
    a. procedures manual
    b. training manual
    c. reference manual
    d. project binder

20. The analyst can best educate users about the ethical use of computer information systems by:
    a. providing URLs to government-enforcement Web sites
    b. distributing a list of ethical "do's and don'ts" with the product documentation
    c. including a discussion of ethics in the training session
    d. setting a good example for users throughout the SDLC

ACTIVITIES

Activity 1

Review LandscapeS' USD (Figure 14-15), hands-on training exercises (Figure 14-16), and ethical issues (Figure 14-17).
    a. Research and recommend hardware resource specifications for LandscapeS' notebook computer.
    b. Develop training exercises to teach users how to generate customer billing output and reports.
    c. Develop three more ethical scenarios for LandscapeS' users to consider.

Activity 2

Use your favorite Internet search engine to research information system ethics. Prepare a one-page user handout titled "Ethics and Your New Information System." Cite your references.

Activity 3

Evaluate the printed documentation or electronic Help feature of one software application in terms of its ease of use, completeness, and effectiveness. Prepare a one-page summary of your analysis.

DISCUSSION QUESTIONS

1. What are the differences between syntax, logic, and runtime errors? Where in the testing sequence would you expect these different error types to appear? To what differing degrees do they impact the system?

2. Why is it impossible to eliminate all errors from information systems?

3. How is 4GL testing different from third-generation testing?

4. In what ways would you expect the user to offer better testing data than that developed by the analyst-programmer?

5. Why do you think the term "analyst" is transformed to "analyst-programmer" in many parts of this section of the book?

6. Describe how you would advise a first-time user to become acquainted with a new desktop microcomputer.

7. Under what circumstances would you recommend shallow detail, broad-coverage training objectives? When would you recommend deeply detailed, narrow-coverage training objectives?

8. Of the three common feedback techniques (instructor observation, student questions, testing), which do you feel is the most effective for CIS training? Explain.

9. To what extent do you believe the average end user needs to know about system software and hardware?

10. People learn in many ways: by reading, listening, watching, discussing, and doing. Why is it important to address several of these learning modalities in your information system training sessions?

PortfolioProject

Team Assignment 14: Testing

This assignment is designed to document your product testing plans, methods, and results (see Figure 14-6). Because testing naturally occurs throughout the design and development phases, much of this assignment requires that you simply document what has already happened. Such "after-the-fact" documentation provides a useful framework for future projects. You are required to submit brief summaries or examples of the following:

a. Submit an outline of your testing plan. This outline should document your testing goals and methods.

b. Summarize your testing results. Include one example of your test data, expected outcome, and achieved outcome, and explain any differences between your expected outcome and achieved outcome. Submit a copy of your summary or example.

c. File a copy of your plan, results summary, and examples in the Portfolio Binder behind the tab labeled "Assignment 14."

Project Deliverable: Training Session

Given the time constraints imposed by the academic calendar and class meeting times, you will not be able to conduct a full-fledged training session. Instead, you are expected to conduct a brief training session, focusing on a small part of your project. Unlike the preliminary, design, and prototype review sessions, the training session is not a slide show presentation. Although you should prepare a few slides to help organize the training activities, most of your time is devoted either to demonstrating your product or to instructing your client on how to use your product.

1. Submit a training session report containing the following:
   a. Cover letter
   b. Training plan and schedule to train users on:
      i. System switchboard
      ii. One subsystem
      iii. System documentation and Help features
   c. Project status report
   d. Project budget
   e. Oral presentation slide show handout (three slides per page)
2. Prepare a slide show and appropriate handouts to support a 20 to 30 minute demonstration of your system and a brief training session covering item 1b.
3. File a copy of your report in the Portfolio Binder behind the tab labeled "Training Session."

conversion,maintenance, **and**review

OVERVIEW

Whether the system replaces an existing system or is entirely new, the analyst is responsible for the orderly transition from a pilot-tested prototype to a completed, working product. This activity is collectively called system conversion. This section begins with a detailed discussion of file preparation and concludes with a recommendation on how to terminate the project with final user acceptance. You will learn how to establish a plan so that time-consuming file preparation efforts do not delay the implementation of your system. You will learn several methods to bring your system online, each with advantages and disadvantages to cost and risk.

Once the information system becomes operational, the maintenance and review phase of the SDLC begins. This chapter describes a systematic review and evaluation technique, which is significantly different in purpose from the activities collectively called system maintenance. The analyst must not only distinguish between review and maintenance, but also between routine maintenance and maintenance upgrades. Simple "wear and tear" contributes to the former; obsolescence causes the latter.

You will learn that different types of system maintenance require different responses from the enterprise and the analyst. Some are more costly than others, but no information system can continue to function properly without systematic maintenance. All types of system maintenance are essential.

Finally, you will learn that managing the maintenance effort is not simple. Much of the methodology employed in the SDLC applies to maintenance work, with one additional variable: as the information system can become functionally obsolete, so can the analyst and other computer professionals if they do not maintain their knowledge of the system and the industry.

- Develop a plan to coordinate the conversion efforts associated with all six information system components.

- Prepare a product review survey form to solicit balanced and informative feedback from the user.

- Advise users about different types of information system maintenance activities.

- Advise users about the purpose and content of periodic reviews and evaluations of the information system.

- Appreciate the need to maintain and expand your skills as a systems analyst.

SYSTEM CONVERSION

System conversion involves all six components of the information system (hardware, software, people, procedures, data, networking), the first four of which we have discussed in some detail. This section focuses on data conversion and the substantial effort required to create or transform data files.

The analyst might find that some of the new system files already exist in machine-readable format, but they require modification to add or delete fields or to change the file medium. In addition, some new system files may not exist at all or exist only within a manual record-keeping system.

Of course, the file preparation activities must commence well before the actual *cutover phase* of the new system, which describes a compressed process of data conversion, testing, changing to the new system, and user training. As with the preparation of training materials, this work can parallel other developmental and implementation activities.

The analyst also should recognize that some of the information in the data files might be confidential or proprietary. The appropriate person in the enterprise should explore this possibility so that precautions can be included in the file preparation procedures.

File Preparations

When no electronic file exists, the analyst must develop a plan and direct the effort to gather source documents, enter data, verify file records, and synchronize interim file maintenance. Except in the most unusual circumstances, the analyst should not create new files. New file creation activities will occur during a very hectic part of the development and implementation stages of the SDLC. The analyst should focus his attention on product development, testing, and training.

Although a variety of people can create new files, the enterprise personnel are perhaps best suited for this task. Their experience can help them interpret source documents, identify errors, and research subtle mistakes. In addition, because they have a large stake in the quality of the file data, they are motivated to do this very tedious job.

To illustrate the nature of this activity, consider the file preparation required to implement the new information system for LandscapeS. Figure 15-1 presents the new system entity-relationship diagram and the relevant existing source documents. Four files need to be created. The customer and plant information must be typed from the Rolodex cards into the appropriate master file. The Customer Account transaction file and Jobs master file can be typed from the original source documents.

FIGURE 15-1   /   *LandscapeS File Preparation*

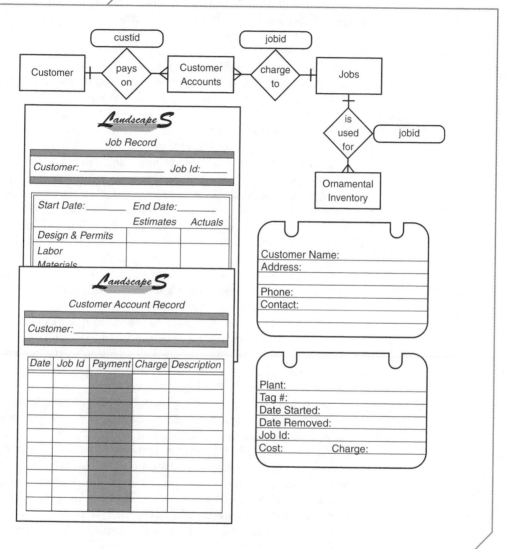

The three master files require unique key-field values. Apparently, the job records and plant records already have unique key-field values, named jobid and tag_num. The existing customer records do not have a unique key-field value, so a new field (custid) is added. The source document information should be verified for completeness and accuracy. Furthermore, the owners need to decide which records they need for the new system. Some of these records may be so old that typing them into the database is useless, whereas others may be old but contain information relevant to future operations.

The Customer Account transaction file is likely to present the most time-consuming task simply because of its size. Once again, the owners can reduce the effort by choosing to enter only selective data. For example, they could decide to omit the records for the completed jobs. Notice that the customer name should be replaced by the unique key-field value (custid). The accuracy and completeness of the data is equally as important in this file as it is in the others. Nevertheless, one added concern is the requirement that the custid and jobid values have corresponding entries in the master files. To ensure such database integrity, the analyst should seriously consider using the new transaction-processing subsystem to build these files, rather than the direct data entry approach taken for the master files.

Unfortunately, source document errors and omissions have no easy remedy. All of LandscapeS' existing manual records are prone to this problem. The job records may show illegible handwriting for a customer's last name. The plant cards may be inconsistent, some with no tag number, others with removed dates earlier than start dates. The phone numbers on the customers' cards may be incomplete. Researching and solving these problems is a tedious, time-consuming task. For the small enterprise, this may be a relatively small problem, but as the size of the enterprise or its files increases, you can expect this activity to require significant resources to complete.

In contrast to the preceding situation, data files may already exist in one electronic form or another. This reduces the file preparation task considerably. However, existing files may need to be modified to add, delete, or expand certain data fields, which requires time for data collection, data entry, and file verification. In some cases, the analyst can automate this transformation by using utilities or creating small, third-generation language programs.

Assume that prior to the new system, LandscapeS maintained a job file on an old Macintosh computer. It seemed natural to the owners to include the job-related data (description, jobid, start and end dates, etc.) and customer data (name, address, phone, etc.) on the same file. This allowed the owners to produce a list of all the jobs and customers. This worked fine until they had repeat customers, wherein they noticed duplicate customer data on the file and, in some cases, inconsistent address and phone data for the same customer name. This problem contributed to their decision to engage a computer professional to provide them with a well-designed information system.

Although LandscapeS' old job file has some problems, much of it can be transformed into the newly designed master files. A simple select operation on the job-related data fields can produce much of the required data for the Job master file. Likewise, the customer data can be copied to a temporary file and sorted on customer name. Then, the analyst can research duplicate customer records to determine which records to delete. Finally, the analyst can add a unique customer identification to the files, plus any other new data called for in the design.

Regardless of the method used to build the new files, special care is required to ensure that they are properly maintained until the new system becomes operational. Thus, in the LandscapeS example, if a new customer comes along after the analyst has created the master file but before the new system is operational, the analyst needs a procedure to capture this information and update the master file. The same is true for changes that involve the other files. *File synchronization* describes this need to keep newly created files maintained during the conversion process. The analyst should clearly document these tasks as a part of the conversion plan.

The accuracy and completeness of the new files is critical to the success of the new information system. The extra work and confusion created by the file preparation activities can put extraordinary pressures on everyone involved in the project, especially those actually doing the file building. Special controls are required to ensure that records are not lost, duplicated, or inadvertently altered. Record counts, summary totals, and carefully inspected sample records are common control techniques that can reduce file errors. Unless the files are very small (fewer than 100 records), the analyst should anticipate some errors in the files during the cutover phase. Therefore, the analyst should build control and correction procedures into the conversion plan.

CONVERSION OPTIONS

The new system becomes operational when it collects data, processes data, and reports information. Over the years, analysts and users have experienced a multitude of problems with newly operational information systems. In this text, you have learned that it is unrealistic to assume that the new system will be error free, even under the best of circumstances. To reduce the anxiety associated with the cutover phase, analysts have developed several conversion options, each with different costs and risks.

Figure 15-2 shows three conversion options for a small-enterprise project: direct, phased, and parallel. A fourth option, called pilot conversion, is omitted because its call for a direct conversion in a single unit or department of the organization is not usually applicable in a small enterprise. You should not confuse the terms pilot conversion and pilot testing. Remember, pilot testing refers to the joint effort of the user and analyst to perform system testing during the development-implementation transition period of the project.

In each of the three options illustrated, the new system perfectly matches the old system from a functional point of view. Thus, as each new subsystem enters the cutover phase, the corresponding portion of the old system disappears. In reality, this perfect match rarely occurs. Instead, portions of the old system might linger on, or portions of the new system might be entirely new. Nevertheless, the figure illustrates that all three options are subject to the same promised conversion date and that the last two options require operational implementation of some or all of the system prior to that promised date.

FIGURE 15-2 / *Conversion Options*

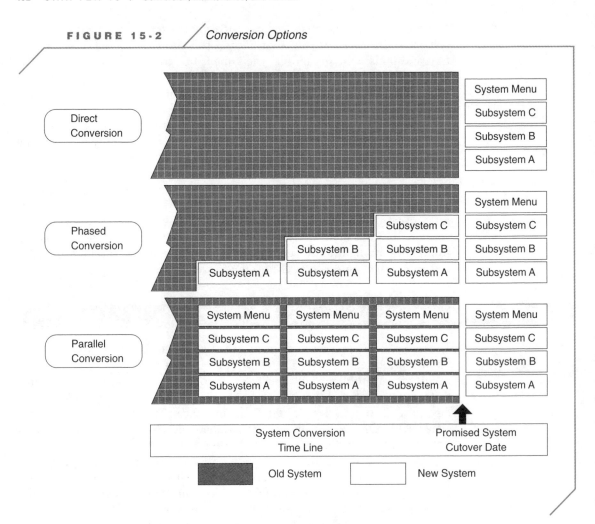

Figure 15-3 shows a series of illustrations portraying the costs and risks associated with different system conversion options. Conventional wisdom holds that a *direct conversion*, in which the new system abruptly replaces the old system, is the least expensive option because no duplication of effort occurs. This is a high-risk option for that same reason. With the old system eliminated, no easy retreat is available if the new system fails.

A *phased conversion* provides for moderation in all aspects. Figure 15-2 shows each subsystem cutover phase in a gradual progression. Although the corresponding new and old subsystems never operate at the same time, costs are somewhat elevated because of the confusion and overhead costs caused by mixing old and new subsystems. Likewise, the risk is moderate if you consider that even if the new subsystem fails, you will at least have some of the old system to rely on.

FIGURE 15-3   /   *Conversion Cost and Risk*

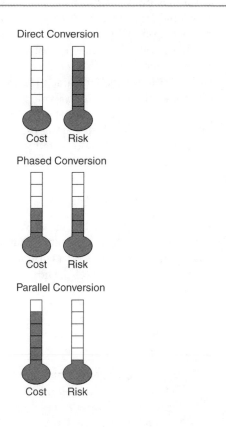

Direct Conversion

Cost   Risk

Phased Conversion

Cost   Risk

Parallel Conversion

Cost   Risk

A *parallel conversion* calls for the new and old systems to operate side by side for a period of time, with careful reconciliation of any differences in the results produced. Costs are high because the enterprise must operate two systems rather than one. The risk is low because you have the complete old system in place should anything go wrong with the new system.

As stated before, the small enterprise is particularly sensitive to changes in the workload of its personnel. Furthermore, the small profit margins it may operate under leave little room for error. Although the new information system may be directed only at a portion of the overall information-processing procedures of the enterprise, small interruptions and distractions in the work environment can often mean the difference between success and failure in meeting enterprise goals. These considerations might convince you that a phased conversion is the best alternative for the small-enterprise project. Other factors, however, may convince you otherwise.

When you select a conversion plan, you should consider the complexity of the information system and the amount of original code incorporated into the design. If, for example, the new system is almost entirely composed of 4GL horizontal or vertical software, or a combination, the direct conversion risk factor may be significantly reduced. Past computer experience of the enterprise personnel may also influence your decision. Someone completely new to computing may be overwhelmed by even the phased conversion, making the parallel conversion the only practical choice. Finally, user resistance to the entire project may justify a direct conversion, even when the risk remains high. Such a "cold turkey" approach leaves employees with only one option, which is to make the new system work.

Whichever the circumstance, the analyst should engage the user in the cutover phase decision-making process. This is consistent with all the previous collaboration efforts and certainly paves the way for the final task—the project review.

The Conversion Plan

As with testing and training, a written plan helps reduce user anxieties and chart the course for a complex sequence of events. A *conversion plan* is a document that details the conversion activities for all six components of the information system. Figure 15-4 shows an outline for a conversion plan.

FIGURE 15-4 / *Conversion Plan Outline*

1. Description of the conversion option
2. Schedule of conversion activities
   a. Site preparation
   b. Hardware installation
   c. System software installation
   d. Data file building and scrubbing
   e. Special conversion procedures
   f. Subsystem installation and cutover schedule
3. Conversion personnel and responsibilities

You cannot separate the conversion activities from other activities within the implementation phase or from testing activities associated with the last part of the development phase. A case in point concerns the installation of the hardware and software. If the system includes pilot testing, which begins during the last part of the development phase, installation should be complete before this activity begins. This is exactly what happens in the real world if the analyst has properly scheduled all of the interconnected activities. Obviously, if the hardware has not arrived early enough, the system cannot be installed in

CONVERSION SUCCESS FACTORS

While there are no serious studies that identify those factors that help predict the success or failure of a conversion, we can speculate on the following three factors and comment on a common practice used by software manufacturers.

Product Stability: Software testing is essential to the development process. Throughout the testing sequence, it is natural for the software to change in response to failures or performance deficiencies. However, it is critical that these changes become infrequent and insignificant as the scheduled conversion draws near. If new versions of software are introduced during conversion, the risk of failure increases.

File Preparation: File creation and conversion activities should be as complete as possible before the conversion occurs to minimize the number of variables the analyst must manage. When the conversion process includes file creation, it is important to stabilize the files as soon as possible.

User Training: Sufficient resources must be applied to prepare the users for the new system. Untrained users compound conversion problems, thereby increasing the chance of failure. Do not expect users to learn the new system on their own.

Beta Release: Off-the-shelf software manufacturers often release a test, or beta, version of their product to complete their development process. A beta tester is an individual or enterprise that agrees to use the new product, with the knowledge that it has not been fully tested and developed. Beta testers report their findings to the software manufacturer. With thousands of worldwide beta testers, representing a wide variety of users and implementations, the software is well tested. Everyone can benefit from this arrangement. However, this is not recommended for small-enterprise information systems development, given the inherent limitations of the project scope. Furthermore, be very cautious about building an information system with beta products—your user may not be very tolerant of software bugs that remain even after extensive beta testing.

time for pilot testing. Gantt charts, as explained in Appendix A, plot activities and time periods during which the activities are performed. The Gantt chart in Figure 15-5 depicts an activity schedule designed to synchronize LandscapeS' project activities.

Notice that this schedule does not match the circular SDLC model or the detailed task lists that appear in this text. Each schedule presents peculiarities, which is a reflection of the varied nature of this type of work. In this case, it may seem odd that syntax, user-interface, and program-module testing commence three weeks before the software and hardware are scheduled to arrive. This is possible because the analysts are using a different computer (their own) to design and develop the prototypes for the project. This allows the work to proceed in parallel fashion, thus making good use of the time between resource ordering and delivery.

FIGURE 15-5 / *LandscapeS Detailed Gantt Chart*

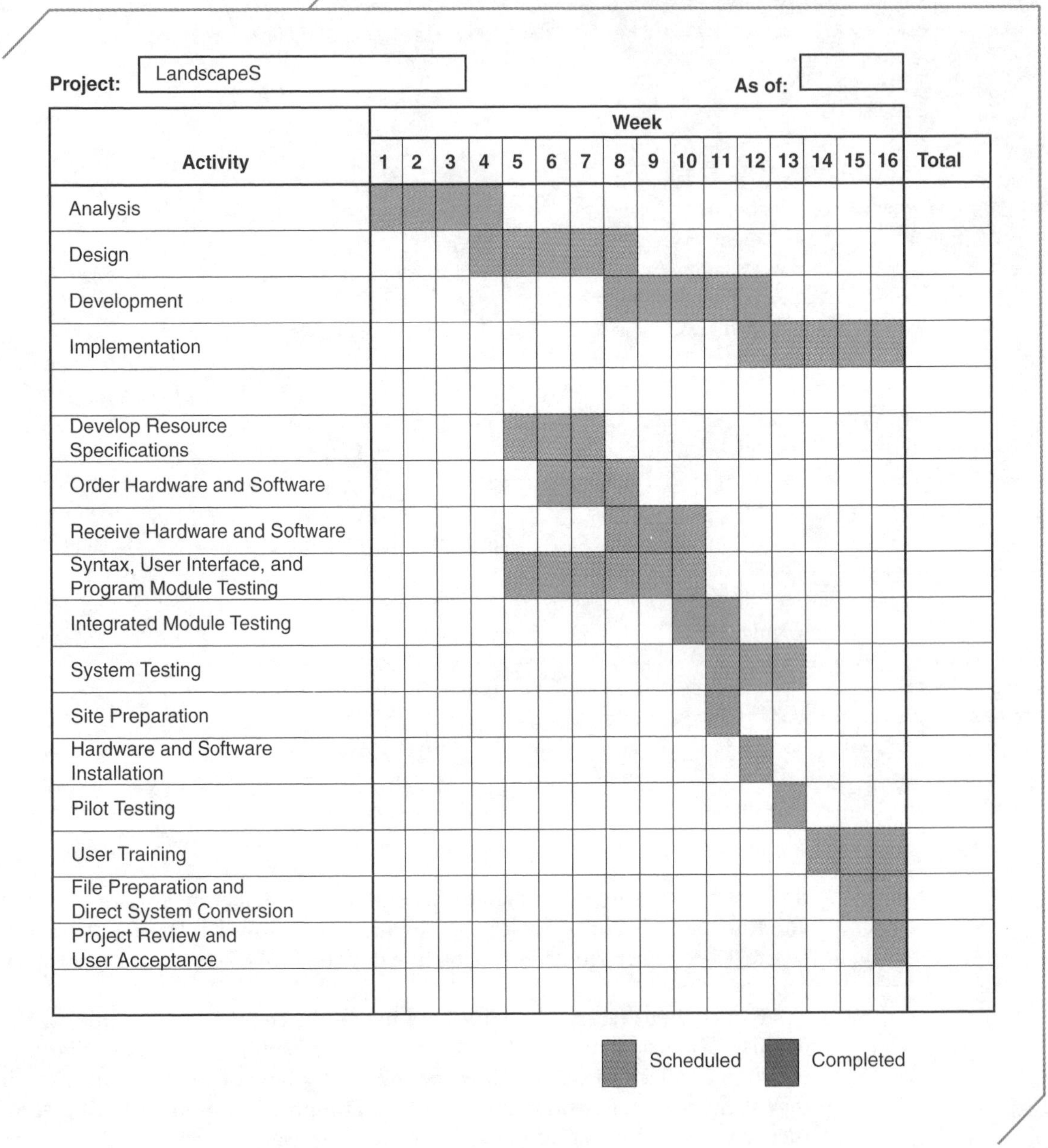

Project: LandscapeS   **As of:**

Activity	Week																Total
	1	2	3	4	5	6	7	8	9	10	11	12	13	14	15	16	
Analysis																	
Design																	
Development																	
Implementation																	
Develop Resource Specifications																	
Order Hardware and Software																	
Receive Hardware and Software																	
Syntax, User Interface, and Program Module Testing																	
Integrated Module Testing																	
System Testing																	
Site Preparation																	
Hardware and Software Installation																	
Pilot Testing																	
User Training																	
File Preparation and Direct System Conversion																	
Project Review and User Acceptance																	

■ Scheduled   ■ Completed

PROJECT REVIEW

In one sense, the SDLC concludes with the project review, leaving maintenance as an ongoing activity. The *project review* provides an opportunity for the analyst and the user to evaluate the project in terms of its success or failure to meet the project specifications. Especially important is that no misunderstanding arises about the continued responsibilities of the user and the analyst. The project review and final user acceptance is perhaps the only clearly distinctive end of one project phase and the beginning of another.

The analyst can use several methods during the project review process. Interviews, data and report sampling, productivity analysis, direct observation, and questionnaires are some popular methods. The summarized results of the project review should be recorded and annotated with any information that will help the analyst and the user with future projects.

A good place to start the project review is with the original project contract introduced in Chapter 2 (Figure 2-7). The four components of this document (problem summary, scope, constraints, and objectives) can serve as the basis for a project review form, as shown in Figure 15-6.

FIGURE 15-6   / *Sample Project Review Form*

Evaluate the following items on a scale of 1 to 5.

	Very Poor				Very Good
	1	2	3	4	5
1. Problem summary					
a. Problem statement	1	2	3	4	5
2. Scope					
a. Accuracy of the					
scope statement	1	2	3	4	5
3. Constraints					
a. Time estimate	1	2	3	4	5
b. Cost estimate	1	2	3	4	5
4. Objectives					
a. Objective 1 product	1	2	3	4	5
b. Objective 2 product	1	2	3	4	5
c.	1	2	3	4	5

The analyst can supplement this form with a detailed survey of each of the key elements of the project, such as input/output interfaces, processing speed, menu selection procedures, training sessions, conversion plan, and so on. With sufficient user-analyst cooperation throughout the project, few surprises should arise during this or any other type of review. Everyone should be well aware of the shortcomings as well as the successes of the information system. The project review is important to document the conclusion of the project's cutover phase.

Product Guarantees

The project contract outlines specific conditions, responsibilities, and objectives. If product performance guarantees are not explicit in this document, future disputes might be difficult to resolve. All parties must attempt to express their expectations and promises as clearly as possible.

In spite of your collective efforts to deliver the product according to contract specifications, at times you will face differences of opinion. Aside from legal counsel, you need to rely on your own ethics and sense of fairness. When the analyst is acting as an independent contractor, this issue may be addressed through special contractual arrangements. For example, a 30-day, free service call agreement is similar to the common hardware warranty. Another postcutover service alternative is the information system maintenance contract, discussed later in the chapter.

User Acceptance

As with an annual vehicle inspection or the signature required on an electrical wiring permit, at the conclusion of an information system project the analyst should secure an acceptance signature from the proper authority within the enterprise. An acceptance signature does not void any product guarantees or warrantees—it merely acknowledges that the analyst has fulfilled the project contract.

In a small-enterprise project, this may seem very formal, but it is necessary. The form signed by the analyst and the user can have both legal and psychological consequences. A lawsuit, filed by either party, is best prevented—and unlikely—if you follow closely the SDLC methodology and obtain an acceptance signature. The psychological impact is probably noticeable right away, as you feel a sense of closure as the project formally ends. In addition, the users may now realize that they can no longer call you at will. Future relationships are open to new negotiations.

PROJECT DELIVERABLE: THE FINAL REPORT

The last of the project deliverables is the final report. While there is no standardized format for this document, it should contain several elements. The report should transmit the working product to the user, provide a brief summary of the overall effort, suggest future enhancements, and describe the need for system maintenance. Do not confuse this report with the system documentation manuals described in Chapter 14. Figure 15-7 presents the contents of the final report.

FIGURE 15-7 / *Contents of the Final Report*

Written Report
1. Summary of project activities
2. Final project status with an analysis
3. Final project budget with an analysis
4. Narrative of potential future enhancements
5. Recommendations for system maintenance

System Documentation
1. Training manual
2. Procedures manual
3. Reference manual

PROGRAMMED REVIEWS

A *programmed review* is a systematic evaluation of all facets of an operational information system. The purpose of the programmed review is to determine how well the system functions when compared to its own, or to industry-established, *performance norms*, which are standards against which the system can be measured. The analyst should schedule programmed reviews for regular intervals and should include representatives from the management, the end users, and the computer-specialists of the enterprise. In some cases, experts from outside the enterprise may also participate in the programmed review process. The team should submit a report at the conclusion of the programmed review.

Do not confuse programmed reviews with the project review or system maintenance activities. Remember, the project review concludes the implementation phase of the project and is designed to evaluate the delivered information system against the project specifications. Programmed reviews occur during the maintenance phase of the product's life cycle and serve to inform the enterprise about how well the information system performs in the real world. Figure 15-8 pinpoints where the programmed review appears on the SDLC model.

FIGURE 15-8   *Information System Programmed Reviews*

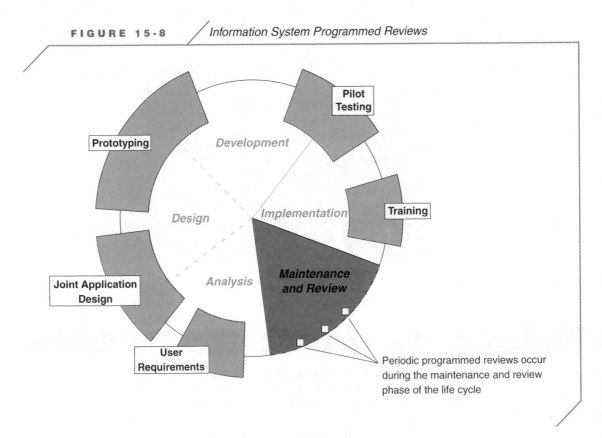

Periodic programmed reviews occur during the maintenance and review phase of the life cycle

Performance Norms

The project objectives provide the initial set of performance norms. They specify how everyone expects the system to perform. Subsequent programmed reviews should measure recent performance against these norms, with the possibility that the norms may be adjusted based on the most recent performance. Of course, this task is much easier when the project objectives are measurable. For example, if an analyst projects that an information system will reduce customer complaints by 10 percent within the first six months of operation, the programmed review team should have little trouble gathering statistics to determine whether this is accurate. On the other hand, qualitative project objectives, such as improved user-friendliness or increased customer satisfaction, are more difficult to evaluate.

Sometimes a portion of the information system's performance can be judged by independent industry standards. Popular computer magazines often publish database access, storage, and retrieval speeds. For example, *PC* magazine often uses a product called WinBench to establish Disk WinMark scores to help the magazine compare computer file-processing performance under Windows. The analyst should use this data with caution, however, because duplicating information system performance under the same test conditions used for the magazine article may be difficult.

Review Report

The *review report* should summarize the programmed review team's findings and, if appropriate, make recommendations for system maintenance or upgrade. Although no standard format exists for this report, it should include information about the performance norms, the circumstances under which the system was tested or the performance data gathered, and the personnel who participated in the programmed review.

Small-enterprise management should expect the review report to comment on the information system's functional life span as well as its operational performance. As the information system approaches its functional obsolescence, the review report should recommend that management consider initiating a new SDLC.

TYPES OF SYSTEM MAINTENANCE

Figure 15-9 shows the SDLC model, with notations that identify several different system maintenance situations. Generally, these situations are different in terms of their positioning within the SDLC model, their associated complexity and cost, and their impact on the small-enterprise information system itself.

Figure 15-10 illustrates how the operating and maintenance costs fall and rise with the different types of maintenance periods. The circle above the graph represents the SDLC, with the shaded part depicting the completed portions, the blackened section indicating the current activity, and the white area depicting those activities still to come. The ongoing system cost line falls during the routine maintenance period and rises as the system ages.

Corrective System Maintenance

As the new system enters the cutover phase, the first operational errors occur. Most of these are user errors. Follow-up training and clarifications to the user manual should correct most of these problems. Some errors, however, may be caused by faulty systems work or program bugs. The analyst can fix these errors under what might be termed a "product warranty." Certainly, both of these error types should diminish in number as time goes on.

FIGURE 15-9 *Types of Information System Maintenance*

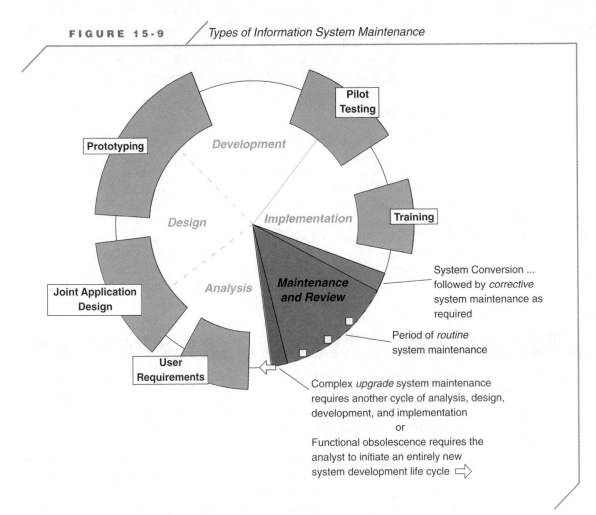

After any initial errors are fixed, the analyst should initiate *corrective system maintenance*, in which a maintenance analyst is available on an on-call basis to the enterprise. Such errors that occur during the corrective system maintenance phase likely fall into one of three categories: cosmetic, nonfatal, or critical. A *cosmetic error* affects system qualities, such as the interface appearance or the smoothness of subsystem transitions. A *nonfatal error* informs the user of a data error or some other minor system malfunction, but permits the system to continue. A *critical error*, such as a file access failure or an infinite processing loop, results in a system- or user-initiated halt. Cosmetic and nonfatal errors can be addressed as part of the normal maintenance process, but critical errors require immediate attention.

FIGURE 15-10 / *Information System Maintenance Costs*

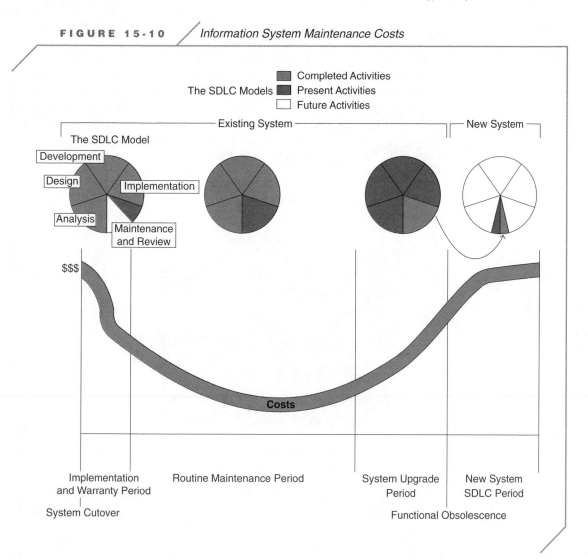

The analyst should document all errors so that the analyst, user, or both can establish performance norms and identify error patterns. For example, if a particular data entry routine continually causes nonfatal errors due to invalid values, the analyst should investigate a more permanent solution rather than merely repairing the immediate damage to the files or output products. One common documentation procedure requires the person who encounters the error to prepare a *system trouble report*, which identifies the circumstances surrounding the error condition and the eventual disposition of the error. Depending on the complexity of the system and the size of the enterprise, such reports can be prepared on anything from a piece of notebook paper to a preprinted, multipart form. This type of system maintenance is the *implementation and warranty period* (Figure 15-10).

Routine System Maintenance

As the information system settles into a long, productive life, two types of routine maintenance are required—one concerns the user, the other the analyst.

Preventive maintenance, as described in the procedures manual, is mostly the user's responsibility in a small-enterprise setting. Figure 15-11 itemizes some of the preventive maintenance tasks. As you look over this list, what items do you think are appropriate for the end user to perform? What type of training do you recommend to cover this material?

FIGURE 15-11 / *Preventive Maintenance Tasks*

1. System backup
2. Software updates for new releases
3. System performance monitoring
     Error rates
     Disk fragmentation
     Indexed file reorganization
     File sizes
     Processing speed
4. Hardware cleaning
5. Printer cartridge replacement

Note that in some cases, the descriptions of day-to-day preventive maintenance tests are collected into a separate operations manual. An *operations manual* may include convenient summaries and checkoff lists to remind the user that system backups are performed at 6:00 p.m. every day, hard disk utilization routines are performed at the end of each week, laser printer cartridges should be removed and gently rocked up and down each month, and so on. Of course, with this approach, the user needs to refer to four manuals (operations, procedures, reference, and training).

Adaptive maintenance is restricted to those system changes that can be introduced without major changes to the hardware and software platform. An example is the periodic requests made by enterprise personnel to change a report format, develop a new query screen to access an existing file, or change the sorting sequence of a file. Traditionally, such changes require the services of someone known as a maintenance programmer. In the small-enterprise environment, the term maintenance analyst may be more appropriate. Once again, this reflects the diverse responsibilities facing today's small-enterprise systems analyst.

Figure 15-10 shows the SDLC model with the entire maintenance and review phase blackened to emphasize that the analysts are still working within the framework of the existing system. The costs of such maintenance should be relatively low and stable.

System Upgrade Maintenance

System upgrade maintenance prolongs the life of the information system. *Upgrade maintenance* refers to any significant changes to any of the six information system components that usually require the services of the analyst. As the needs of the enterprise change over time, the adaptive maintenance requests begin to require more analyst time and stretch the limits of the existing hardware and software. For example, when a user decides there is a real need to incorporate multimedia, or even simple video, into the information system, the analyst may find that the hard disk storage capacity is too small to accommodate the new database file sizes. A second hard disk may be the answer.

Figure 15-12 lists some of the hardware and software upgrades that can help to delay functional obsolescence.

FIGURE 15-12   / *System Upgrade Opportunities*

1. New software versions or upgrades
2. CPU upgrades
3. Hard disk additions
4. RAM additions
5. Expansion board additions

Once the hardware and software upgrades are in place, the analyst can analyze, design, develop, and implement new subsystems. Sometimes existing subsystems can be modified to take advantage of the upgrades. These modifications are often so complex that they, too, are subject to the phases of the life cycle. Therefore, the corresponding SDLC model in Figure 15-10 shows all but the maintenance phase blackened. In a sense, the analyst is going through the SDLC again, but still within the essential boundaries of what is still the existing information system.

To illustrate, consider the following Request for System Services prepared by one of LandscapeS' owners (Figure 15-13). Many events, both external and internal to the enterprise, influence the decisions about information system upgrades. LandscapeS expanded its office staff to accommodate a steady growth in its business operations. The office staff needs a desktop computer to run the information system, while the field staff needs to retain its mobility. The owner suggests a new desktop and wireless LAN for the office and an upgrade to make it possible to connect the notebook to the LAN.

FIGURE 15-13 / *LandscapeS Request for System Services*

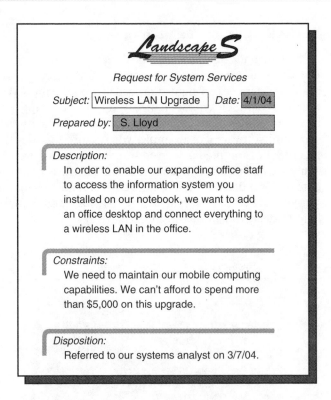

LandscapeS

Request for System Services

Subject: Wireless LAN Upgrade   *Date:* 4/1/04

Prepared by: S. Lloyd

Description:
In order to enable our expanding office staff
to access the information system you
installed on our notebook, we want to add
an office desktop and connect everything to
a wireless LAN in the office.

Constraints:
We need to maintain our mobile computing
capabilities. We can't afford to spend more
than $5,000 on this upgrade.

Disposition:
Referred to our systems analyst on 3/7/04.

INFORMATION SYSTEM OBSOLESCENCE

The far right panel of Figure 15-10 illustrates what happens when the existing system becomes functionally obsolete. The costs, which have been rising steadily, begin to level off as the analyst stops trying to upgrade the existing system. No longer able to continue its metamorphosis, the existing system is retired, as an entirely new SDLC is initiated. The illustration shows that the analyst is engaged in the preliminary stages of the analysis phase.

The computer lab at your school has probably gone through several systems development life cycles: first replacing a mainframe computer with a minicomputer, then adding 8- or 16-bit microcomputers, followed closely with the 32-bit version, and finally a local area network. Computer professionals have come to expect such upheavals, but small-enterprise users may not anticipate the accelerating technology of information processing. The analyst can help the user prepare for future system changes by instructing the user to look for two symptoms that foretell functional obsolescence: changing cost/benefit ratios and changing user needs.

System Costs versus System Benefits

When system costs exceed system benefits, it is time to consider a change. Chapter 11 explains that identifying and quantifying information system costs and benefits are difficult tasks that often lead to imprecise results. Qualitative project objectives, such as customer goodwill, employee morale, and opportunity costs, defy the analyst's attempt to chart costs and benefits on a timeline similar to the one presented in Figure 15-10. As with the formal cost/benefit analysis, however, the analyst and the user need to estimate values for the qualitative project objectives and proceed with a perceptive eye for the reasonableness of the overall trends the new system brings. Figure 15-14 offers a sample of the cost and benefit categories that may apply in different situations.

FIGURE 15-14 / *Costs and Benefits of an Operational System*

Costs	**Benefits**
1. Operations	1. Operational Information
2. Maintenance	2. Decision Support Information
3. Error Correction	3. Accuracy and Consistency
4. Training	4. Access Speed
5. Opportunity (space, time, money)	5. Increased Productivity
6. Employee Frustration	6. Employee Satisfaction
7. Customer Frustration	7. Customer Satisfaction

Replace or Upgrade?

As the maintenance analyst for Wind Surf Now magazine, you have identified the very early signs of information system obsolescence—maintenance costs are slowly rising and the users' system upgrade requests are becoming more difficult to satisfy. Your preliminary analysis reveals that although you can accommodate the latest user request, an entirely new SDLC is required within the next 18 to 24 months.

Although maintenance work is profitable, you make much more money on systems projects that involve all phases of the SDLC. In addition, new systems work almost guarantees another long-term maintenance contract.

What should you propose to Wind Surf Now magazine?

Thinking Critically

Changing Information Needs of the Enterprise

When the existing system can no longer be upgraded to accommodate the changing information needs of the enterprise, this also signals the time for a change. Fortunately, this circumstance is much easier to identify than cost/benefit analysis. The failure of any of the six CIS components to adapt to new user requirements should trigger a close look at the entire system.

Hardware upgrades have physical limits: processing speed, disk speed, memory size, expansion slots, and so forth. New software products and upgrades sometimes require hardware features that are beyond those of the present system. Data entry and information distribution can tax the limits of the system as the enterprise grows. Procedures can easily become a tangled web as they are revised in response to more processing complexities. Finally, there are limits to the end user's ability to work efficiently with what inevitably becomes a patchwork of hardware, software, and procedure upgrades.

MAINTENANCE CONTRACTING

Maintaining an existing system as it ages may be a thankless job for a computer professional. The wide range of technical expertise required is often overwhelming, depending on the generational overlap of the system hardware and software. The system documentation and procedures manuals may be inaccurate, incomplete, and illegible if they were not updated during the life of the system. The system design may not conform to the structured, modularized standards of today's systems. Finally, all the above impediments may result in a product that does not satisfy the user or the sensibilities of the analyst.

Everyone involved in operating the system needs to understand the difficulty of system maintenance, whether enterprise personnel or an outside consultant performs that maintenance. Some latitude should be built into the time, cost, and quality parameters of the maintenance work. As with a new system, a written statement of the problem and the objectives is essential.

Preparing for a New System

An ancillary value of well-organized programmed review documentation, error reports, routine maintenance records, and system upgrade histories is that they can provide supporting evidence to justify a new system study. Not only does this information point to the deficiencies of the existing system, but it also might highlight some areas on which the new system should improve, serving as the beginning of the new system performance specifications.

Maintaining Professional Skills

Currency training describes formal and informal training on new CIS products, methodologies, and techniques. It applies to the analyst and enterprise personnel who are responsible for information system operations and is an important part of the fast-paced field of information processing.

End users also need to pay some attention to developments in technology—if not to operate the current system, then to prepare for the next one. As part of the maintenance package, the analyst should recommend a subscription to a computing magazine and an industry-specific publication that incorporates computer applications for the field in question.

The analyst should not focus on a particular product or system, which is a common consequence of accepting responsibilities for specific projects and applications. At times, the analyst may feel the need to devote all of their energies on the current assignment. However, it is important that the analyst find time for self-study, exploration, and an occasional class. Analysts must maintain their own natural resource as well as any other.

THE CORNUCOPIA CASE

From the beginning of this project, the analysts and the owner enjoyed a good working relationship. The analysts were sensitive to the owner's caution and reluctance to introduce too much change into her enterprise operations. The owner devoted the time required to answer questions, evaluate product designs, and participate in problem-solving sessions. At the end of the project, these roles reverse to the extent that the user sees many possibilities for new system features, whereas the analysts argue that it might be better to let the new system operate as is for a while.

The project formally concludes soon after the system becomes operational and the initial problems associated with the cutover phase are resolved. However, this is not the end of the SDLC. Assuming that system maintenance and upgrade needs are attended to, the analysts expect this information system to have a functional life of three to five years. The procedures manual specifies how the information system should be maintained to help guarantee its performance.

File Preparation

The Customer master file already exists as an ASCII file on the owner's old Macintosh computer. The analysts easily import this file into Access. However, several new fields have been created (CustomerID, Status, etc.), and they need to be populated with data. The alphabetized file holds 74 records. The CustomerID field is assigned as five uppercase letters based on first and last name. The Status field is automatically set to "Active." The entire task is estimated to take not more than one hour of the analyst's time.

The CD master file does not exist. With more than 3,000 CDs in the current inventory and thousands more available through catalog sales, this appears to be a considerable task—especially given the potentially nonstandard values for several fields (e.g., there may be several artists and several compositions on a single disk). Furthermore, almost all of the 12 hours of enterprise personnel time contracted for training and file creation is scheduled for training.

Through their vertical software research, the analysts discovered a company that provides current and past CD data on disk. The one-time cost does not exceed the surplus budget amount indicated in the project status reports. The analysts decide to purchase the disk.

Conversion Option

The owner agrees that a direct conversion plan is the best alternative for this project. It will certainly minimize the cost. The risk is low because most of the new system is entirely new, and the existing manual system can be easily resurrected. The owner shares the analysts' feeling that the system is not too complex and that the current personnel will welcome the system into the workplace.

Project Review

The owner is very happy with the new system. However, she is eager to upgrade to a Web-based system so that she can download her disk file updates. The analysts agree that these changes would greatly improve the information system, but they caution against making any decision without a careful study of the costs involved. They recommend that the owner document these requests on the new "Request for Services" form they have included in the maintenance section of the procedures manual.

Information System Programmed Reviews

The idea of a periodic review of the information system seems too formal for the owner. Although she appreciates the need to evaluate all of her enterprise operations, she feels that she can accomplish this by simply observing the day-to-day operations. Under no circumstances is she interested in forming a programmed review team.

The analysts suggest that the project objectives itemized in the project contract might serve as a set of performance standards for a very informal review. They argue that it is important to establish some historical reference points for comparison during future reviews. Reluctantly, the owner agrees to schedule a programmed review during the month following the first-year anniversary of system implementation.

Information System Maintenance

The analysts offer the owner a $100-per-month maintenance contract that includes two service call-outs per month, and one regularly scheduled service call per month. The owner rejects this offer based on her recent experience with other small-enterprise computer users, who tell her that their computers never break down. In addition, she feels that the 30-day free call-out service included with the normal product warranty is sufficient to get the bugs out of the system.

As an alternative, the analysts offer to work on a pure call-out basis, which costs $50 per hour with a one-hour minimum charge. Phone calls will be billed at $25 per incident. The owner accepts this offer.

Time and Money

The final project status report (Figure 15-15) and budget (Figure 15-16) reveal that the project was delivered on time, but over budget. As previously discussed, the analysts underestimated the labor cost category by almost 25 percent. This was somewhat offset by lower actual costs for software. The fact that the overall budget is within the original estimate should not obscure the issue. The money saved on hardware and software purchases could have been used to subscribe to an online CD disk service. Alternatively, the savings could have simply reduced the total cost to the user. Furthermore, if the actual labor hour data is a true indication of the effort required to do the work, then the analysts must consider this factor in all future estimates.

In their attempt to explain the labor charges, the analysts argue that the ninth and tenth weeks of the project, during which they were involved with prototyping and programming activities, presented special problems. Regardless of their reasons, they adjust their bill to the original total cost estimate of $10,000.

FIGURE 15-15 / *Cornucopia Project Status — Week 16*

Activity	% Comp.	Status	1	2	3	4	5	6	7	8	9	10	11	12	13	14	15	16	Total
Analysis - Estimate	100%		4	5	5	4													18
Actual	100%	ok	2	4	2	3													11
Design - Estimate	100%					4	4	5	5	4									22
Actual	100%	ok			1	3	8	5	7	6									30
Develop - Estimate	100%								4	4	5	5	4						22
Actual	100%	ok						1	8	20	4	6	4	1					44
Impl. - Estimate	100%											4	5	5	5	5			24
Actual	100%	ok									1	2	4	5	6	4			22
Total - Estimate			4	5	5	8	4	5	5	8	4	5	5	8	5	5	5	5	86
Actual			2	4	3	6	8	5	7	7	8	20	5	8	8	6	6	4	107
Contract	100%	ok	C																
Prelim. Present.	100%	ok				C													
Design Review	100%	ok							C										
Prototype Review	100%	ok										C							
Training Session	100%	ok															C		
Final Report	100%	ok																C	

FIGURE 15-16 / *Cornucopia Project Budget — Week 16*

A	1	2	3	4	5	6	7	8	9	10	11	12	13	14	15	16	Total
Date: Cornucopia Project Budget As of: Week 16																	
Estimates Hardware						2500	500	500	500								4000
Software						1000	250	250									1500
Labor	200	250	250	400	200	250	250	400	200	250	250	400	250	250	250	250	4300
Total	200	250	250	400	200	3750	1000	1150	700	250	250	400	250	250	250	250	9800
Actuals Hardware						4250											4250
Software						225		425									650
Labor	100	200	150	300	400	250	350	350	400	1000	250	400	400	300	300	200	5350
Total	100	200	150	300	400	4725	350	775	400	1000	250	400	400	300	300	200	10250
Weekly +/- Hardware	0	0	0	0	0	1750	-500	-500	-500	0	0	0	0	0	0	0	0
Software	0	0	0	0	0	-775	-250	175	0	0	0	0	0	0	0	0	-850
Labor	-100	-50	-100	-100	200	0	100	-50	200	750	0	0	150	50	50	-50	1050
Total	-100	-50	-100	-100	200	975	-650	-375	-300	750	0	0	150	50	50	-50	450
Cum +/- Hardware	0	0	0	0	0	1750	1250	750	250	250	250	250	250	250	250	250	
Software	0	0	0	0	0	-775	-1025	-850	-850	-850	-850	-850	-850	-850	-850	-850	
Labor	-100	-150	-250	-350	-150	-150	-50	-100	100	850	850	850	1000	1050	1100	1050	
Total	-100	-150	-250	-350	-150	825	175	-200	-500	250	250	250	400	450	500	450	

SUMMARY

The cutover phase is a major task that must be coordinated with installation, the delivery of platform hardware and software, the integration of horizontal, vertical, and custom software, testing, file preparation, and training. The analyst is responsible for this orchestration. Some activities must begin well before the system becomes operational, whereas some can commence only during or even at the end of the cutover phase.

File preparation can present a significant workload to enterprise personnel. The analyst must impress upon the workers that file accuracy is critical to the smooth operation of their system. Thorough file-building efforts result in efficient system operation.

The actual conversion from the existing system to the new system brings the unique dynamics of every enterprise and its new system together with two important project parameters one last time: cost and risk. This delicate balancing act should involve both the analyst and the principal end user.

This expedites the resolution of the inevitable conflicts that can arise during this transitional period.

An information system project should be evaluated and brought to a definitive close by signing a user acceptance form. Such a form can protect the analyst in legal situations and may help the end user understand that the analyst's services are complete.

Historically, system maintenance has been the least favored task associated with systems analysis and information system operations. In terms of time, system maintenance is the longest period of the SDLC and it can have a tremendous influence on the overall success of a product.

There are varieties of maintenance tasks that occur during the SDLC. From corrective to routine to upgrade, maintenance work takes on a different sense of urgency. Corrective maintenance is a high priority and potentially highly stressful activity. Routine maintenance is aptly named. Upgrade maintenance becomes more challenging as the system ages and its inevitable replacement becomes more likely.

Obsolescence is a natural part of the SDLC. The decision to initiate a new SDLC requires a long-term view of costs and benefits. Once a small-enterprise system is in place, the analyst should help the user understand that advancements in technology may require future system upgrades.

TEST YOURSELF

1. What are the major elements of an information system conversion plan?

2. What survey questions can best solicit balanced and informative feedback from the user?

3. What are the different types of information system maintenance activities?

4. What is the purpose and content of periodic information system review and evaluation?

5. Why is it necessary for you to maintain and expand your skills as a systems analyst?

Chapter Content

1. Conversion activities involve all six of the information system components. True or False?

2. Because there are very few file conversion utilities, most files have to be created from scratch. True or False?

3. The analyst should take full responsibility for file creation, including the actual data entry, file conversion, and data validation activities. True or False?

4. The conversion plan is greatly affected by the conversion plan employed. True or False?

5. Programmed reviews are regularly scheduled events during the information system's maintenance phase. True or False?

6. By establishing measurable project objectives, the analyst can more easily assess a system's performance over time. True or False?

7. Corrective system maintenance is usually performed under a product warranty contract. True or False?

8. Routine system maintenance includes both preventive and adaptive maintenance. True or False?

9. Upgrade maintenance is designed to prolong the life of the information system by adapting one or more of the six components of the information system. True or False?

10. Functional obsolescence occurs when it is no longer physically possible or economically feasible to upgrade a system to meet the changing information needs of the enterprise. True or False?

11. The need for a full-scale SDLC is eliminated when an enterprise is merely retiring one information system and moving on to another. True or False?

12. Maintenance contracting requires good system documentation and regular performance review reports. True or False?

13. Currency training is designed to prevent analyst obsolescence. True or False?

14. The end user should be responsible for only the simplest preventive maintenance tasks. True or False?

15. A signed user acceptance form provides evidence that the analyst fulfilled the project contract. True or False?

16. Programmed reviews evaluate the system's performance against which of the following standards?
    a. the project specifications
    b. government-established performance norms for the industry
    c. industry-established performance norms or past performance of the system
    d. user-supplied goals and objectives

17. The programmed review report is composed by:
    a. government agents
    b. efficiency experts
    c. programmer trainees
    d. computer specialists, management, and end users

18. Which of the following conversion options have high cost and low risk?
    a. direct conversion
    b. phased conversion
    c. parallel conversion

19. Which of the following conversion options have moderate cost and moderate risk?
    a. direct conversion
    b. phased conversion
    c. parallel conversion

20. Which of the following conversion options have low cost and high risk?
    a. direct conversion
    b. phased conversion
    c. parallel conversion

ACTIVITIES

Activity 1

a. Review the request for system services (Figure 15-13) and your hardware resource specifications for LandscapeS' notebook computer (see Activity 1 in Chapter 14). Evaluate the system services request in terms of your previous hardware specifications. Use the Internet to research wireless LAN options from at least two vendors. Prepare a one-page summary of your findings.

b. How would you classify the request in Figure 15-13? Is this adaptive system maintenance, a system upgrade, or the first signs of functional obsolescence? Draft a response to LandscapeS describing your recommended course of action.

c. After reviewing LandscapeS' file preparation image (Figure 15-1), create five to ten sample records for each of the four files. Be sure to coordinate the key field values where appropriate.

Activity 2

Estimate the time required to build a customer master file using the following set of assumptions. Each record contains 17 fields, totaling 183 bytes per record. Four of these fields (62 bytes worth) already exist in an old disk file, which now has 735 records. About 30 new customers are not on the old file. All of the customer information required for the new file is on old invoice source documents that are stored by transaction date in a file cabinet—a stack of invoices about two feet tall. A simple data entry screen form is already available. The person who will do the research and data entry is an experienced processing clerk who can keyboard at 50 wpm with a 95 percent accuracy rating. Show your work and any further assumptions you make about the task.

Activity 3

Research the current maintenance contract options and prices offered by two of your local computer stores. Prepare a one-page summary of your findings.

DISCUSSION QUESTIONS

1. If a new system involves only the personnel records of an enterprise, how might this affect the risk associated with a direct conversion plan?

2. How does the phased conversion plan affect the training schedule?

3. Explain how a parallel conversion plan affects the effective conversion date from the analyst's point of view. In other words, when do all of the programs, files, procedures, and so forth have to be ready?

4. Would you consider a 30-day money-back guarantee prudent for a CIS project? Explain your answer.

5. What justification would you offer a client who questions the need to retain your services for a monthly maintenance fee?

6. How do the system performance monitoring tasks provide information that is useful to the overall routine maintenance activity?

7. What type of training would you recommend to train users on preventive maintenance tasks?

8. What are some of the possible uses for a hardware-software platform that becomes functionally obsolete?

9. Explain why the maintenance costs of the existing system may flatten, but not decrease, during the period identified in Figure 15-10 as the "New System SDLC Period."

10. What is the functional life span of your education?

PortfolioProject

There is no team assignment for this chapter. The cover letter for the final report should include a statement about the delivery and user acceptance of the information system. Your status report and budget analysis should address situations where your actual performance differed from your original estimates.

Your system documentation should include a disk or CD-ROM containing all of your project files. You have already created most of the content for the system documentation, but there may be some sections where substantial work remains. For example, in Chapter 14 you were required to provide a brief summary of your testing plan and prepare training materials for a small portion of the information system. Figure 14-10 provides the table of contents for the training manual, procedures manual, and reference manual. You should discuss with your instructor the degree of completeness required for each section of the system documentation.

Project Deliverable: Final Project Report and System Documentation

1. Submit the final report to your instructor containing the following:
   a. Cover letter
   b. Summary of project activities
   c. Final project status report with an analysis
   d. Final project budget with an analysis
   e. Narrative of potential future enhancements
   f. Recommendations for system maintenance
2. Submit system documentation, as directed by your instructor, assembled in the following documents:
   a. Training manual
   b. Procedures manual
   c. Reference manual
   d. Disk or CD-ROM containing all project files
3. File a copy of your report in the Portfolio Binder behind the tab labeled "Final Report."

Appendix A
projectmanagement

Chapters 2 and 6 introduce project management as part of the Cornucopia case and the Portfolio Project assignments. This appendix expands on that brief introduction.

The Silhouette Sea Charter and CIS Lab examples introduced in Chapter 2 illustrate this discussion on project management. You will learn how to identify and estimate the major budget elements of the SDLC, how to develop and maintain a status report, how to coordinate the individual efforts of project team members, and how to catalog the various data elements and procedures used in your Portfolio Project. Appendix A also discusses Gantt charts and PERT charts and briefly introduces automated project management tools.

Your Data Disk contains the following files to help you implement the project management tools described in this appendix:

- status.xls (project status)
- budget.xls (project budget)
- hours.xls (analyst hours log)

PROJECT BUDGETS

Although the project contract lists broad cost constraints, detailed costs begin to accrue the moment you agree to the project. As an analyst, you must have a plan, or *project budget*, that forecasts cost components covering the life of the project. This will serve as a critical control mechanism as your work progresses.

Several of the costs of the SDLC are observable and measurable, such as hardware and software purchases or your labor costs per hour. However, other costs are less obvious and less tangible. The enterprise may experience some interruptions in its normal operations, some staff hours may be diverted toward interactions with you as the analyst, and there may be some lost opportunities to spend project money on alternative activities. All these costs are of concern, but to focus your attention on the budgeting process, this discussion is confined to the hardware, software, and labor costs.

Hardware and Software Costs

You can obtain hardware and software costs from several sources. Manufacturers, advertising, the Internet, magazine reviews, mail-order catalogs, and retail outlets provide ample product pricing. Still, there are several reasons why you may need to shop extensively for the best hardware and software price. Vertical software is often bundled with hardware as a "turnkey solution." Likewise, horizontal software is often included in hardware purchase prices. Furthermore, if the new design calls for a software upgrade, the user may qualify for special upgrade pricing. Finally, as the analyst, you provide a value-added service, as you work to adapt, modify, and integrate hardware and software products into a new system. This may qualify you for special discounts that can lower costs to the user.

In the early stages of the SDLC, these costs are estimated in a very general way. Later, as each design element is added, detailed hardware and software specifications allow you to develop more accurate cost estimates. For your purposes, you estimate that the average small enterprise needs a high-end desktop computer and several horizontal software packages to form the basic resources necessary for a simple computer information system. This works out to roughly $4,000 in hardware and $1,500 in software. As the design phase nears completion, specific costs associated with specific products replace these numbers.

Labor Costs

Labor costs, which are a function of time and labor rate, present a different problem. To illustrate, if you ask your mechanic how long it will take to change the water pump on your car, the answer is likely to come from a book or an electronic database. Over many years, a mechanic's experience in replacing water pumps in a variety of vehicles provides sufficient data to establish an average labor cost. However, similar averages are impossible to establish in systems analysis and design because computing procedures change so dramatically and so frequently. Simply put, no standards exist to tell you how long it should take to customize software or create a data flow diagram. Furthermore, even if you could establish such time standards, labor rates vary considerably, depending upon geographic location and system complexity.

In the absence of extensive personal experience, you should consider adapting an established technique for estimating the labor rate required for different analyst tasks. General building contractors, for example, use two approaches: time-and-materials and two-times-materials. In the time-and-materials approach, the contractor is not restricted to a labor estimate. The final bill is fully dependent on the actual hours and material costs required to complete the work. This open-ended cost arrangement is very rare in systems analysis and design. In the two-times-materials approach, the labor rate or time estimates are based on a more easily determined cost component: materials. As the name implies, the labor is twice the cost of the materials, because materials such as lumber, concrete, windows, and so on require labor to transform them into the final product. You could apply a variation on the two-times-materials approach to your computing work. For example, you might base your labor costs estimate on the hardware and software costs as follows: labor costs equal the total hardware costs, plus one-third of the software costs. Using the numbers in the previous section, this works out to $4,500 ($4,000 + $500), yielding a total project cost of $10,000 for a modest, small-enterprise computer information system.

An alternative approach, which is called the *sum-of-the-tasks* approach, builds the labor costs estimate by evaluating each of the SDLC tasks. For example, you might determine that each of the analysis phase tasks (preliminary investigation, contract preparation, data flow diagram creation, and so on) requires *x* number of hours. By totaling these individual estimates, and those of the other phases, you arrive at an overall estimate. This approach is perhaps more easily justified to the user, but it requires more advanced planning and experience with the detailed tasks of the SDLC.

Project Cost Estimates versus Actual Costs

Whatever method you use to establish your original project cost estimates, you must track your *actual costs* and compare them to the *estimated costs*. This not only provides a basis for job-cost accounting, but it also helps you build a body of reliable experience for future project cost estimating.

The following budgets for Silhouette Sea Charter and the CIS Lab illustrate the two cost-estimating approaches. You use the top-down budgeting approach for the Silhouette Sea Charter project because you expect the new system design to include the standard small-enterprise hardware and software previously discussed. This assumes a four-month project schedule and a cost estimate of $10,000 for the entire project.

Figure A-1 illustrates a project budget that provides estimated and actual costs for the primary components. The estimates portion of the spreadsheet is completed during the analysis phase. The "actuals" portion is updated periodically to reflect the hardware and software purchases and the billable labor hours you report, assuming a labor rate of $50 per hour. The spreadsheet formulas embedded in the template automatically calculate the Monthly and Cum sections of the budget. The formulas are set up to show positive values if the actual costs exceed the estimated costs. The bottom two sections provide important project monitoring data.

Of course, this simple spreadsheet implementation can be changed to accommodate projects with greater or fewer budget elements and with projects of longer or shorter duration. Figure A-2 presents the budget for the CIS Lab project. As you can see, this project involves only labor costs and requires only one month to complete.

FIGURE A-1   *Silhouette Sea Charter Project Budget*

	A	B	C	D	E	F	G	H	I	J	K	L	M	N	O	P	Q	R
		SilhouetteBudget.xls																
1	Date:	Silhouette Project Budget								As of: Month		2						
2																		
3		1	2	3	4	5	6	7	8	9	10	11	12	13	14	15	16	Total
4	Estimates Hardware		2000	2000														4000
5	Software	500	500	500														1500
6	Labor	1000	1500	1000	1000													4500
7	Total	1500	4000	3500	1000	0	0	0	0	0	0	0	0	0	0	0	0	10000
8	Actuals Hardware		3000															3000
9	Software	400	400															800
10	Labor	1200	1200															2400
11	Total	1600	4600	0	0	0	0	0	0	0	0	0	0	0	0	0	0	6200
12	Weekly +/- Hardware	0	1000	0	0	0	0	0	0	0	0	0	0	0	0	0	0	1000
13	Software	-100	-100	0	0	0	0	0	0	0	0	0	0	0	0	0	0	-200
14	Labor	200	-300	0	0	0	0	0	0	0	0	0	0	0	0	0	0	-100
15	Total	100	600	0	0	0	0	0	0	0	0	0	0	0	0	0	0	700
16	Cum +/- Hardware	0	1000	0	0	0	0	0	0	0	0	0	0	0	0	0		
17	Software	-100	-200	0	0	0	0	0	0	0	0	0	0	0	0	0		
18	Labor	200	-100	0	0	0	0	0	0	0	0	0	0	0	0	0		
19	Total	100	700	0	0	0	0	0	0	0	0	0	0	0	0	0		

Budget / Sheet2 / Sheet3 /

FIGURE A-2   *CIS Lab Project Budget*

	A	B	C	D	E	F	G	H	I	J	K	L	M	N	O	P	Q	R
		CISLabBudget.xls																
1	Date:	CIS Lab Project Budget								As of: Week		2						
2																		
3		1	2	3	4	5	6	7	8	9	10	11	12	13	14	15	16	Total
4	Estimates Hardware	0	0	0	0													0
5	Software	0	0	0	0													0
6	Labor	50	75	75	75													275
7	Total	50	75	75	75	0	0	0	0	0	0	0	0	0	0	0	0	275
8	Actuals Hardware	0	0															0
9	Software	0	0															0
10	Labor	50	100															150
11	Total	50	100	0	0	0	0	0	0	0	0	0	0	0	0	0	0	150
12	Weekly +/- Hardware	0	0	0	0	0	0	0	0	0	0	0	0	0	0	0	0	0
13	Software	0	0	0	0	0	0	0	0	0	0	0	0	0	0	0	0	0
14	Labor	0	25	0	0	0	0	0	0	0	0	0	0	0	0	0	0	25
15	Total	0	25	0	0	0	0	0	0	0	0	0	0	0	0	0	0	25
16	Cum +/- Hardware	0	0	0	0	0	0	0	0	0	0	0	0	0	0	0		
17	Software	0	0	0	0	0	0	0	0	0	0	0	0	0	0	0		
18	Labor	0	25	0	0	0	0	0	0	0	0	0	0	0	0	0		
19	Total	0	25	0	0	0	0	0	0	0	0	0	0	0	0	0		

Budget / Sheet2 / Sheet3 /

The bottom-up approach was used to develop this budget for several reasons. First, no hardware or software is needed because the lab manager has already purchased the TKSystem software ($250) and installed it on a spare computer. Remember, the revised project description requires you, as the analyst, only to "document the system and prepare training materials for the lab personnel." Second, there are a limited number of clearly identifiable tasks and labor-hour estimates required for this project, as shown in Figure A-3.

FIGURE A-3 / *CIS Lab Project Labor Estimate*

Project Tasks	Labor Estimate
1. Preliminary investigation	2 hours
2. Develop DFD	2 hours
3. Develop USD	1 hour
4. Develop menu tree	1 hour
5. Develop system flowchart	1 hour
6. Develop documentation	2 hours
7. Develop training materials	2 hours

A charge-back accounting system requires you to bill each department for services at $25 per hour. This works out to $275 spread over the four-week period. The CIS Lab budget is sufficient to cover this charge. These two examples illustrate that (1) cost-estimating techniques may vary from project to project, and (2) cost estimating is not an exact science. Therefore, actual costs and completion progress reports are required to monitor the status of the project.

PROJECT STATUS REPORTS

Status reports serve many purposes. The project leader must track progress toward the goals of the project to determine whether adjustments are required to preserve the schedule. Users should remain informed of the project status in order to prevent last-minute surprises—especially when, or if, any changes directly involve them or affect the budget. Finally, you can use status reports as a means to improve your budget-estimating skills for the future.

In the beginning, the project is considered to be one large job, but this view will soon change to one in which you see the job as a collection of tasks, each task with its own purpose. Thus, you must collect project-tracking data regularly to reflect the work performed on each task. Keep in mind, however, that because you work on more than one task at a time, the record-keeping process can be very tedious.

Gantt Charts

The *Gantt chart* provides a visual summary of the project tasks, including the start and stop date, percent complete, and status condition. Figure A-4 illustrates a basic Gantt chart for the Silhouette Sea Charter project. The time is changed to show the project throughout 16 weeks rather than four months. This worksheet itemizes four activities and six events that are common to the case studies in this text. The Gantt chart clearly shows that the end of each activity slightly overlaps the beginning of the next activity. In addition, by comparing the blackened areas to the shaded areas, you can see that the project is slightly behind schedule, as indicated by the "- -" in the status column. Finally, the Gantt chart provides an overall view of how the major events of the project are scheduled in relation to the project activities. You should notice two important events associated with the analysis phase. Finalization of the project contract and presentation of the preliminary project overview are closely related events. The preliminary presentation recasts the project contract into oral and written components describing your understanding of the project and basic approach to the work that lies ahead.

FIGURE A-4 / *Silhouette Sea Charter Gantt Chart*

	A	B	C	D	E	F	G	H	I	J	K	L	M	N	O	P	Q	R	S	T
1	Silhouette Project																			
2	Date:				Scheduled			Completed							As of: Week 8					
3																				
4	Activity	% Comp.	Status	1	2	3	4	5	6	7	8	9	10	11	12	13	14	15	16	Total
5	Analysis - Estimate	95%	ok																	
6	Design - Estimate	80%	--																	
7	Develop - Estimate	5%	--																	
8	Impl. - Estimate	0%	ok																	
9																				
10	Contract	100%	ok																	
11	Prelim. Present.	100%	ok																	
12	Design Review	80%	--																	
13	Prototype Review	0%	ok																	
14	Training Session	0%	ok																	
15	Final Report	0%	ok																	
16																				

Spreadsheet Implementation

As is, the Gantt chart only displays the dates a given activity starts and stops, with no mention of the labor hours required for each activity or how those hours might be spread over the time period. However, project leaders need all this information to manage the allocation of analyst resources. These deficiencies are addressed in the simple spreadsheet implementation, shown in Figure A-5, where the Silhouette Sea Charter project plan is displayed in much more detail. First, the project status is shown on a weekly rather than a monthly basis. Second, the shaded blocks of the Gantt chart are replaced by the number of labor hours associated with that time period. Third, the worksheet provides space to record both hour estimates and actual hours worked, labeled "Actuals" in the spreadsheet. This is consistent with the way the project budget is assembled. Fourth, the event status lines now reflect the scheduled time (S) and the completed time (C). This document allows the reader to determine quickly whether activities or events are progressing as planned.

FIGURE A-5   /   *Silhouette Sea Charter Project Status Report*

SilhouetteStatus.xls

Date: | Silhouette Project Status | | As of: Week 8

	A	B	C	D	E	F	G	H	I	J	K	L	M	N	O	P	Q	R	S	T
	Activity	% Comp.	Status	1	2	3	4	5	6	7	8	9	10	11	12	13	14	15	16	Total
4	Analysis - Estimate	100%		4	4	4	4													16
5	Actual	95%	ok	5	5	5	5													20
6	Design - Estimate	100%					4	6	6	6	6									28
7	Actual	80%	--				4	5	5	6	6									26
8	Develop - Estimate	20%								6	4	4	4	4						22
9	Actual	5%	--							2										2
10	Impl. - Estimate	0%											4	5	5	5	5		24
11	Actual	0%	ok																0
12	Total - Estimate			4	4	4	8	6	6	6	12	4	4	4	8	5	5	5	5	90
13	Actual			5	5	5	9	5	5	6	8	0	0	0	0	0	0	0	0	48
14																				
15	Contract	100%	ok	C																
16	Prelim. Present.	100%	ok			C														
17	Design Review	80%	ok								S									
18	Prototype Review	0%	ok											S						
19	Training Session	0%	ok															S		
20	Final Report	0%	ok																S	
21																				

Status / Sheet2 / Sheet3 /

Task Tracking

Despite the improvement it offers in tracking the project, Silhouette Sea Charter's project status report does not show the detailed tasks required to complete each phase of the project. In this respect, it is similar to the top-down cost-estimating approach previously discussed. The labor cost component on the budget should match the total labor hours on the status worksheet, both for estimates and for actual hours worked. However, there is still no way to associate labor hours with any particular task. One remedy is to make the status report much more detailed, showing every task for every activity. Figure A-6 presents the status report in this detailed form for the CIS Lab project.

FIGURE A-6   CIS Lab Project Status Report

CISLabStatus.xls

	Activity	% Comp.	Status	1	2	3	4	5	6	7	8	9	10	11	12	13	14	15	16	Total
1	Date:	CIS Lab Project Status						As of: Week 2												
2																				
3	Activity	% Comp.	Status	1	2	3	4	5	6	7	8	9	10	11	12	13	14	15	16	Total
4	Prim Inv - Estimate	100%		2																2
5	Actual	100%	ok	2																2
6	DFD - Estimate	100%			2															2
7	Actual	100%	ok		2															2
8	USD - Estimate	100%			1															1
9	Actual	100%	ok		1															1
10	Menu - Estimate	0%				1														1
11	Actual	100%	++		1															1
12	SysFlow - Estimate	0%				1														1
13	Actual	100%	++																	0
14	UserDoc - Estimate	0%				1														1
15	Actual	100%	++																	0
16	OpsDoc - Estimate	0%					1													1
17	Actual	100%	++																	0
18	TrnMan - Estimate	0%					2													2
19	Actual	100%	++																	0
20	Total - Estimate			2	3	3	3	0	0	0	0	0	0	0	0	0	0	0	0	11
21	Actual			2	4	0	0	0	0	0	0	0	0	0	0	0	0	0	0	6

Status / Sheet2 / Sheet3 /

This report was developed from the CIS Lab project task list in Figure A-7. These entries illustrate the bottom-up approach to cost estimating. To itemize and estimate labor hours for each task, you must have a great deal of experience. Notice that the budget (Figure A-2) indicates that you are over budget by $25. However, a close inspection of the status report reveals that this overage is due to your early start on the menu. The project is slightly ahead of schedule and the labor charges are not a cause for alarm. This exercise demonstrates how a project leader might use this data.

FIGURE A-7 *CIS Lab Project Task List*

Tasks	Description	Analyst	Hours	Scheduled Start	Scheduled Stop
1	Preliminary investigation	Martha	2	Week 1	Week 1
2	Develop DFD	Martha	2	Week 2	Week 2
3	Develop USD	George	1	Week 2	Week 2
4	Develop menu tree	George	1	Week 3	Week 3
5	Develop system flowchart	Martha	1	Week 3	Week 3
6a	Develop user documentation	George	1	Week 3	Week 3
6b	Develop operations documentation	Martha	1	Week 4	Week 4
7	Develop training manuals	George	1	Week 4	Week 4

To support the periodic budget and status report updating process, you must report the actual hours you spend on each task, or "billable" hours. One relatively easy solution to this problem is for you to log your hours on a simple ***analyst hours log***, such as the one illustrated in Figure A-8. This spreadsheet allows each analyst to record hours worked, along with a summary worksheet for total hours worked. This spreadsheet can be easily adapted to allow you to report hours worked for each of the SDLC tasks itemized in Figures 6-14, 6-15, and 6-16.

FIGURE A-8 *CIS Lab Project Analyst Hours Log*

CISLabHours.xls

CIS Lab Project Analyst Hours Log

Activity	1	2	3	4	5	6	7	8	9	10	11	12	13	14	15	16
Analysis	2	2														
Design																
Development																
Implementation																
Total	2	2	0	0	0	0	0	0	0	0	0	0	0	0	0	0

Week

Martha / George / Total

The task list, analyst hours log, project status report, and project budget are products of an integrated information system designed to generate information that helps you control the project. As such, they are prepared according to a set of procedures. Figure A-9 summarizes these procedures. More experienced spreadsheet users should consider combining all three project management spreadsheets (analyst hours log, project budget, and project status report) into one spreadsheet with multiple worksheets. By cross-referencing the appropriate cells, it is possible to automatically update actual hours worked on the project budget and project status report worksheets as hours worked are entered on the analyst hours log worksheet.

FIGURE A-9 / *Building the Project Budget and Project Status Report*

1. Develop gross estimates of the hardware and software costs based on your experience and the cost constraints in the project contract.

2. Use either the top-down or bottom-up approach to estimate the labor costs of the project.

3. If you use the bottom-up approach in Step 2, you can compute the labor costs per time period directly from the task list. If you use the top-down approach, you must at least break down these costs into the major phases of the SDLC.

4. Spread the costs over the project time span specified in the project contract, remembering to allow for delivery time, installation, and testing. Note: Because these topics are not specifically covered until the last section of the text, you might simply follow the general examples presented in this chapter.

5. Develop the project budget.

6. Develop the project status sheet. Although this requires you to focus on activities rather than on cost components, be careful to match the labor hour estimates between the task list, budget, and status sheet.

7. On a regular basis, collect the billable hours forms and post the actual hours to the appropriate activity or task on the status sheet.

8. Post the status sheet actual hours to the budget, being careful to make any adjustments for differing time periods on the two documents.

PROJECT RESOURCE MANAGEMENT

Many project tasks share a linear relationship, while other tasks can be worked on simultaneously. A PERT chart illustrates such task dependencies. PERT is an acronym for program evaluation and review technique. Although especially useful in large systems projects that involve thousands of tasks and dozens of workers, the PERT chart can help the small-enterprise analyst allocate and coordinate the resources necessary to keep the project on schedule.

In some cases, you may have access to off-the-shelf project management software. The following discussion on project resource management provides background information so that you can better understand how to use such a product.

PERT Charts

Because the PERT chart shows which tasks must be completed before others can begin, it can help the project leader identify the tasks that must be pursued in sequence, as well as tasks that can be worked on simultaneously. This chart can reveal how to distribute the work efficiently across the available workforce.

Figure A-10 illustrates a PERT chart for the CIS Lab project. This chart reveals additional information about the CIS Lab project tasks. You already know which tasks need to be performed and how much time they require. With the help of the PERT chart, you notice that completion of the DFD is critical to the rest of the project. This may not be surprising, given the earlier discussion on the importance of this model. However, you also notice that Tasks 3, 4, and 5 can be worked on simultaneously, which permits the project leader to assign another analyst to two of these activities. Figure A-11 summarizes the steps to follow to create a PERT chart.

FIGURE A-10 / *CIS Lab Project PERT Chart*

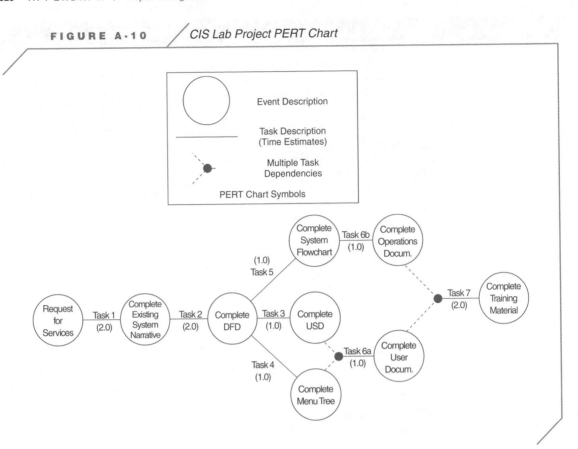

FIGURE A-11 / *Building the Project PERT Chart*

1. Develop a PERT worksheet.
   a. List all the tasks along with the estimated time to complete each task. This information should be available on the task list.
   b. For each task, identify the task or tasks whose completion must immediately precede or succeed the current task. This documents the dependencies between and among the tasks.
   c. Identify events to describe the beginning and ending of each task.

2. Draw the PERT chart from left to right. Each event is represented by a circle. Each task is represented by a line connecting its associated beginning and ending events. The estimated time required to complete a task appears below the task line.

Although the task dependencies for this project may seem obvious, with most projects, you cannot expect task dependencies to be self-evident. The use of these various tools can reveal many hidden tasks or aspects of the various projects you encounter. The CIS Lab PERT worksheets (Figures A-12a and A-12b) illustrate the method used to document the task dependencies for the CIS Lab project.

FIGURE A-12a / *CIS Lab Project PERT Worksheet — Task Dependencies*

Task Number	Estimated Time to Complete	Preceding Task Number	Succeeding Task Number
1	2 hours	none	2
2	1 hour	1	3, 4, 5
3	1 hour	2	6a
4	1 hour	2	6a
5	1 hour	2	6b
6a	1 hour	3, 4	7
6b	1 hour	5	7
7	2 hours	6a, 6b	none

FIGURE A-12b / *CIS Lab Project PERT Worksheet — Event/Task Relationships*

Event Description	Begins Task	Ends Task
Request for services	1	none
Complete existing system narrative	2	1
Complete DFD	3, 4, 5	2
Complete USD	6a	3
Complete menu tree	6a	4
Complete system flowchart	6b	5
Complete user documentation	7	6a
Complete operations documentation	7	6b
Complete training manual	none	7

The Critical Path

The *critical path* is the path from start to finish along which no task can be delayed without delaying the completion date of the entire project. In other words, no "padding" exists for any of the critical-path tasks. In large projects, analysts pay close attention to the critical-path tasks they have been assigned. High-visibility work of that nature can make or break a career. With small-enterprise projects, however, the critical path looks much different. With tight budgets and rigid time constraints, every task takes on critical-path importance. Whether working alone or in a project team, you have little padding in the schedule, which makes effective project management all the more important.

Figure A-13 illustrates the critical path for the CIS Lab project. The time requirements for Tasks 4 and 5 are reduced to .5 hours to make this illustration effective. The critical path is determined by computing the total time required for every possible start-to-finish path, and then identifying the path with the highest total time. In this case, Tasks 1, 2, 3, 6a, and 7 total 8.0 hours. If any of these tasks is not completed on time, the project completion date must be delayed. Systematic procedures are available to compute the critical path when the visual method is not as easy as in this simple example.

FIGURE A-13 / *CIS Lab Project Critical Path*

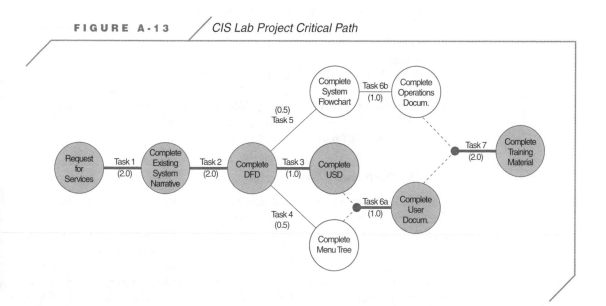

PROJECT DICTIONARY

The project dictionary provides a systematized catalog, definition, and cross-reference of the project products. During the earliest stages of the project, you create many narrations, diagrams, charts, budgets, and status reports as part of the analysis phase. You will soon find that the proliferation of such documents can overwhelm all but the best management systems. This problem is compounded by the use of project teams and user-based methodologies. Quite simply, the more people involved in the process, the more difficult it is to keep track of the various pieces of the project.

Scope of the Project Dictionary

The project dictionary functions much like a data dictionary in that it provides a quick reference to important project items. In this case, however, you are not concerned with external entities, data stores, or data flows. Rather, the concentration is on the essential project documents needed to create, modify, and share with other analysts. To illustrate, suppose that Martha completes the CIS Lab project DFD, saves it to a hard disk with the name "MyDFD," and leaves for a weekend of backpacking in the wilderness. George comes in on Saturday expecting to work on the ERD, but cannot find Martha's file. This is not a major crisis, but it does demonstrate that an organized cataloging procedure can make project teams more efficient.

Given such a broad requirement, entering everything associated with the project into the project dictionary can present a formidable data management problem in and of itself. This can quickly consume the limited resources available for small-enterprise computing. You do need a reliable reference tool, but the tool cannot become the focus of the project. One remedy is to establish a project standard for naming and storing files.

Project Dictionary Standards

Figure A-14 illustrates one approach to tackling this data management task. The approach must be consistent, understandable, and workable. In this case, a simple file naming convention and Windows Explorer are used to implement the project dictionary. The folder, subfolder, and file names correspond to the major elements of the project. Files with multiple versions are named with a suffix to identify the version number. Windows Explorer conveniently provides file size, creation date, and other documentation options to help identify the document.

FIGURE A-14 / *Disk Directory as Project Dictionary*

```
Drive Root:
Silhouette Project
      Project Management
                  StatusWk-01.xls
                  BudgetWk-01.xls
                  AnalystHrWk-01.xls
General Correspondence
                  ReqServices.doc
                  QandA-01.doc
Project Deliverables
      Contract
                  Contract-v01.doc
                  Contract-v02.doc
      Preliminary Presentation
                  PrelimRpt.doc
                  PrelimShow.ppt
      Design Review
      Prototype Review Session
      Training Session
      Final Report
      Existing System Models
                  ExDFD-v01.vsd
                  ExDFD-v02.vsd
                  ExERD-v01.vsd
                  ExUSD-v01.vsd
      New System Models
                  USD-v01.vsd
      Web Site
                  HomePg-v01.htm
```

AUTOMATED PROJECT MANAGEMENT TOOLS

Project management software is available that provides many project management capabilities. Some are stand-alone products, while others are built into CASE products. Figure A-15 illustrates the Silhouette Sea Charter Gantt chart implemented in Microsoft Visio. Visio, which also includes templates for constructing PERT charts, timelines, and calendars, is a relatively simple project management tool. Microsoft offers a much more sophisticated product called Project.

FIGURE A-15 / *Silhouette Sea Charter Gantt Chart with Visio*

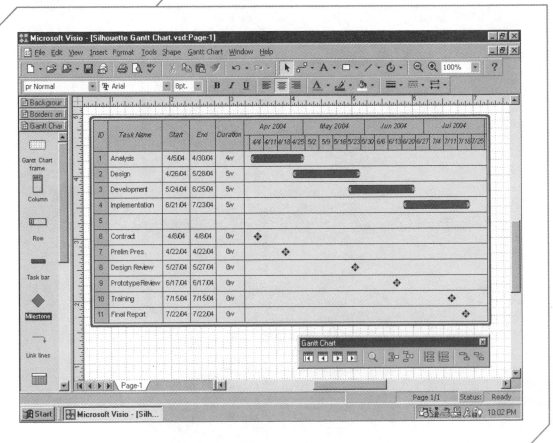

Another approach is to use popular horizontal software to design your own management tools. As discussed previously in this appendix, the project budget, status report, and analyst hours log are implemented with a spreadsheet program. One option that is no longer practical, however, is the paper-and-pencil approach to this task.

Although the project management techniques presented in this appendix are based on relatively simple projects, you should not infer that all projects, big or small, require such extensive project management. Just as with other tools and techniques, project management should not become the primary focus of your work. As an analyst, you must decide on an appropriate level of project management for each project. For example, the CIS Lab project does not really need a PERT chart to complete the work successfully. Sometimes a feasibility study, status report, and project budget are sufficient. Other times, you need sophisticated project management software, such as Microsoft Project. In general, you should expect project management efforts to grow with the size and complexity of your work.

Appendix B
studentportfolio
projects

The systems analyst rarely works alone. Consultations with managers, users, vendors, and technical support personnel are routine. To some degree, you will experience this type of interaction during your Portfolio Project as you work with instructors and computer lab technicians. For the most part, these dealings are purposeful and productive. However, your work as a member of a project team is never as straightforward. Inevitably, team members approach their assignments with different skills, attitudes, and work habits. This appendix presents some guidelines to help you and your team work together more effectively.

TEAM FORMATION AND LEADERSHIP

There are many ways to form and organize project teams. Teams can be self-selected, appointed by an outside party (such as your instructor), or chosen by a team leader. In some cases, teams are formed simply based on member availability. Given that the completion of an information system project requires many different skills, it is reasonable to assume that team membership is also based on the skills of potential members. Thus, a four-member team appointed by your instructor might be composed of one person skilled in database processing, another person with programming experience, a third person well trained in spreadsheet design and development, and a fourth person talented in word processing and graphic communications. In some respects, the team's organizational structure is obvious in this example—members are assigned duties based on their skills. However, who coordinates these efforts? Who assigns tasks that require multiple skills? In short, who provides project management?

Project management responsibilities can be filled in many ways. A traditional approach is to designate one member to assume all leadership duties throughout the project. Shared leadership is another approach, which requires each member of the team to fill the leadership role at some time during the project, depending on the activity. For example, the database person might serve as the database administrator, directing other team members to research user needs, develop test files, and write operating instructions. The programmer, on the other hand, might control the detailed processing design activities for all subsystems. Moreover, if the word processing and graphics expert is responsible for all user interface design, he or she might assume the leadership role for all meetings and design review sessions between the team and the user. Regardless of the approach, it is important to clearly define the roles and responsibilities of each team member.

ESTABLISHING TEAM GROUND RULES

One common complaint about team projects is that some members do not "pull their own weight." This situation becomes worse when all members of the team are treated equally when a grade is based on the team's performance. The following strategies can help reduce this problem:

- **Work to each member's strength.** In line with the assumption that team members possess different skills, you should assign tasks based on those skills. By experiencing success in one area, team members who might otherwise be inclined to underachieve may in fact overachieve.

- **Respect each member's differences.** Often quick to identify a problem and design a remedy, computer professionals are sometimes slow to recognize the value in other, more methodical or introspective approaches. Similarly, team members who do not necessarily leap to a solution should not stop thinking as soon as one is suggested. The best solutions are usually the result of collaboration. In this way, what may have once seemed to be either overly passive or aggressive behavior can be a valued piece of the process.

- **Communicate with other team members.** Misunderstandings are often caused by an inability to communicate, especially with respect to individual work habits. Behavior that appears to demonstrate a total lack of concern may be attributable to something completely different. Do not trust nonverbal communications to inform you about how your team members approach a task. Asking "How are you doing on this task?" can encourage the most recalcitrant team member to respond.

- **Establish a team protocol.** The operational dynamics of the team project greatly influence your level of success with the items above. Two important operational concerns involve the dynamics of team meetings and the agreed upon standards of conduct for team members. To help your project team work effectively, Figure B-1 presents a sample agenda for your first team meeting. While subsequent meetings might focus on specific project activities, they should also include agenda items that address standards of conduct. For example, if your group agrees that team members can and should ask for help when they fall behind schedule, you might add a regular agenda item called "Task Status and Help Request." Alternatively, if your group decides that it will discipline habitually unacceptable behavior, you might include an agenda item on performance review. The group should openly evaluate its own performance and, when necessary, initiate action to bring an individual member or the entire team back on course.

FIGURE B-1   *Portfolio Project Team Meeting Agenda*

1. Introductions (10 minutes)
   a. Computer skills
   b. Personal schedules, phone numbers, e-mail
      addresses
2. Selection of a team meeting facilitator (5 minutes)
3. Discussion of portfolio project handout (15 minutes)
4. Discussion of team protocol (15 minutes)
   a. Meeting time and place
   b. Agenda format for future meetings
   c. Standards of conduct
5. Distribution of initial task assignments (10 minutes)
   a. Task description
   b. Deliverable format and content
   c. Responsible team member or members
   d. Due date and time
6. Adjourn

SUGGESTED PORTFOLIO PROJECT ENTERPRISES

The sample Portfolio Project enterprises described below are based on actual small enterprises. The information from each enterprise has been tailored to accommodate the constraints imposed by the academic setting in which you are working. If you received a packet from your instructor for your Portfolio Project, it ideally includes a sampling of written materials from the enterprise, such as sample pamphlets, customer literature, forms, advertisements, and so on. You should use these materials to familiarize yourself with the enterprise. Web addresses for the actual enterprises are also included to get you started on your research into the enterprise.

The following descriptions do not provide sufficient detail for you to complete your Portfolio Project. To develop a complete understanding of the nature of your Portfolio Project and to clarify what your client wants, you must engage in a series of question-and-answer sessions with your client, as specified in Portfolio Project Team Assignment 2, at the end of Chapter 2. In addition, you should conduct some background research into the general business or service area associated with your client's enterprise. There are several ways to approach this. First, you can investigate a local enterprise similar to your client's enterprise. Second, you can search the Internet for sites devoted to similar enterprises. Third, you can search your school library for trade journals, magazines, or scholarly publications that target the type of enterprise under study.

Your Portfolio Project status and budget reports require that you keep track of the hours spent on your project. Given the academic setting in which you are working, there are two important points to remember about the hours posted to these reports. First, the hours that your team reports should reflect the *average* number of hours that team members devoted to the project during the period. Thus, if three team members worked a total of 14 hours during the week, you should post 4.6 hours to the status and budget reports. Second, you must use your best judgment in determining how much time you spend actually working on the Portfolio Project, as opposed to how much time you spend learning about systems analysis and design. For example, it is not appropriate to charge the project for time you spend reading about data flow diagrams in Chapter 3. However, it is appropriate to charge the project for time you spend creating and revising the data flow diagrams for the project.

Ski Park Project

This sample Portfolio Project is patterned after a local ski park (*www.skipark.com*). During the analysis phase, you should add to your understanding of the enterprise through Internet research and a simulated client/analyst dialog. Assume that the enterprise currently relies on a manual information-processing system. Your task is to design, develop, and deliver a modest computer information system based on standard desktop or workstation hardware and off-the-shelf software. Your system may incorporate a LAN or some other architecture appropriate to the small enterprise. You are operating under two important constraints: (1) your budget is $10,000, and (2) your Portfolio Project is due by the end of the term.

Your client wants a user-friendly system that includes the following information products:

* Customer database and correspondence subsystem
* Lift ticket sales and equipment rental subsystem
* Facility usage statistical analysis subsystem
* Web site

The park opens as soon as there is enough snow (usually mid-November) and closes when the snow is gone (usually early May). The park is open every day, including holidays, from one hour after sunrise to one hour before sunset. There are 10 lifts (numbered 1 to 10). Lift ticket prices are as follows:

	Adult	Jr/Sr	Pup
	(age 13–64)	(8–12, 65+)	(7 and under)
All Day	$33	$17	$5
Half Day (after 1PM)	$26	$13	$3
Special 2 of 3 Day Pass	$58	$32	$7
Special 3 of 4 Day Pass	$84	$44	$9

Season lift ticket passes are available as follows:

- Purchase before December 1: $450
- Purchase between December 1 and January 1: $400
- Purchase between January 2 and March 1: $300
- Purchase after March 1: $200

Ski rental pricing is as follows:

	Adult	**Jr/Sr**	**Pup**
	(age 13–64)	(8–12, 65+)	(7 and under)
All Day	$21	$17	$10
Half Day (after 1PM)	$17	$14	$8

Snowboard rental prices are as follows:

	Adult	**Jr/Sr**	**Pup**
	(age 13–64)	(8–12, 65+)	(7 and under)
All Day	$28	$25	$20
Half Day (after 1PM)	$22	$19	$15

Anyone can become a member of the park conservancy by donating $50 (or more) per season. Members are invited to special events throughout the season. The customer database includes members, even if they do not buy a pass, plus anyone who has purchased a season pass or a special pass. Persons who are in the database become inactive if they do not purchase another pass or renew their membership within twelve months of the time their pass or membership expires. Every database record receives correspondence (U.S. mail and e-mail) of some type. Examples are:

- Early November: Invitation to buy a pass for the new season
- Early December: Reminder to buy a pass (if they haven't already)
- Throughout the season: Announcement of special promotional rates and events
- Early January: Invitation to "return to the park" at a discount price (for inactive persons only)
- End of November: Thank you for tax deductible contribution of $xx (for members only)

The ticket manager maintains records as follows:

- Season tickets sold
- All day tickets sold for each day

- Half day tickets sold for each day
- Special 2 of 3 day passes sold for each day
- Special 3 of 4 day passes sold for each day

The lift manager maintains records by sampling the number of persons (separating skiers and snowboarders) using the lift during the following times:

- 10:00 AM – 10:10 AM
- 1:00 PM – 1:10 PM
- 3:00 PM – 3:10 PM

These samples are extrapolated to provide data for morning, midday, and afternoon. Managers use the data to analyze overall daily attendance and lift loads.

Repertory Theatre Project

This sample Portfolio Project is patterned after a local repertory theatre (*www.redwoodcurtain.com*). During the analysis phase, you should add to your understanding of the enterprise through Internet research and a simulated client/analyst dialog.

Assume that the enterprise currently relies on a manual information-processing system. Your task is to design, develop, and deliver a modest computer information system based on standard desktop or workstation hardware and off-the-shelf software. Your system may incorporate a LAN or some other architecture appropriate to the small enterprise. You are operating under two important constraints: (1) your budget is $10,000, and (2) your project is due by the end of the term. Your client wants a user-friendly system that includes the following information products:

- Subscriber database and correspondence subsystem
- Ticket sales and reservation subsystem
- Performance statistical analysis subsystem
- Web site

The theatre stages five plays during the season, which runs from late February to November. Each play runs for four weeks, with four performances each week (Thursday, Friday, and Saturday nights; Sunday matinee). The first Thursday and Friday night performances are "preview nights." The first Saturday night performance is "opening night." The shows close on Saturday night, with no Sunday matinee the fourth week. The theatre is "dark" between shows. The current season schedule is as follows:

- Play 1: *The Cripple of Inishmaan* by Martin McDonagh (February – March)
- Play 2: *Amy's View* by David Hare (April – May)
- Play 3: *Fuddy Meers* by David Lindsay-Abaire (July – August)
- Play 4: *Crumbs from the Table of Joy* by Lynn Nottage (September)
- Play 5: *Dinner with Friends* by Donald Margulies (October – November)

The theatre has open seating, with 130 seats available for each performance. Ticket prices are as follows:

Season Ticket Plans
- Gala: Opening night – $50
- Premiere: Friday or Saturday nights (not available the first week) – $45
- Value: Thursday night or Sunday matinee (Thursday not available the first week) – $35
- Economy: Thursday or Friday night previews – $20

Single Ticket
- Thursday, Friday previews – $6
- Saturday opening night – $20
- Thursday (nonpreview) or Sunday matinee – $10
- Friday or Saturday (nonpreview or opening night) – $15

Anyone can become a member of the theatre by donating $50 (or more) per season. Members are invited to special events throughout the season. The subscriber database includes members, even if they are not subscribers. Persons who are in the database become inactive if they do not renew their subscription or membership within twelve months of the time their subscription or membership expires. Every database record receives correspondence (U.S. mail and e-mail) of some type. Examples are:

- Early January: Invitation to subscribe for the new season
- Early February: Reminder to subscribe (if they haven't already)
- Two weeks prior to opening night: Announcement of each play (even if they have subscribed)
- Two weeks before the second play of the season: Invitation to "return to the theatre" at a discount price (for inactive persons only)
- End of November: Thank you for tax deductible contribution of $xx (for members only)

The ticket manager maintains records for each performance as follows:

- Tickets sold to subscribers
- Tickets sold as singles
- Comp tickets distributed
- House count (how many seats are filled)

This data is used to analyze attendance by play, performance time slot, performance day, and time of year.

Community Center Project

This sample Portfolio Project is patterned after a local community center (*www.fortunariverlodge.com*). During the analysis phase, you should add to your understanding of the enterprise through Internet research and a simulated client/analyst dialog.

Assume that the enterprise currently relies on a manual information-processing system. Your task is to design, develop, and deliver a modest computer information system based on standard desktop or workstation hardware and off-the-shelf software. Your system may incorporate a LAN or some other architecture appropriate to the small enterprise. You are operating under two important constraints: (1) your budget is $10,000, and (2) your project is due by the end of the term. Your client wants a user-friendly system that includes the following information products:

- Customer database and correspondence subsystem
- Facility reservation subsystem
- Facility usage statistical analysis subsystem
- Web site

The local nature conservancy owns and operates the community center. It provides meeting facilities for small and medium-size events: weddings, concerts, conventions, retreats, and so forth. There are four rooms and a kitchen. Several catering businesses provide food and beverage service. Room set-up capacities and rates are as follows:

	Iris Room	Fern Room	Rose Room	Peach Room
	$510	$285	$200	$115
Reception	400 persons	270 persons	170 persons	70 persons
Classroom	300 persons	200 persons	100 persons	50 persons
Banquet	200 persons	135 persons	65 persons	35 persons

The entire facility can be rented for $665. Kitchen rental is additional, with a minimum charge of $115 plus a per person charge of $1.75. Rental rates are for a 12-hour period. The fee includes tables and chairs, setup, takedown, and cleanup.

Anyone can become a member of the center conservancy by donating $50 (or more) per year. Members are invited to special events throughout the season. The customer database includes members, even if they do not rent the facility, plus anyone who has rented the facility. Persons who are in the database become inactive if they do not rent the facility with 24 months of their last rental or renew their membership within twelve months of the time their membership expires. Every database record receives correspondence (U.S. mail and e-mail) of some type. Examples are:

- Two weeks after event: Quality control (customer satisfaction) survey
- Early February: Invitation to rent the facility
- Throughout the year: Announcement of special promotional rates and events
- Early January: Invitation to "return to the center" at a discount price (for inactive persons only)
- End of November: Thank you for tax deductible contribution of $*xx* (for members only)

The center manager maintains records as follows:

- Facility rental revenues by date and type of event
- Customer satisfaction (on a letter-grade scale) by date and event type
- Facility vacancy rates

Managers use the data to analyze overall staff performance and revenue streams.

Health Club Project

This sample Portfolio Project is patterned after a local health club (*www.healthsport.com*). During the analysis phase, you should add to your understanding of the enterprise through Internet research and a simulated client/analyst dialog.

Assume that the enterprise currently relies on a manual information-processing system. Your task is to design, develop, and deliver a modest computer information system based on standard desktop or workstation hardware and off-the-shelf software. Your system may incorporate a LAN or some other architecture appropriate to the small enterprise. You are operating under two important constraints: (1) your budget is $10,000, and (2) your project is due by the end of the term. Your client wants a user-friendly system that includes the following information products:

- Membership database subsystem
- Membership correspondence subsystem
- Membership usage statistics subsystem
- Web site

The health club is open to the public on a pay-for-use basis, but it is primarily a membership-based operation. Membership initiation fees are $75 for an individual and $100 for a couple or family. Monthly fees are as follows: individual ($40), couple ($55), or family ($55 plus $10 for each child between 10 and 18 years old). Monthly fees are paid through automatic debit or credit card. Members are given the option of having up to $5 of their monthly fee donated to a scholarship fund, which the club matches. This fund is used to help pay some, or all, of the monthly fee for qualifying members.

Every member has a member card with a unique seven-digit ID. Each family member receives their own card, with the first five digits of the membership ID the same and the last two digits different. Each time a member uses the club, he or she must drop off the card at the front desk. A club assistant (1) records the date and member ID, and (2) places the card in a return slot at the front desk. Members retrieve their card when they leave the club.

The membership database includes every member, past and present. Persons who are in the database become suspended if their automatic payments are discontinued. Once these payments are resumed, the member moves back to active status. After six months on suspended status, the member moves to inactive status. Members automatically become inactive if they notify the club that they are canceling their membership. Every database record receives correspondence (U.S. mail and e-mail) of some type. Examples are:

- Throughout the year: Invitations to enroll in special classes or activities
- As needed: Notification of movement to suspended or inactive status
- Mid-January: Invitations to inactive members to rejoin
- End of November: Thank you for tax deductible contribution of $xx (for members who choose to donate to the scholarship fund)

Membership statistics (absolute numbers and relative percentages) include month-by-month information on the following:

- Age: youth (10–18), adult1 (19–25), adult2 (26–50), adult3 (51 and older)
- Gender
- City of residence
- Active members
- Suspended members
- Inactive members
- Member usage (grand total by day, member total by month)
- Usage forecast based on past six months

Appendix C
technicalwriting
andpresentations

INTRODUCTION

This appendix presents summary instruction on technical writing and presentations to help you effectively prepare the documents, illustrations, and personal reports required to complete information system projects. During the SDLC, many tasks require the analyst to prepare and deliver written, visual, and oral communications to management, vendors, users, and other analysts. Many courses teach communications skills, such as composition, speech, and debate. Although the details of technical writing and presentations demand special consideration, the basic principles of communication presented in these courses are universal, regardless of the subject matter or the audience. You should develop communications skills as your computing skills progress.

TECHNICAL WRITING

Technical writing should be direct and to the point. Its purpose is to inform, instruct, or influence the reader. To distinguish technical writing from other types of writing, consider one of Shakespeare's most famous lines: "To be or not to be, that is the question." Certainly, Hamlet's introspection is meant to inform the reader, but it does so indirectly and unpredictably by soliciting many interpretations, emotions, and, indeed, further questions. Conversely, technical writing is meant to inform the reader directly, and with predictable results.

As an analyst, you will work with several categories of technical material. The following list differentiates these broad categories, which describe specific deliverables associated with the SDLC.

- Policies—Broad guidelines
- Procedures—How to carry out policies
- Narratives—Informal descriptions
- Specifications—Detailed references
- Manuals—Instructional "how tos"
- Reports—Summarizations

Technical Writing and the SDLC

Within the SDLC, many occasions arise that require technical writing. The following list summarizes those project deliverables that involve technical writing.

- Analysis phase: Feasibility report, project contract
- Design phase: Preliminary presentation report, requests for proposals, design review session report
- Development phase: Prototype review session report, testing plan, system reference manual
- Implementation phase: Conversion plan, training session handouts, training manual
- Maintenance phase: Procedures manual, maintenance contract

Writing to the Audience

Always consider the intended audience when preparing such materials. For example, the training session handouts, which instruct users, should be much more detailed than the design review session report, which informs management about the general design of the system. On the other hand, the feasibility report reflects a balance between generalities and specifics, as it informs the reader about the broad project parameters (generalities) and influences management's decision about a potential project contract (specifics). Thus, each document is written to suit its intended purpose and audience.

The following samples illustrate different writing styles used by Microsoft to describe .NET to three different audiences.

- The first excerpt, taken from the .NET homepage (*microsoft.com/net/basics/whatis*), explains the product in general terms. This material is written for all audiences.

  > Microsoft .NET is a set of Microsoft software technologies for connecting your world of information, people, systems, and devices. It enables an unprecedented level of software integration through the use of XML Web services: small, discrete, building-block applications that connect to each other—as well as to other, larger applications—via the Internet.

- The second excerpt is written for IT professionals (*microsoft.com/net/business/it_pros*).

  > The .NET Framework is the programming model underlying .NET for developing, deploying, and running XML Web services and applications. XML Web services are units of code that allow programs written in different programming languages and on different platforms to communicate and share data through standard Internet protocols such as XML, SOAP, Web Services Description Language (WSDL), and Universal Description, Discovery, and Integration (UDDI).

- The third excerpt, written for developers, is from the .NET Framework Developer's Guide (*msdn.microsoft.com/library*).

  > The .NET Framework has two main components: the common language runtime and the .NET Framework class library. The common language runtime is the foundation of the .NET Framework. You can think of the runtime as an agent that manages code at execution time, providing core services such as memory management, thread management, and remoting, while also enforcing strict type safety and other forms of code accuracy that ensure security and robustness. In fact, the concept of code management is a fundamental principle of the runtime. Code that targets the runtime is known as managed code, while code that does not target the runtime is known as unmanaged code. The class library, the other main component of the .NET Framework, is a comprehensive, object-oriented collection of reusable types that you can use to develop applications ranging from traditional command-line or graphical user interface (GUI) applications to applications based on the latest innovations provided by ASP.NET, such as Web Forms and XML Web services.

In addition to informing, instructing, and influencing readers, written material helps focus attention on the project. This is necessary because the level of audience interest varies over the SDLC. Even with increased and sustained user involvement in the project, it is unrealistic to assume that participants will maintain the same project focus throughout the entire process. Figure C-1 illustrates participant interest levels during the SDLC. Consider how the writing style, amount of detail, and overall tone of the project deliverables itemized above influences your audience.

FIGURE C-1   *Participant Interest Levels and the SDLC*

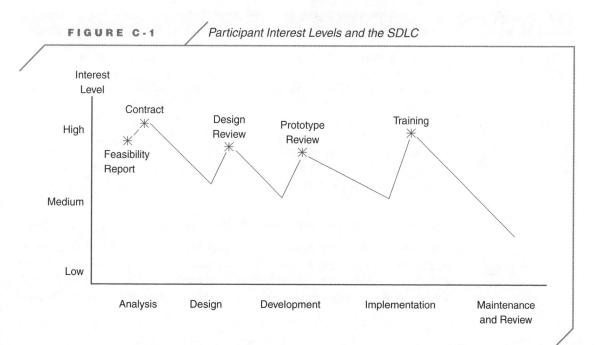

Technical Writing: Step by Step

Many of the written project deliverables follow an established format. When using a standard document format, such as the project contract, you can concentrate on content. Even free-form reports and manuals are structured to include standard items such as budgets, status reports, interface designs, and so on. However, at some point in the project, you need to prepare an original, customized technical document. The following guidelines can help you with this task.

1. Define the purpose of the document and the action, if any, that you expect to promote.
2. Identify the audience and the appropriate writing style, level of detail, and overall tone.
3. Outline the important topics to address in the document and organize their appearance coherently.
4. Where necessary, research the topics and apply your own experience and expertise to formulate your essential thoughts.
5. Prepare a rough draft.
6. Review the draft with a colleague if possible, and revise the document repeatedly until you are satisfied that it will meet your objectives.
7. Edit the document for proper grammar, spelling, punctuation, and so on.
8. Decide on the appropriate combination of written, visual, and audio elements for the document.
9. Carefully plan the desired visual effect of the document.
10. Carefully incorporate mechanisms for future document revisions, such as the appearance of content changes, the placement of revision dates, and so on.

TECHNICAL PRESENTATIONS

Communications with your client are about you and the product. Each contact with the user helps to create a lasting impression about the quality of your work. Although written materials leave an obvious physical presence, presentations leave impressions that may be even more permanent.

Presentation hardware and software allow you to deliver information in many exciting formats. In addition to traditional mediums, computer-based products allow you to create professional-looking presentations that include images and sounds. Your audiences may be just as interested in your delivery as they are with the content.

Characteristics of Effective Presentations

Effective presentations have a clear purpose. Imagine how your audience might respond to the question, "So, what happened at the meeting?" If they cannot answer this question about your presentation in one or two simple sentences, the chances are great that your message was not effective. To avoid this problem, it is best to limit the number of major points covered during a presentation to no more than one each 15 minutes. Thus, an hour-long presentation should present no more than four major points to remember. This gives you ample time to support, illustrate, demonstrate, and summarize each point as you proceed.

Effective presentations are honest. Never underestimate the intelligence of your audience. If they leave with a feeling that you have manipulated the presentation by artful or unfair means to suit your own advantage, they will neither respect nor remember your message, even if they agree with you. If you do not know the answer to a question, say so and promise to return with the answer at some other time.

Effective presentations are entertaining. Most audiences expect that even the simplest presentation should hold their attention. Your appearance should be clean and professional, and your presentation style should hold the interest of your audience Remember, this is the age of marketing and media, and you are selling your services as an analyst.

Effective presentations are tailored to fit the audience. Audience interest levels vary considerably during your presentation. Understanding this, you should pace the presentation so that participants who mentally wander in and out of the meeting can easily recognize the major point under discussion. Figure C-2 presents a typical pattern of audience interest during a technical presentation.

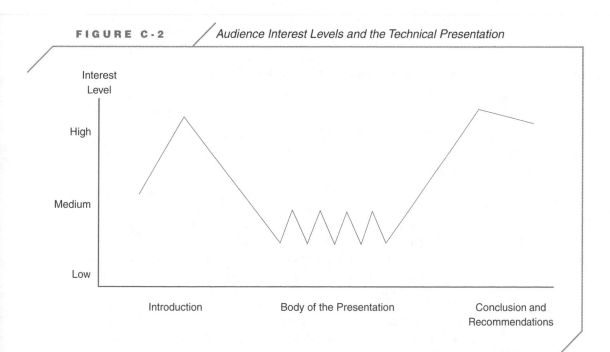

FIGURE C-2 / *Audience Interest Levels and the Technical Presentation*

Orchestrating the Technical Presentation

You should prepare a plan to guide the presentation. The plan should outline each segment, noting the major points, the principal presenters, the supporting illustrations and demonstrations, and the necessary presentation hardware and software. Although the actual presentation is likely to include some impromptu comments and an occasional verbatim technical reading, you should work from this plan extemporaneously. This assures the audience that the activity has some structure, while, at the same time, it encourages you to choose the words that best communicate with the listeners.

You should pay special attention to the visual dimensions of the presentation. As much as visual aids can clarify points or focus audience attention, they can also distract when employed excessively or improperly. For example, when a presenter arrives with a three-inch stack of printed slide handouts, some people may silently groan at the prospect of an extremely long presentation. In addition, a long presentation with overly stylized fade-ins, clip art, and unusual colors may distract the audience.

Finally, you should exercise responsible time management throughout the presentation. Start on time and end on time. Do not linger on one topic or rush through another. The timing cues you dispense are powerful communicators. Beginning on time suggests that you have important information to present. Ending on time acknowledges the value of the audience's time. However, disciplined time management does not mean that you should be completely unyielding about the schedule. You should try to reserve some time for unexpected questions or technical problems.

Technical Presentations: Step by Step

Some tactical points to consider as you prepare and deliver a technical presentation are contained in the following list. You may choose to revise this list to accommodate your own experience.

Preparation

1. Prepare a concise statement of the purpose of the presentation.
2. Prepare an outline of the major points you want to cover.
3. Annotate the outline with references to the supporting technical documentation and illustrations for each point.
4. Develop a plan for the visuals you intend to include in the presentation.
5. Develop a time management plan, leaving up to 10 percent of the time unscheduled.
6. Create and assemble a folder or notebook containing everything you need to conduct the presentation.
7. Rehearse the presentation, making adjustments as required.
8. Arrange for the production and assembly of presentation handouts.

Delivery

1. Arrive at the presentation site early enough to test any required presentation hardware and software.
2. Make a mental note of any physical surroundings that may facilitate or restrict your movements during the presentation.
3. As the audience arrives, locate at least one person to whom you intend to direct your very first sentence. Then, scan the room from left to right to practice eye movements to use as the presentation progresses.
4. Throughout the presentation, continually search the audience for any signs that suggest a change of pace is in order.
5. Make yourself available at the end of the presentation so that members of the audience can approach you with individual questions and comments.

Glossary

Active Server Pages (ASP) A Microsoft protocol to implement server-side programming. (Ch. 13)

actual costs Dollar amounts associated with project purchases and labor charges. (Appendix A)

adaptive maintenance System maintenance restricted to those system changes that can be introduced without major changes to the hardware and software platform. (Ch. 15)

aggregation A type of object relationship in which one class is composed of one or more instances of another class. (Ch. 5)

4GL products A loosely defined term describing the collection of PC-based, user-friendly software products. (Ch. 1)

4GL An acronym for fourth-generation language. (Ch. 10)

4GL file types File types associated with 4GL products, such as .txt, .dbf, .wks, and .bmp files. (Ch. 8)

analyst hours log A log of hours analysts devote to specific tasks during the SDLC. See **billable-hours log**. (Appendix A)

application builder A software program that automates much of the design and development processes, permitting the analyst to create systems from specifications rather than engaging in detailed computer programming. (Ch. 12)

application service provider (ASP) An independent enterprise that offers a variety of application and information system services and solutions via the Internet. (Ch. 13)

assembly language A low-level computer programming language tailored to fit a specific central processing unit architecture and characterized by its symbolic form. (Ch. 12)

attributes Characteristics of data file records. Also referred to as *data fields.* (Chs. 4,12)

back-end CASE A CASE tool designed to prototype and develop the system. See **lower CASE.** (Ch. 11)

bandwidth Describes the type and speed of data transmission possible over a communication medium. (Ch. 13)

baseline measurements A set of initial performance measurements against which future performance can be compared. (Ch. 2)

basic training cycle Training that includes student feedback so that the analyst can adjust training materials and procedures before the cycle repeats itself. (Ch. 14)

batch processing A system or subsystem that gathers transaction file data for processing at a later time. (Ch. 4)

billable-hours log A form used to record analyst hours spent on various project activities. (Ch. 7)

bit depth Each image pixel is assigned a color according to its bit value. The larger the bit depth, the greater the range of possible values. A pixel with a bit depth of 4 can assume 16 different color values. A bit depth of 8 yields 256 color values. A 32-bit image can assume millions of different color values. (Ch. 9)

blurred SDLC An expression used to describe the overlapping activities of the five phases of the system development life cycle. (Ch. 1)

bottom-up DFD creation Building a data flow diagram by identifying the inputs and outputs associated with individual tasks. (Ch. 3)

build versus buy decision The situation faced by analysts when there is a feasible, prepackaged solution alternative to the custom-built solution. (Ch. 2)

bugs A term used to describe errors in computer programs. (Ch. 14)

bus width The size of the electronic pathways connecting the various computer components. (Ch. 13)

cardinality Describes the type of relationship between two files as one-to-one, one-to-many, or many-to-many. (Ch. 4)

case sensitivity Refers to how the information system reacts to upper- and lowercase letters. (Ch. 14)

central repository A well-coordinated, disk-based collection of project files, supporting documents, and data dictionary. (Ch. 14)

class A blueprint for an object, which includes definitions for data attributes and methods. (Ch. 5)

class identification process A method by which object classes may be initially identified by isolating the data files on the entity-relationship diagram. (Ch. 5)

client A computer that receives network services. (Ch. 13)

client pull An HTML technique in which the original transmission includes instructions that tell the client to reestablish the connection to the server and request update information. (Ch. 13)

class diagram This UML diagram shows the set of classes and class relationships associated with a class. (Ch. 5)

class relationship Describes the various ways in which class members are associated with one another. (Ch. 5)

client-server database model Describes an information system in which database services are provided by one computer (the server) to another (the client). (Ch. 13)

client/service software Software that facilitates the distribution of network programs and data throughout the network. (Ch. 13)

client-side programming Programs that reside and execute on the client computer, thus offloading some of the client-server processing functions to the client computer. (Ch. 13)

clipboard An electronic buffer equipped with sophisticated automatic file conversion capabilities. (Ch. 10)

CMOS A semiconductor fabrication technology used to implement processor instructions on read-only memory chips. (Ch. 13)

code of conduct A set of rules that define and govern the behavior of all individuals associated with computer information systems. (Ch. 14)

code of ethics A set of standards concerning the way in which computer users and professionals should approach the use of computer resources and information systems. (Ch. 14)

cohesive Describes the degree to which a module is well focused in the function it performs. (Ch. 7)

cohesive task A group of closely related actions that transform data into information. (Ch. 3)

collaborative relationship A type of object relationship in which one class provides information to another class in a collabotative fashion. (Ch. 5)

command interpreter An operating system module that receives user commands, determines their validity, and associates them with a set of microcode. (Ch. 13)

common gateway interface (CGI) An Internet technology standard that specifies how a client's browser can pass requests to and receive responses from a Web-based database. (Ch. 13)

compatibility A term used to describe a computer information system that can interact with other computer information systems. (Ch. 2)

composite system flowchart A system flowchart that reduces several DFD subsystems or menu options into one processing rectangle. (Ch. 10)

computer-assisted system engineering (CASE) A software tool designed to help the analyst perform the activities associated with the system development life cycle. (Ch. 1)

context diagram The highest level of data flow diagram, showing one process symbol for the entire information system, external entities, and data flows to and from the external entities. (Ch. 3)

context switching Describes the computer operating system's ability to transition from one active program environment to another. (Ch. 9)

conversion plan A well-organized document detailing the conversion activities for all five components of the computer information system. (Ch. 15)

corrective system maintenance System maintenance to correct cosmetic, nonfatal, or critical errors that occur immediately after conversion. (Ch. 15)

cosmetic error Superficial system interface error that does not affect processing. (Ch. 15)

cost and delivery parameters Two of the most important project constraints. (Ch. 2)

cost/benefit analysis An activity that plots project costs and benefits over time to determine if, or when, the system becomes economically justifiable. (Ch. 11)

coupled Describes the degree to which a module interacts with other modules within the program or system. (Ch. 7)

critical error Information system error that results in an abnormal system halt. (Ch. 15)

critical path The sequence of PERT chart activities that must be completed on time if the project is to complete on time. (Appendix A)

currency training Formal and informal analyst training on new computer information system products, methodologies, and techniques. (Ch. 15)

customized application Generally refers to an information system product that is adapted from a single horizontal software package. (Ch. 12)

cutover phase A collection of activities, such as file building, that occur at the end of project testing and eventually lead to full implementation. (Ch. 15)

data attributes Specific data elements associated with a class, defined by their name and data type. (Ch. 5)

data dictionary An ordered catalog of all the data elements in a computer information system. (Ch. 5)

data driven Web site A Web site capable of providing user interactive access to an enterprise database. (Ch. 13)

data flow Depicted on a data flow diagram as an identifiable data element that moves among external entities, processes, and data stores. (Ch. 3)

data flow diagram (DFD) A picture of the processes, external entities, data flows, and data stores of an information system. (Ch. 3)

data integrity Describes the level of accuracy and consistency maintained throughout the data files of a computer information system. (Ch. 8)

data model A graphical representation that focuses on data file relationships. (Chs. 3,4)

data store Depicted on a data flow diagram as a repository of data. (Ch. 3)

data stream A collection of individual characters that make up data elements. (Ch. 4)

data structure A specific organizational strategy for associating individual data elements. Common data structures are records, files, and arrays. (Ch. 4)

data structure hierarchy The sequence of data structures ordered from the most elementary (characters) to the most complex (knowledge base). (Ch. 8)

data type Specifies how the computer stores data, and how the data can be used. Common data types are numeric, character, and date. (Ch. 4)

data warehouse A collection of files containing information about the past operations of an enterprise. (Ch. 4)

database A collection of files containing data pertinent to the current and immediate operations of an enterprise. (Ch. 4)

database schema A detailed layout of the database file structures and relationships. (Ch. 9)

database server An Internet technology in which the database and much of the database processing logic is stored on the server. (Ch. 13)

debug tool A program development feature that allows the programmer to observe the processing

results and data values at the time the computer reports an error condition. (Ch. 14)

debugging The processing of detecting and correcting errors in a computer information system. (Ch. 14)

decision support system (DSS) A system that allows the user to develop information as the need arises. (Ch. 7)

dedicated server A computer totally dedicated to providing network services. (Ch. 13)

design prototype Quickly developed, expendable stand-in screen forms, reports, and simple processing routines that are used to facilitate user participation in the design process. (Ch. 7)

design review session A formal oral and written presentation that includes an overview of the new system design and resource requirements as well as a description of how the system meets the objectives set forth in the project contract. (Ch. 11)

detailed system flowchart A system flowchart that shows the subsystems annotated with the implementing 4GL application and associated file types. (Ch. 10)

DFD decomposition An activity in which the analyst breaks down information system processes and subprocesses into more detailed components. (Ch. 3)

direct conversion A conversion strategy in which the new information system abruptly replaces the old information system. (Ch. 15)

direct memory access (DMA) A special, high-speed transfer of data directly between peripheral devices and memory, without going through the central processing unit. (Ch. 13)

disk cache Special random access memory areas reserved to serve as high-speed buffers for hard disk data transfer. (Ch. 13)

document processing Describes a system in which outputs can be composed from several different source files through the use of sophisticated file-sharing techniques. (Chs. 9,10)

documentation review A fact-finding and diagnostic activity in which the analyst reads through the existing information system operations, procedures, and maintenance manuals or instructions. (Ch. 2)

efficient file sharing Describes software products that are capable of sharing files without a lot of user intervention, data transformations, and other inconveniences. (Ch. 10)

electronic commerce The use of the Internet to facilitate enterprise transactions from consumer-to-business, business-to-business, and business-within-business. Also referred to as e-commerce. (Ch. 10)

e-mail Electronic messages sent to and from computers connected through network technologies. (Ch.1)

embedded object An object that is actually copied into the destination document. (Ch. 10)

encapsulation Object data attribute values and behaviors are only accessible to other program elements via special service behaviors associated with the object. (Ch. 5)

enhanced SDLC A term used to describe a systems development life cycle that consistently incorporates user participation and modern 4GL products to deliver computer information systems. (Ch. 1)

enterprise object A person, place, thing, or event associated with an activity of the enterprise. (Ch. 3)

entity Describes a data file in the entity-relationship diagram. (Ch. 4)

entity-relationship diagram (ERD) A picture of the data file relationships in an information system. (Chs. 3,4)

entrepreneurial systems analyst Describes an independent systems analyst who concentrates on servicing the expanding small-enterprise market. (Ch. 1)

error threshold An error limit that, when exceeded, signals a serious problem with an information system. (Ch. 2)

estimated costs Anticipated costs associated with project purchases and labor charges. (Appendix A)

event A distinguishable action or point in time that identifies the end of one task and the beginning of another. (Ch. 3)

event-driven applications Computer systems designed to react to user-initiated events, such as a mouse operation or a menu selection. Also referred to as *event-driven computing*. (Ch. 12)

event-driven computing Computer systems designed to react to user-initiated events, such as a mouse operation or a menu selection. Also referred to as *event-driven applications*. (Chs. 7,9)

expandability The ability of a computer information system to be enhanced by the addition of more resources. (Ch. 2)

expansion bus Specialized circuitry that permits peripheral devices to connect to the microprocessor. (Ch. 13)

expert system information networks A system that applies diagnostic analysis to problems. (Ch. 7)

extensible markup language (XML) An extension to standard HTML designed to permit users to structure and define complex data structures and information for Web pages. (Ch. 13)

external entity A person or system that either provides data to or receives information from an information system but does not directly process the data within the information system. (Ch. 3)

extranet A network with access limited to one or more enterprise partners, typically implemented via the Internet. (Ch. 13)

fact finding and diagnosis An analysis phase activity in which the analyst performs industry research, interviews, observations, and documentation review in an effort to identify and define information system problems. (Ch. 2)

feasibility analysis A process by which the analyst studies the information needs and possible solutions to determine whether the project can be completed within the project constraints. Also referred to as a *preliminary investigation*. (Ch. 2)

feasibility report A document detailing the study of the project needs, constraints, and possible solutions and that contains a recommendation to either continue or abandon the project. (Ch. 2)

file A collection of related data records. Also referred to as a *data store*. (Ch. 4)

file design fundamentals A collection of rules and techniques used to describe the various file types found in computer information systems. (Ch. 8)

file import/export File sharing in which the importing program actually translates the incoming file into its own native format. (Ch. 10)

file manager The segment of the operating system that handles file storage and access duties. (Ch. 13)

file organization The relationship between the physical file storage characteristics and the logical file processing algorithm. (Ch. 8)

file retention period The period of time over which a file must be retained to satisfy internal or external control procedures. (Ch. 9)

file server A LAN computer dedicated to providing client services. (Ch. 13)

file structure A description of a file's field names, sizes, and types. (Ch. 8)

file synchronization Describes the need to keep newly created files maintained during the conversion process. (Ch. 15)

file transfer protocol (FTP) Communications software standards used to transfer entire files from one computer to another. (Ch. 10)

firewall software Software designed to protect unauthorized access across a Web server to a LAN server. (Ch. 13)

firmware Computer programs that are permanently encoded onto a chip. (Ch. 1)

first normal form (1NF) A database in which there are no repeating groups. (Ch. 5)

first-level DFD A data flow diagram showing the major subsystems of an information system. (Ch. 4)

fourth-generation programming language (4GL) A high-level, nonprocedural computer language requiring the user or programmer to specify what procedures to perform. (Ch. 1)

fourth-generation language (4GL) programming Computer programming using high-level, code-generating software. (Ch. 12)

front-end CASE CASE tools designed to help the analyst with analysis and design activities of the project. See **upper CASE.** (Ch. 11)

front-ends Describes software programs used to create screen forms associated with database applications. (Ch. 7)

full CASE A CASE tool that incorporates all upper and lower CASE features. (Ch. 11)

functional obsolescence A situation in which the existing information system can no longer be modified to satisfy user needs. (Ch. 1)

Gantt chart A tabular form used to plot activities and time periods during which the activities are performed. (Appendix A)

go, no-go decision A decision on whether to continue, suspend, or abandon the project. (Ch. 11)

graphical user interface dialog (GUID) A sequence of screen forms that not only instruct the user on how to proceed with system operations but also accept user selections and specifications. (Ch. 7)

groupware Describes software products that facilitate multiuser access to the same programs and data files. (Ch. 13)

hierarchical password system A security system in which users must supply a series of passwords to gain access to different levels of information. (Ch. 9)

high-tech file sharing Describes object embedded and linked file-sharing methods. (Ch. 10)

high-level language A computer programming language that is highly portable among different computer architectures and characterized by its human language syntax and vocabulary. (Ch. 12)

history file A file of data records reflecting historical information, usually at the transaction level of detail. (Ch. 9)

horizontal software A classification of application software designed for use across a broad spectrum of users. (Ch. 1)

hypertext markup language (HTML) A computer language that formats documents for transmission over the Internet. (Ch. 12)

I/O bottleneck A situation in which computer system performance degrades because of wait times associated with input and output operations. (Ch. 2)

I/O manager An operating system module that secures access to peripheral devices. (Ch. 13)

implementation and warranty period A short period of time immediately following system conversion, during which the analyst is obliged to correct any system errors free of charge. (Ch. 15)

indexed file A file organization scheme that provides for logically ordered access to physically unordered files. (Ch. 8)

industrial-strength information system A computer information system designed to withstand inconsistent or exceptional data entries, improper or illegal user input requests, and unusual or erratic processing demands. (Ch. 11)

industry research A fact-finding and diagnostic activity in which the analyst studies the information system problems and solutions of similar enterprises. (Ch. 2)

information accuracy A goal by which to judge the performance of an information system. (Ch. 2)

information adaptability A goal by which to judge the performance of an information system. (Ch. 2)

information affordability A goal by which to judge the performance of an information system. (Ch. 2)

information processing requirements A collection of goals used to evaluate the performance of an information system. (Ch. 2)

information relevancy A goal by which to judge the performance of an information system. (Ch. 2)

information system A collection of hardware, software, procedures, data, and people coordinated to produce information products and services. (Ch. 1)

information system hierarchy Associating different information system products and services with different enterprise audiences. (Ch. 1)

information timeliness A goal by which to judge the performance of an information system. (Ch. 2)

information usability A goal by which to judge the performance of an information system. (Ch. 2)

inheritance Classes defined as subsets of other classes automatically inherit data attributes and behaviors. (Ch. 5)

input/output prototype A prototype limited to the user interfaces of the system. (Ch. 11)

instance A specific object, defined by a class. For example, the Customer class provides the blueprint from which many specific customers, or instances of Customer, are created for a particular enterprise. (Ch. 5)

instructor-directed learning A learning environment in which the teacher provides the situations and questions experienced by the student. (Ch. 14)

intangible benefits Project benefit elements for which there is no easily established dollar value. (Ch. 11)

intangible costs Project cost elements for which there is no easily established dollar value. (Ch. 11)

integrated CASE An extension of the full CASE approach to include project management features. (Ch. 11)

integrated development environment (IDE) A well-coordinated set of computer software tools designed to help the programmer design, test, debug, document, and revise programs. Sometimes called a *programmer's workbench*. (Chs. 1,14)

integrated software Horizontal software that combines the three major applications (word processing, spreadsheet, and database) into one package with a common user interface. (Ch. 10)

interaction diagram This UML diagram shows how several objects work together to provide information services. (Ch. 5)

internal controls System security measures that are built into the software to prevent illegal, illogical, or inconsistent hardware, software, and data manipulation. (Ch. 8)

Internet A global network of computer networks. (Ch. 1)

interrupt management An operating system strategy whereby the currently active operation is interrupted to serve another, higher-priority request. (Ch. 13)

intranet An enterprise network that uses Internet technology as the distributing medium. (Ch. 1)

IPO chart A tabular listing of the input data and output information associated with a specific processing task. IPO is an acronym for input-process-output. (Ch. 3)

Java An object-oriented computer language used to create cross-platform applets and applications that are transmitted over the Internet. (Ch. 12)

joint application design (JAD) An analysis-design method that brings the analyst and user together in a series of well-focused workshops. (Ch. 7)

key field A data element used to distinguish a record within a file. (Ch. 4)

legacy system An ongoing, existing system that uses older technologies that are incompatible with newer 4GL software. (Ch. 10)

linkage A term describing the situation in which files contain common fields. (Ch. 4)

linked object A situation in which an object is linked by reference, rather than copied, to the destination document. (Ch. 10)

local area network (LAN) A system that connects computers located within a small geographic area. (Ch. 1)

local-bus A data transmission technology in which the connection between a device and memory is managed separately from the main system bus. (Ch. 13)

logic error Error caused when the formula used to solve a problem contains a flaw. (Ch. 14)

logical DFD A data flow diagram that removes all reference to the physical implementation of an information system. (Ch. 3)

low-tech file sharing Describes clipboard-based and import-export file-sharing methods. (Ch. 10)

lower CASE A CASE tool designed to prototype and develop the system. See **back-end CASE**. (Ch. 12)

machine language A low-level computer programming language requiring special knowledge of the computer's internal architecture. (Ch. 12)

macro language A computer language composed of application software commands. (Ch. 12)

management information system (MIS) A system that integrates information collected from different parts of the enterprise. (Ch. 7)

management system See **management information system (MIS)**. (Ch. 1)

many-to-many A file relationship in which records in each file may be associated with one or more records in the other file. (Ch. 4)

master file A data file containing record values that seldom change. (Ch. 4)

memory cache Specialized memory designed to service the needs of the microprocessor quickly. (Ch. 13)

memory manager The segment of the operating system that handles memory assignment and access duties. (Ch. 13)

menu tree A hierarchical display of the operational choices available to a user. (Ch. 6)

menuing software Software that helps the analyst construct menu sequences and attach application software to the menu options. (Ch. 10)

messages Describes the data that moves to and from objects via the service behaviors associated with objects. (Ch. 5)

methodology A comprehensive strategy designed to accomplish a well-defined goal. (Ch. 1)

methods Programming code segments that provide functionality to objects in an object-oriented environment. (Ch. 5,12)

microcode Simple, single-purpose instructions tailored to work on specific microprocessing circuitry. (Ch. 13)

middleware The term used to describe a diverse set of software that functions as a bridge between incompatible technologies. (Ch. 10)

mobile computing A computer information system that includes portable computers that use networking facilities to interface with the system. (Ch. 13)

modeling methodologies A collection of graphical representations, or abstractions, of an information system. (Ch. 4)

module A collection of instructions that perform a specific function. (Ch. 10)

modular design A programming design process in which functional modules are defined, coded, and tested before being assembled into larger, more complex programs. (Ch. 7)

Moore's Law An informal axiom that predicts microprocessor performance will double every eighteen months. (Ch. 1)

multimedia communications An information system output composed of text, video, audio, animation, and interactive elements. (Ch. 9)

multisourced output An output document composed of information from several different application packages. (Ch. 9)

multitasking An operating system resource-sharing strategy that permits more than one program to operate on the computer in the same time frame. (Ch. 13)

multithreading An operating system strategy that allows the CPU to work simultaneously on multiple processing instruction segments of a program. (Ch. 13)

NetPC A microcomputer specifically designed and configured to connect to a network, minimizing costly features found in stand-alone microcomputers. (Ch. 13)

.NET platform Microsoft's Internet-based application development and distribution environment. (Ch. 13)

network administrator An individual who is responsible for the design, installation, and maintenance of network hardware, software, and procedures. (Ch. 13)

network interface card (NIC) A circuit board installed in a network and attached to the network cabling. It facilitates communications between the network node and the network. (Ch. 13)

network operating system (NOS) An operating system specially designed to manage the transmissions between the network server and nodes. (Ch. 13)

networking Describes a collection of computers, connected together in a fashion so that they can share hardware, software, and data resources. (Ch. 13)

node A computer that is serviced by a network. (Ch. 13)

nonfatal error Information system error that permits system operations to continue without interruption. (Ch. 15)

nonprocedural programming Computer programming in which the programmer need only specify the goal of the procedure, with the computer language supplying all the detailed code necessary to carry out the instruction. (Ch. 12)

normalization An activity in which the analyst commonly removes all repeating groups and many-to-many file relationships, although some less common actions may also be required. (Ch. 4)

object An information element (person, place, thing, event, or activity) relevant to the operations of an enterprise. Usually objects are defined by their attributes and behaviors. (Ch. 5)

objectlike language A programming language that provides many, but not all, features of an object-oriented programming language. (Ch. 12)

object linking and embedding (OLE) A high-tech file-sharing method in which a copy of the source data is placed into the destination file. (Ch. 10)

object model A graphical representation of an information system in which the focus is on object classes and their relationships. (Ch. 5)

object modeling Describes the process of identifying enterprise objects and the relationships between those objects. (Ch. 3)

object-oriented analysis (OOA) A process through which an information system is defined by enterprise objects and their relationship to one another. (Ch. 5)

object-oriented application software Describes a class of software that employs object-oriented principles, making it easier for users to create and maintain applications. (Ch. 1)

object-oriented database (ODBMS) A database product in which all the data files, queries, reports, and screen forms are defined as objects. (Chs. 8,12)

object-oriented programming language A computer language that permits the programmer to define classes by their data attributes and behaviors. (Ch. 1)

object-oriented programming Computer programming using high-level languages that provide large libraries of reusable code and define objects by their attributes and behavior. (Ch. 12)

object relational database management system (ORDBMS) An extension of the relational database model in which individual database vendors incorporate object-oriented capabilities, such as object inheritance and polymorphism. (Ch. 8)

object-oriented systems analysis and design (OOSAD) An information system analysis and design methodology that focuses on objects rather than data and/or processes as separate elements. (Ch. 5)

officeware Describes a collection of horizontal software designed to improve office worker productivity. (Ch. 13)

off-the-shelf application software A prepackaged, ready-to-use, application-specific computer program, such as a word processor or spreadsheet. (Ch. 1)

online research An investigative or discovery process that relies on access to electronic media via the Internet. (Ch. 2)

online system An information system that processes data immediately or very soon after the event occurs. (Ch. 7)

online transaction processing (OLTP) A system or subsystem that captures and processes transaction file data at the time of the event. (Chs. 4,7)

on-demand reports Information system outputs that are produced on an irregular basis, usually at the specific direction of the user. (Ch. 9)

on-site observations A fact-finding and diagnostic activity in which the analyst observes the enterprise firsthand. (Ch. 2)

one-to-many A file relationship in which records in one file may be associated with one or more records in the other file, but not vice-versa. (Ch. 4)

one-to-one A file relationship in which each record in one file may be associated with one and only one record in the other file. (Ch. 4)

open architecture A microcomputer hardware strategy that made it possible for third-party manufacturers to produce peripherals for use with otherwise proprietary microprocessors. (Ch. 13)

operating system software A computer program designed to monitor and operate computer hardware, manage file access and memory, and allow the user to launch other computer programs and applications. (Ch. 1)

operational system An information system designed to serve everyday workers with detailed, job-specific information. (Ch. 1)

operational obsolescence A situation in which a product can no longer operate. (Ch. 1)

operations manual A well-organized document that describes how to operate the system and perform routine system maintenance. (Ch. 15)

opportunity costs Costs, both tangible and intangible, associated with activities not performed because resources were occupied in another fashion. (Ch. 2)

PAPA An acronym for the four major issues of information ethics: privacy, accuracy, property, and accessibility. (Ch. 14)

parallel conversion A conversion strategy in which the new and old information systems operate side by side until the old system is finally abandoned. (Ch. 15)

PCMCIA slot A notebook computer expansion port designed to accept a PC card for a modem, memory, network card, or storage device. (Ch. 13)

peer-to-peer network A network in which each node can serve as client or server to another node. (Ch. 1,13)

performance norms Performance standards against which currently operating systems can be measured. (Ch. 15)

periodic reports Information system outputs that are produced on a regularly scheduled basis. (Ch. 9)

personal contacts A fact-finding and diagnostic activity in which the analyst visits the enterprise to conduct interviews, administer questionnaires, and perform on-site observations. (Ch. 2)

personal interviews A fact-finding and diagnostic activity in which the analyst talks with enterprise personnel. (Ch. 2)

PERT chart A graph that shows the dependency relationships among project activities, the time estimated to complete an activity, and the events that mark the beginning and end of activities. (Ch. 6)

phased conversion A conversion strategy in which the new information system gradually replaces the old information system. (Ch. 15)

physical DFD A data flow diagram that shows the physical implementation characteristics of an information system. (Ch. 3)

pilot test A testing procedure in which the user interacts with the information system to determine how well the system works without analyst intervention. (Ch. 14)

polymorphism Describes the ability of a single system function to perform differently, depending upon the circumstances under which it is invoked. (Ch. 5)

preliminary investigation A process by which the analyst studies the information needs and possible solutions to determine whether the project can be completed within the project constraints. Also referred to as a feasibility analysis. (Ch. 2)

preliminary presentation A formal report, both oral and written, that describes the project's major objectives, general design, and present progress towards completion. (Ch. 7)

preventive maintenance Routine system maintenance to clean and adjust hardware components, install minor software updates, back up data files, and so on. (Ch. 15)

print queue A software routine that temporarily stores files to be printed until the printer is free. (Ch. 9)

problem summary A brief statement of the information system problems and needs of the enterprise. (Ch. 2)

procedural programming Computer programming in which the programmer must specify exactly how a procedure is to be performed by supplying detailed, high-level instructions. (Ch. 12)

procedures manual A collection of instructions describing how to use a computer information system. (Ch. 14)

process An activity that transforms data into information. (Ch. 3)

process model A graphical representation of the activities that transform data inputs into information outputs. (Ch. 3)

processing prototypes A prototype with basic file maintenance and transaction processing included, along with the user interfaces to the system. (Ch. 11)

program flowchart A two-dimensional view of the program logic. (Ch. 10)

programmed review A systematic, regularly scheduled evaluation of all facets of an operational information system. This review occurs during the maintenance and review phase of the SDLC. (Ch. 15)

programmer's workbench Describes a collection of software development tools and utilities that accompany many 4GL products. Sometimes called an *integrated development environment*. (Ch. 14)

project binder A well-organized file of all the system documentation. (Ch. 14)

project budget A systematic financial plan of all costs associated with the project. (Appendix A)

project constraints The portion of the project contract that specifies when the project is due, how much the project should cost, and any other requirements that govern the work. (Ch. 2)

project contract A document detailing a project's problem summary, scope, constraints, and objectives. (Ch. 2)

project deliverables A collection of information system products, including everything from project contracts and software to user acceptance and review evaluations. (Ch. 2)

project dictionary A systematized catalog, definition, and cross-reference of the project products, such as reports, software, data files, status reports, budgets, and so on. (Ch. 6)

project documentation The complete collection of procedural, operational, and reference manuals that describe a project. (Ch. 2)

project management A collection of activities that include project planning, supervision, and cost control. (Ch. 6)

project management software A computer program designed to help the analyst keep track of the many and varied resources, activities, and events associated with the systems development life cycle. (Chs. 1,15)

project objectives The portion of the project contract that sets out the measurable goals of the information system. (Ch. 2)

project review A user and analyst review, conducted at the end of the conversion activities, of how well the information system meets the project specifications. (Ch. 15)

project scope The portion of the project contract that defines the boundaries of the information system. (Ch. 2)

protocol A set of rules that governs data communications. (Ch. 13)

prototype USD The user's system diagram with shaded areas to denote the prototyped parts of the full system. (Ch. 11)

prototype A simplified model of a computer information system, usually presenting the user interfaces and some simple file-processing features of the real system to come. (Ch. 11)

query-by-example (QBE) A visually oriented database access method in which users create information outputs based on special criteria. (Ch. 9)

questionnaires A fact-finding and diagnostic activity in which the analyst administers questionnaires to enterprise personnel. (Ch. 2)

rapid application development (RAD) A process that combines joint application design, 4GL product expertise, CASE tools, and prototyping into a unified approach to the SDLC. (Chs. 11,12)

reference manual A project documentation manual that contains the technical material, such as program listings, error codes, hardware specifications, and so on. (Ch. 14)

referential integrity The ability of the information system to prevent inconsistent data manipulations across the various subsystems. (Ch. 14)

register width The number of bits in a CPU register. (Ch. 13)

relational database management system (RDBMS) A database product that enables the user to relate multiple files within the database. (Ch. 12)

relational database An easily accessible collection of related records, each described by attribute values. (Ch. 4)

relationship A term used to describe the occurrence of a common field between two files. (Ch. 4)

remote I/O A user interface separated by some distance from the processing computer system. (Ch. 13)

repeating group A set of fields that occur an undetermined number of times within a record. (Ch. 4)

request for bids (RFB) A formal notice that specifies intent to purchase specific products and invites vendors to submit price quotes and contract terms that meet the specifications. (Ch. 11)

request for proposals (RFP) A formal notice that declares intent to purchase products described by the function they must perform and invites vendors to submit product specifications, price quotes, and contract terms that will meet the performance requirements. (Ch. 11)

request for system services A document describing the information needs of the enterprise. (Ch. 2)

resident operating system The portion of the operating system that is always loaded into memory. (Ch. 13)

resolution Screen resolution is expressed as the number of horizontal and vertical pixels used to represent an image. Generally, greater pixel densities produce better pictures. Thus, the 1280 by 1024 screen resolution yields a sharper image than does the 800 by 600 screen resolution. (Ch. 9)

reusable prototypes Prototypes that are designed to be transformed into a full-featured, working system. (Ch. 11)

reverse engineering The process of disassembling a system in order to discover how it is put together. (Ch. 7)

review report A report describing the performance of an operational information system and recommending maintenance, upgrade, or replacement. (Ch. 15)

ROM-BIOS A basic set of input/output instructions, stored on a read-only memory chip, that load key elements of the operating system when the user turns on the computer. (Ch. 10)

runtime errors Errors caused when input data is inconsistent with data definitions or processing procedures. (Ch. 14)

schedule manager An operating system module that arbitrates multiple requests for system resources. (Ch. 13)

scope The conditions that define a subset of data selected for access. (Ch. 9)

screen form An electronic form designed to inform the user and facilitate data entry. (Ch. 7)

second normal form (2NF) A 1NF database in which every attribute in each file is dependent only on the key field of that file. (Ch. 4)

secure hypertext transmission protocol (SHTTP) A data transmission encryption and decryption protocol designed to protect Internet communications from invasion. (Ch. 13)

secure sockets layer (SSL) A data transmission encryption and decryption protocol designed to protect Internet communications from invasion. (Ch. 13)

sequential file A file whose physical order matches its logical order. (Ch. 8)

server A computer that provides network services to client computers attached to the network. (Ch. 13)

server push An HTML feature in which the connection between the client and server is never broken. (Ch. 13)

server-side programming A program that resides on the server computer and provides client access to an underlying database. (Ch. 13)

single stepping Executing each instruction separately, so the programmer can observe the flow of the macro and the results of each action, and isolate any action that causes an error or produces unwanted results. (Ch. 14)

sinks External entities that receive information from an information system. (Ch. 3)

small enterprise Any organization designed to serve a specific function that is characterized by its limited personnel, budget, and scope of operation. (Ch. 1)

small enterprise computing Describes an environment in which microcomputers and off-the-shelf software form the foundation for computer information systems. (Ch. 1)

softcopy Information system outputs presented on a screen, as opposed to printed matter. (Ch. 9)

software suite A collection of otherwise distinct horizontal software packages that are sold as a unit. (Ch. 1,10)

source data input inefficiencies A situation in which input procedures are awkward, confusing, or antiquated. (Ch. 2)

source document The document from which original data values are captured for entry to the information system. (Ch. 7)

sources External entities that provide data to an information system. (Ch. 3)

specialization A class relationship in which one class is a subclass of another, thus inheriting all of the parent class' data attributes and methods. (Ch. 5)

statechart diagram This UML diagram shows how objects react to a variety of events and stimulations from other objects. (Ch. 5)

state-transition diagram (STD) A picture of the dialog-dependent actions of an information system, where the dialogs (computer prompts and user replies) cause the system to change from one process to another. (Ch. 12)

status report A periodic oral or written summary of progress on project activities. (Appendix A)

strategic information system An information system designed to serve executive-level personnel with enterprise-wide information products; a system that reaches beyond the enterprise to focus on external data. (Ch. 1,7)

structure chart A two-dimensional view of the relationships among the different functional components of a software product. (Ch. 10)

structured design A design process in which large problems are broken down into smaller, more manageable components, or modules. (Ch. 7)

structured query language Often abbreviated as SQL, this computer language provides English-like commands and syntax used to access databases. (Ch. 9)

subsystem integration breakdown A situation in which one or more information subsystems do not interface properly with the rest of the information system. (Ch. 3)

sum-of-the-tasks A term describing the budget preparation activity in which the costs of the individual tasks are added together to arrive at the overall project budget. (Appendix A)

swap file A portion of the hard disk reserved for operating system use. (Ch. 13)

SWAT team A team of advanced tool specialists used to develop a project quickly. (Ch. 11)

syntax errors Errors caused by invalid or incorrect usage of computer program or application command formation rules. (Ch. 14)

syntax rules The rules that govern the formation of computer commands. (Ch. 9)

system architecture The combination of hardware and software that supports the information system. (Ch. 13)

system documentation A collection of system documents that provide a detailed history of the information system. (Ch. 14)

system-critical error An error that prevents the information system from operating in even the most basic fashion. (Ch. 14)

system environment A short-hand reference to the combination of hardware, system software, 4GL software, and customized software used to deliver information system services. (Ch. 13)

system flowchart A picture of the relationships among inputs, outputs, and software. (Ch. 6)

system integrity The ability of the information system to perform its functions accurately and consistently across all subsystems. (Ch. 14)

system model A graphical representation that focuses on the relationships among data files and computer software. (Ch. 3)

system prototype A prototype of the complete system. (Ch. 11)

system resource requirements A list of hardware, software, data-handling, procedure-handling, and personnel resource needs associated with the SDLC. (Ch. 11)

system security testing Hardware and software testing to determine how easy or difficult it is to use the system without knowing the access passwords and procedures. (Ch. 14)

system testing Testing that involves all five components of the computer information system. (Ch. 14)

system trouble report A document used to report information system problems. (Ch. 15)

systems development life cycle (SDLC) A framework used to describe a circular process designed to analyze, design, develop, implement, and maintain and review computer information systems. (Ch. 1)

tags Special HTML notations that mark the beginning and ending of Web page elements. (Ch. 12)

task An activity that transforms data into information. (Ch. 3)

technique A style or means of applying a tool to a specific problem. (Ch. 1)

telecommunications Describes a host of electronic data transmission technologies designed to facilitate the movement of data from one computer to another. (Ch. 1)

testing plan A detailed description of the systematic testing procedures for the entire information system. (Ch. 14)

testing scope Describes the functional and operational boundaries or limits of a testing procedure. (Ch. 14)

thin client A microcomputer specifically designed and configured to connect to a server, minimizing costly features found in stand-alone microcomputers. (Ch. 13)

third normal form (3NF) A 2NF database in which all the dependent relationships are contained with the individual files. (Ch. 4)

third-generation language A high-level, procedural computer language requiring the programmer to instruct the computer on how to perform every procedure. (Ch. 1)

three-stage concept A sequenced view of enterprise computing in which information system complexity grows as one system is retired and another is implemented. (Ch. 7)

throwaway prototypes Prototypes that are designed to be discarded when the real system is developed. (Ch. 11)

tool A concept or object which, when used properly, is designed to perform a specific function. (Ch. 1)

top-down DFD Building a data flow diagram by asking the user a series of questions about how the information system works. (Ch. 3)

total cost of ownership (TCO) A term used to define the combination of costs, from purchase through retirement, associated with computing technology. (Ch. 13)

total-quality management (TQM) A management philosophy stressing the need for every team member to participate actively in collaborative efforts to produce quality products. (Ch. 1)

traditional file types File types associated with third-generation programming languages, such as master, transaction, and backup files. (Ch. 8)

training manual A well-organized collection of all training materials. (Ch. 14)

transaction file A data file that records information about an event. (Ch. 4)

transaction processing system A computer information system that captures, organizes, stores, and reports information about enterprise events. (Ch. 7)

transient operating system The portion of the operating system that is only loaded into memory when needed. (Ch. 13)

turnkey system A prepackaged computer information system designed to serve a specific audience. (Ch. 1)

unified modeling language (UML) A widely accepted standard for object-oriented information system analysis. UML provides specifications for several diagrams, each of which focuses on a slightly different aspect of the information system. (Ch. 5)

upgrade maintenance System maintenance characterized by significant changes to any of the five components of an information system, most often requiring the services of a systems analyst. (Ch. 15)

upper CASE A CASE tool designed to help the analyst with the analysis and design activities of the project. See **front-end CASE.** (Ch. 11)

use case diagram This UML diagram shows the functions (uses) and users (actors) of the information system. (Ch. 5)

use case modeling A use case is a sequence of steps designed to complete a single business task. Use case modeling expands this to include all activities of an enterprise. (Ch. 5)

user acceptance testing Activities designed to determine the user's level of satisfaction with the new information system. (Ch. 14)

user inquiry Extemporaneous information output requests that are conceived, planned, and executed by users. (Ch. 9)

user interface testing Testing to determine how well the user input screens, output screens and reports, dialog sequences, menus, and so on perform. (Ch. 14)

user's system diagram (USD) A user-friendly version of the data flow diagram, using familiar icons for processes, data stores, and external entities as well as simplified data flows. (Ch. 6)

user-directed learning A learning situation in which the student is allowed to explore the information system, asking questions of the system documentation or instructor only as questions arise. (Ch. 14)

user-driven design A design process in which the focus is on sustained user participation. (Ch. 7)

utility software Special-purpose computer programs designed to complement or add new functions to the operating system software. (Ch. 1)

vertical software A classification of application software designed for use by a narrowly defined set of users. (Ch. 1)

video bus A data transmission technology that allows the high-demand graphical user interfaces direct access to the high-speed CPU bus. (Ch. 13)

virtual extranet A special implementation of a virtual private network (VPN) where access extends to several partnered enterprises. (Ch. 13)

virtual intranet A special implementation of a virtual private network (VPN) where access is limited to users within a single enterprise. (Ch. 13)

virtual memory A memory management strategy that expands the effective, logical amount of memory beyond its physical limitations. (Ch. 13)

virtual private network (VPN) A private network implemented via the public Internet. (Ch. 13)

walk-through A problem solving strategy in which the analyst simulates the processes using some sample data. (Ch. 3)

Web hosting An independent enterprise that provides all the hardware, software, and support required to implement and maintain a database-driven Web site. (Ch. 13)

Web server A computer dedicated to servicing Internet requests. (Ch. 13)

wide area network (WAN) A system that expands the principles of local area networking to a wider geographic area. (Ch. 1)

WWW The world wide web is a network of networks, with content presented in a variety of multimedia formats. (Ch. 1)

Selected References

Andrews, Jean. 2001. *I-Net+ Guide to Internet Technologies*, Boston, MA: Thompson Learning.

Arinze, Bay. 1994. *Microcomputers for Managers*. Belmont, CA: Wadsworth.

Baum, David. 1992. "Go Totally RAD and Build Apps Faster." *Datamation* (September 15, 1992).

Behan, Maria. 1998. "Selling from Your Website." *Beyond Computing* (March 1998).

Blaha, Michael, and Premerlani, William. 1998. *Object-Oriented Modeling and Design for Database Applications*. Upper Saddle River, NJ: Prentice Hall.

Booch, Grady. 1994. *Object-Oriented Analysis and Design with Applications*, 2nd Edition. Redwood City, CA: The Benjamin/Cummings Publishing Company.

Booch, Grady, et al. 1999. *The Unified Modeling Language User Guide*, Menlo Park, CA: Addison-Wesley.

Callaway, Erin. 1998. "TCO Workout." *PC Week* (March 30, 1998).

Dejoie, R., G. Fowler, and D. Paradice. 1991. *Ethical Issues in Information Systems*. Boston: Boyd & Fraser.

DelRossi, Robert. 1994. "Rapid Application Development." *Infoworld* (February 14, 1994).

DeMarco, Tom. 1979. *Structured Analysis and System Specification*. Englewood Cliffs, NJ: Prentice Hall.

Dewitz, Sandra. 1996. *Systems Analysis and Design and the Transition to Objects*. New York: McGraw-Hill.

Dicarlo, Lisa. 1998. "Two Roads to Thinner Clients." *PC Week* (March 23, 1998).

Gore, M., and J. Stubbe. 1994. *Contemporary Systems Analysis*, 5th Edition. Dubuque, IA: Business and Education Technologies.

Gregory, W., and W. Wojtkowski. 1990. *Applications Software Programming with Fourth-Generation Languages*. Boston: Boyd & Fraser.

Harrington, Jan. 1987. *Relational Database Management for Microcomputers: Design and Implementation*. New York: Holt, Rinehart and Winston.

Horner, Donald. 1989. *Operating Systems Concepts and Applications*. Glenview, IL: Scott, Foresman and Company.

Kador, John. 1998. "The Business Uses of Push Technology." *Beyond Computing* (January/February 1998).

Kallman, Ernest, and Grillo, John. 1996. *Ethical Decision Making and Information Technology*. New York: McGraw-HillCompanies, Inc.

Lehman, John. 1991. *Systems Design in the Fourth Generation*. New York: John Wiley and Sons.

Leinfuss, Emily. 1993. "GUI Application Development." *Infoworld* (June 21, 1993).

Lorents, Alden, and Morgan, James. 1998. *Database Systems - Concepts, Management, and Applications*. Fort Worth, Texas: The Dryden Press.

Luce, Thom. 1989. *Computer Hardware, System Software, and Architecture*. Watsonville, CA: Mitchell Publishing.

Martin, James, 1989. *Information Engineering*. Englewood Cliffs, NJ: Prentice Hall.

Martin, Merle. 1991. *Analysis and Design of Business Information Systems*. New York: Macmillan.

McKie, Stewart. 1997. "Integrating Electronic Commerce." *DBMS Magazine* (September 1997).

Miller, Michael. 1994. "Applications Integration: Getting it Together." *PC Magazine* (February 8, 1994).

Musciano, Chuck, and Kennedy, Bill. 1996. *HTML The Definitive Guide*, Sebastopol, CA: O'Reilly & Associates.

Niederst, Jennifer. 1996. *Designing for the Web*. Sebastopol, CA: O'Reilly & Associates.

Norman, Ronald. 1996. *Object-Oriented Systems Analysis and Design*. Englewood Cliffs, NJ: Prentice Hall.

Norton, Peter. 1990. *Inside the IBM PC and PS/2*, 3rd Edition. New York: Brady.

Padwick, Gordon. 1996. *Building Integrated Office Applications*. Indianappolis, IN: Que Corporation.

Peng, C., et al. 1998. "Accessing Existing Business Data from the World Wide Web." *IBM Systems Journal* (Vol. 37, No. 1, 1998).

Petreley, N., and D. van Kirk. 1993. "Avoid the Application Development Maze." *Infoworld* (February 8, 1993).

Pratt, Philip. 1992. *Microcomputer Database Management Using dBASE IV*. Boston: Boyd & Fraser.

Satzinger, John W., et al. 2002. *Systems Analysis and Design in a Changing World*, 2/e Cambridge, MA,: Thompson Learning.

Sharp, Brad. 1998. "Creating an E-Comm Architecture." *Unix Review* (February 1998).

Simpson, Alan. 1996. *HTML Publishing Bible*. Foster City, CA: IDG Books Worldwide, Inc.

Salpeter, Judy. 1992. "Are You Obeying Copyright Law?" *Technology & Learning* (May/June 1992).

Taylor, Sandra. 1997. "4GLs Gear Up for Full-Cycle Development." *Application Development Trends* (December 1997).

Whiting, Rick. 1998. "The Embedded Legacy." *Software Magazine - Year 2000 Survival Guide* (April 15, 1998).

Whitten, J., and L. Bentley. 1998. *Systems Analysis and Design Methods*, 4th Edition. Boston: Irwin.

Yourdon, Edward. 1989. *Modern Structured Analysis.* Englewood Cliffs, NJ: Prentice Hall.

Yourdon, E., and L. Constantine. 1979. *Structured Design.* Englewood Cliffs, NJ: Prentice Hall.

Index

abstractions
  described, 65
  process modeling and, 65
Accelerated Graphics Port
  (AGP), 404
access control, 410–411. *See also*
  security
Access (Microsoft)
  4GL programming and, 353,
    374, 360, 367, 379, 385
  code generators and, 368
  database implementation with,
    218–219
  file types, 212
  networking and, 418, 426
  process design and, 293–294
  queries and, 251, 385
  security and, 412–413
  testing and, 447
accessibility, 36, 37
accuracy, 34–35, 37
ACM (Association for Computing
  Machinery), 468
activity diagrams, described, 126
adaptability, 36, 37
Add mode, 339
Add New Customer button, 298
Add Record icon button, 389
ADDEM command, 446
Adjust Hours option, 146
administrator account type, 410
Adobe
  PageMaker, 244
  Photoshop, 244, 435
affordability, 35–36, 37
aggregation, 124
AGP (Accelerated Graphics
  Port), 404
AITP (Association of Information
  Technology Professionals),
  468–469
American Standard Code for
  Information Interchange
  (ASCII), 213
Analysis phase
  described, 23, 440–44
  DFDs and, 77

analyst(s). *See also* Analysis phase
  as agents of change, 20
  as group facilitators, 21–22
  hours, validation of, 130
  hours log, 525–526
  as problem-solving strategists,
    20–21
  role of, 18–20
animations, 249
application builders, described, 362
application(s)
  beta releases, 495
  customized, 17, 372
  described, 11, 17
  development of, without
    programming languages, 372
  horizontal, 17, 275, 518
  maturation of, 7
  service providers (ASPs), 427
  suites, 7, 11, 276
  turnkey, 17, 408, 518
  types of, 17
  vertical, 17, 46, 274–275, 518
architecture, 402–404, 429
ASCII (American Standard
  Code for Information
  Interchange), 213
ASP (Microsoft Active Server
  Pages), 426
ASPs (application service
  providers), 427
assembly language, described, 352
Association for Computing
  Machinery (ACM), 468
Association of Information
  Technology Professionals
  (AITP), 468–469
ATMApplet, 363
attributes
  4GL programming and,
    353–354, 361, 378
  classes and, 121–122
  described, 103, 353–354
  ERDs and, 103
  inheritance of, 354
  linkages and, 103

object modeling and, 120
  programming languages and,
    353–354
audience
  effective presentations to,
    548–549
  writing to, 546–547
audio
  devices, 244
  files, sharing, 285
  softcopy design and, 249
  synchronization, 244
Authorware, 244
automated input, 194

B2B (business-to-business)
  e-commerce, 277, 423
B2C (business-to-customer)
  e-commerce, 277
B2E (business-to-employee)
  e-commerce, 277
backup files
  database design and, 212
  described, 212
  networking and, 416
  output design, 245, 265
  system maintenance and, 504
bandwidth, 414
bar code readers, 194
baseline measurements, 48
BASIC
  classification of, as a third-
    generation programming
    language, 17
  described, 352
bat file type, 213
batch
  files, 212–213, 281
  processing, 98, 177
behaviors, 120
benefits, intangible, 314–315
beta releases, 495
bin file type, 213
bit depth, 244
.bmp file type, 213
boot process, 408